Recent Titles in This Series

(Continued in the back of this publication)

The Penrose Transform and
Analytic Cohomology
in Representation Theory

CONTEMPORARY MATHEMATICS

154

The Penrose Transform and Analytic Cohomology in Representation Theory

AMS-IMS-SIAM Summer Research Conference
June 27 to July 3, 1992
Mount Holyoke College,
South Hadley, Massachusetts

Michael Eastwood
Joseph Wolf
Roger Zierau
Editors

American Mathematical Society
Providence, Rhode Island

The AMS-IMS-SIAM Summer Research Conference in the Mathematical Sciences on The Penrose Transform and Analytic Cohomology in Representation Theory was held at Mount Holyoke College, South Hadley, Massachusetts, from June 27 to July 3, 1992, with support from the National Science Foundation, Grant DMS-8918200 02.

1991 *Mathematics Subject Classification.* Primary 22E46;
Secondary 22E45, 32L25, 14F05, 53C65.

Library of Congress Cataloging-in-Publication Data

The Penrose transform and analytic cohomology in representation theory: AMS-IMS-SIAM Joint Summer Research Conference, June 27 to July 3, 1992, supported by the National Science Foundation / Michael Eastwood, Joseph Wolf, Roger Zierau, editors.

    p. cm. – (Contemporary mathematics; 154)

    "The AMS-IMS-SIAM Joint Summer Research Conference in the Mathematical Sciences on the Penrose Transform and Analytic Cohomology in Representation Theory was held at Mount Holyoke College, South Hadley, Massachusetts from June 27 to July 3, 1992"–Galley.

    Includes bibliographical references.

    ISBN 0-8218-5176-4

    1. Semisimple Lie groups–Congresses. 2. Representations of groups–Congresses. 3. Penrose transform–Congresses. I. Eastwood, Michael G. II. Wolf, Joseph Albert, 1936–. III. Zierau, Roger, 1956–. IV. American Mathematical Society. V. Institute of Mathematical Statistics. VI. Society for Industrial and Applied Mathematics. VII. AMS-IMS-SIAM Joint Summer Research Conference in the Mathematical Sciences on the Penrose Transform and Analytic Cohomology in Representation Theory (1992: Mount Holyoke College) VIII. Series: Contemporary mathematics (American Mathematical Society); v. 154.

QA387.P46   1993

$512'.55$–dc20

                                             93-27398
    CIP

CONTENTS

PREFACE

This volume contains papers presented at the conference "The Penrose Transform and Analytic Cohomology in Representation Theory," held 27th June to 3rd July, 1992, at Mount Holyoke College, South Hadley, Massachusetts. All articles have been refereed. The conference was one of a series of Summer Research Conferences sponsored jointly by the American Mathematical Society, the Institute of Mathematical Statistics, and the Society for Industrial and Applied Mathematics, and supported by a grant from the National Science Foundation. The organizing committee consisted of Robert J. Baston (Co-chair), Michael G. Eastwood (Co-chair), Victor W. Guillemin, Joseph A. Wolf, and Roger Zierau.

The main idea of this conference was to bring together representation theorists and differential geometers in order that they may profitably interact. In particular, it seemed that various integral transforms from representation theory, complex integral geometry, and mathematical physics were instances of the same general construction sometimes called the "Penrose transform". Speakers were asked to give talks bearing the diverse and non-specialist nature of their audience in mind. In this respect and also in informal discussion, the conference was very successful. There is considerable scope for further research and we hope that this conference and its proceedings will be seen as a useful catalyst.

Generally speaking, the programme was arranged with the more expository talks early on so that later speakers could refer to topics already discussed. The order has been maintained in these proceedings. Unfortunately, Victor Guillemin was unable to attend but has contributed a joint article with Meng-Kiat Chauh. Lisa Mantini kindly agreed to speak in his place. The programme and list of participants is given overleaf.

Special thanks are due to Carole Kohanski as Conference Coördinator and Donna Harmon for Editorial Assistance.

Michael Eastwood
Joseph Wolf
Roger Zierau

Programme

A.W. Knapp	"Introduction to Representations in Analytic Cohomology"
J.A. Wolf	"Geometric Quantization for Semisimple Lie Groups"
D.A. Vogan, Jr.	"Unipotent Representations and Cohomological Induction"
B. Kostant	"Nilpotent Coadjoint Orbits and Unipotent Representations of Semisimple Lie Groups"
M.G. Eastwood	"Introduction to Penrose Transform"
L. Barchini	"Intertwining Operators into Dolbeault Cohomology Representations"
C.R. LeBrun	"Holonomy Groups, the Penrose Transform, and Algebraic Geometry"
S. Gindikin	"Realizations of $\overline{\partial}$-cohomology on Pseudo-Hermitian Symmetric Manifolds"
E.G. Dunne	"Twistor Theory for Indefinite Kähler Symmetric Spaces"
D. Miličić	"$\mathcal{D}$-modules and the Classification of Harish-Chandra Modules"
T.N. Bailey	"Parabolic Invariant Theory and Geometry"
R.J. Baston	"Multiplicity One for Zuckerman's Functors"
L. Mantini	"A Penrose-Radon Transform in L^2-cohomology for Massless Field Equations"
J.E. Rice	"Cousin Complexes and Resolutions of Representations"
H.W. Wong	"Dolbeault Cohomologies and Zuckerman Modules"
W. Schmid	"$G_{\mathbb{R}}$-Equivariant Sheaves and Cohomology"
D.M. Barbasch	"Unipotent Representations and Derived Functor Modules"
R. Howe	"Some Examples of the Infinitesimal Structure of Degenerate Principal Series"
R. Zierau	"Unitarity of Certain Dolbeault Cohomology Representations"

Participants

T.N. Bailey, M.W. Baldoni, D.M. Barbasch, L. Barchini, R.J. Baston, A.O. Brega, R. Donley, E.G. Dunne, M.G. Eastwood, E. Galina, S. Gindikin, A.R. Gover, R. Howe, J. Huang, R. Keown, A.W. Knapp, T. Kobayashi, L. Koch, B. Kostant, R. Kunze, C.R. LeBrun, J.D. Lorch, L. Mantini, W. McGovern, R.C. Mclean, D. Milicic, J. Novak, J.W. Rice, W. Schmid, R.J. Stanton, R.W. Stokes, J.A. Tirao, D.A. Vogan, J.A. Wolf, H.W. Wong, H. Zheng, R. Zierau, Y.M. Zou.

Contemporary Mathematics
Volume **154**, 1993

Introduction to Representations in Analytic Cohomology

A. W. KNAPP

ABSTRACT. This is a survey of background and old results concerning representations in cohomology sections of vector bundles. The base space is a homogeneous space G/L, where G is a connected reductive Lie group and L is the centralizer of a torus. When G is compact, the representations in question are the subject of the Bott-Borel-Weil Theorem. When G is noncompact and L is compact, the representations are identified by the Langlands Conjecture, which was proved by Schmid. For noncompact L, difficult analytic problems blocked progress initially. To avoid these difficulties, Zuckerman and Vogan developed an algebraic analog, cohomological induction, that gave a construction of identifiable representations that were often irreducible unitary. Recent progress has related the analytic representations and their algebraic analogs in various ways.

1. Sections of homogeneous vector bundles

This paper gives some background from representation theory for understanding the connection between the Penrose transform and analytic realizations of group representations. It is assumed that the reader is acquainted with elementary facts about holomorphic vector bundles and the elementary structure theory of semisimple groups. Discussions of these two topics may be found in Wells [**19**, Chapter I] and Knapp [**8**, Chapter V], respectively. The results in this paper largely are not new, and, for the most part, references will be given in place of proofs.

1991 *Mathematics Subject Classification.* Primary 20G05, 22E45, 32L10; Secondary 55R91, 83C60.

The author was supported by NSF Grant DMS 91 00367.

This paper is in final form and no version of it will be submitted for publication elsewhere.

1.1. Setting.

Throughout the paper we work with the following situation, sometimes limiting ourselves to special cases: G is a connected linear reductive Lie group with complexification $G^{\mathbb{C}}$, T is a torus subgroup, and $L = Z_G(T)$ is the centralizer of T in G. It is known that L is connected; a proof may be constructed by combining [8, Corollary 4.22] with the style of argument at the top of p. 126 of that book. Therefore the complexification $L^{\mathbb{C}}$ is meaningful. Let Q be a parabolic subgroup of $G^{\mathbb{C}}$ with Levi factor $L^{\mathbb{C}}$.

We denote Lie algebras of Lie groups A, B, etc., by $\mathfrak{a}_0$, $\mathfrak{b}_0$, etc., and we denote their complexifications by $\mathfrak{a}$, $\mathfrak{b}$, etc. The complex Lie algebras of complex Lie groups $G^{\mathbb{C}}$, $L^{\mathbb{C}}$, Q are denoted $\mathfrak{g}$, $\mathfrak{l}$, $\mathfrak{q}$. We use an overbar to denote the conjugation of $\mathfrak{g}$ with respect to $\mathfrak{g}_0$.

We can decompose the Lie algebra $\mathfrak{q}$ of Q as a vector space direct sum $\mathfrak{q} = \mathfrak{l} \oplus \mathfrak{u}$, where $\mathfrak{u}$ is the nilradical. Then $\mathfrak{u}$ and $\bar{\mathfrak{u}}$ are both nilpotent complex Lie algebras, and we have $[\mathfrak{l}, \mathfrak{u}] \subseteq \mathfrak{u}$ and $[\mathfrak{l}, \bar{\mathfrak{u}}] \subseteq \bar{\mathfrak{u}}$.

We assume that $\mathfrak{q}$ is a θ-**stable parabolic**; this condition means that

$$(1.1a) \qquad\qquad \mathfrak{g}_0 \cap \mathfrak{q} = \mathfrak{l}_0.$$

It is equivalent to assume a vector space sum decomposition

$$(1.1b) \qquad\qquad \mathfrak{g} = \bar{\mathfrak{u}} \oplus \mathfrak{l} \oplus \mathfrak{u}.$$

Under the condition (1.1), the natural mapping $G/L \to G^{\mathbb{C}}/Q$ is an inclusion, and the image is an open set. Thus the choice of Q has made G/L into a complex manifold with G operating holomorphically.

An example to keep in mind is the group $G = U(m, n)$ of complex matrices that preserve an indefinite Hermitian form. Here $G^{\mathbb{C}} = GL(m+n, \mathbb{C})$. If we take T to be any closed connected subgroup of the diagonal, then L will be a block diagonal subgroup within G, necessarily connected. We can choose $\mathfrak{u}$ to be the complex Lie algebra of corresponding block upper triangular matrices and $\bar{\mathfrak{u}}$ to consist of the corresponding block lower triangular matrices.

1.2. Associated bundles.

It is well known (see [16]) that

$$(1.2) \qquad\qquad p : G \to G/L$$

is a C^{∞} principal fiber bundle with structure group L. Let V be a finite-dimensional real or complex vector space, let $GL(V)$ be its general linear group, and let $\rho : L \to GL(V)$ be a C^{∞} homomorphism. The **associated vector bundle**

$$(1.3a) \qquad\qquad p_V : G \times_L V \to G/L$$

is a vector bundle with structure group $GL(V)$ whose bundle space is given by

$$(1.3b) \qquad G \times_L V = \{(g, v)/\sim\} \qquad \text{with} \qquad (gl, v) \sim (g, \rho(l)v)$$

for $g \in G$, $l \in L$, and $v \in V$; the bundle structure will now be described.

The bundle (1.2) can be given in terms of transition functions. Namely for a suitably fine open cover $\{U\}$ of G/L, there are fiber preserving C^∞ maps $h_U : p^{-1}(U) \to U \times L$ that specify the local product structure on G, and the assumption is that, for $x \in U \cap V$, $h_{VU}(x) = h_V \circ h_U^{-1}|_{p^{-1}(x)}$ is a member of L and depends in C^∞ fashion on x. The functions $\{\rho(h_{VU}(x))\}$ are the transition functions for (1.3). These determine a vector bundle structure for (1.3) by Steenrod [16, Theorem I.3.2].

The space of C^∞ sections of (1.3) is denoted $\mathcal{E}(G \times_L V)$. The group G acts on $G \times_L V$ by left translation: $g_0(g, v) = (g_0 g, v)$ in the notation of (1.3b), and this action induces an action of G on $\mathcal{E}(G \times_L V)$ by $(g_0 \gamma)(g, v) = \gamma(g_0^{-1} g, v)$ for $\gamma \in \mathcal{E}(G \times_L V)$. When V is complex, this construction yields a representation of G (understood to be on a complex vector space) with a natural continuity property: $(g_0, \gamma) \mapsto g_0 \gamma$ is continuous from $G \times \mathcal{E}(G \times_L V)$ to $\mathcal{E}(G \times_L V)$ if $\mathcal{E}(G \times_L V)$ is given its usual C^∞ topology.

Similarly

$$(1.4) \qquad p : G^{\mathbb{C}} \to G^{\mathbb{C}}/Q$$

is a holomorphic principal fiber bundle with structure group Q. In the above situation if V is complex and if ρ extends to a holomorphic homomorphism $\rho : Q \to GL(V)$, then we can construct an associated vector bundle

$$(1.5a) \qquad p_V : G^{\mathbb{C}} \times_{\mathbb{C}} V \to G^{\mathbb{C}}/Q$$

with bundle space given by

$$(1.5b) \qquad G^{\mathbb{C}} \times_Q V = \{(g^{\mathbb{C}}, v)/ \sim\} \qquad \text{with} \qquad (g^{\mathbb{C}} q, v) \sim (g^{\mathbb{C}}, \rho(q)v).$$

The bundle (1.5) is a holomorphic vector bundle.

The inclusion $G/L \hookrightarrow G^{\mathbb{C}}/Q$ induces via pullback from (1.5a) a bundle map

$$(1.6) \qquad G \times_L V \hookrightarrow G^{\mathbb{C}} \times_Q V.$$

In terms of (1.3b) and (1.5b), this map is given simply by $(g, v) \mapsto (g, v)$. The result is that the C^∞ complex vector bundle $G \times_L V$ acquires the structure of a holomorphic vector bundle. We can regard the space of holomorphic sections $\mathcal{O}(G \times_L V)$ of $G \times_L V$ as a vector subspace of $\mathcal{E}(G \times_L V)$.

In applications it is important to be able to relax the assumptions on the original $\rho : L \to GL(V)$ and still be able to use Q to impose the structure of a holomorphic vector bundle on $G \times_L V$. See Tirao-Wolf [17] for this generalization.

To any section γ of $G \times_L V$ we can associate a function $\varphi_\gamma : G \to V$ by the definition

$$(1.7a) \qquad \gamma(gL) = (g, \varphi_\gamma(g)) \in G \times_L V.$$

Under this correspondence, C^∞ sections γ go to C^∞ functions φ_γ, and we obtain an isomorphism

(1.7b)
$$\mathcal{E}(G \times_L V) \cong \left\{ \varphi : G \to V \,\middle|\, \begin{array}{l} \varphi \text{ of class } C^\infty, \\ \varphi(gl) = \rho(l)^{-1}\varphi(g) \text{ for } l \in L,\, g \in G \end{array} \right\}.$$

The usual C^∞ topology on $\mathcal{E}(G \times_L V)$ corresponds to the C^∞ topology on the space of φ's.

The correspondence $\gamma \leftrightarrow \varphi_\gamma$ works locally as well, with sections over an open set $U \subseteq G/L$ corresponding to functions φ on the open subset $p^{-1}(U)$ of G transforming as in (1.7b). Again γ of class C^∞ corresponds to φ_γ of class C^∞. Let $\mathcal{E}(U)$ be the space of C^∞ sections over U.

In the special case that $G \times_L V$ admits the structure of a holomorphic vector bundle because of (1.6) and (1.5), we can speak of the space of holomorphic sections $\mathcal{O}(U)$ over an open set $U \subseteq G/L$. The first proposition tells how to use φ_γ to decide whether γ is holomorphic.

PROPOSITION 1. *Suppose that ρ extends to a holomorphic homomorphism $\rho : Q \to GL(V)$ and thereby makes $G \times_L V$ into a holomorphic vector bundle. Let $U \subseteq G/L$ be open, let γ be in $\mathcal{E}(U)$, and let φ_γ be the corresponding function from G to V given by (1.7). Then γ is holomorphic if and only if*

(1.8)
$$(Z\varphi_\gamma)(g) = -\rho(Z)(\varphi_\gamma(g))$$

for all $g \in p^{-1}(U)$ and $Z \in \mathfrak{q}$, with Z acting on φ_γ as a complex left-invariant vector field.

PROOF. Suppose γ is in $\mathcal{E}(U)$. We can regard U as open in $G^{\mathbb{C}}/Q$ and use the formula
$$\gamma(g^{\mathbb{C}}Q) = (g^{\mathbb{C}}, \tilde{\varphi}_\gamma(g^{\mathbb{C}})) \in G^{\mathbb{C}} \times_Q V$$
to define $\tilde{\varphi}_\gamma$ on the open set $p^{-1}(U) \subseteq G^{\mathbb{C}}$. The function $\tilde{\varphi}_\gamma$ satisfies

(1.9a)
$$\tilde{\varphi}_\gamma(g^{\mathbb{C}}q) = \rho(q)^{-1}(\tilde{\varphi}_\gamma(g^{\mathbb{C}}))$$

for $g^{\mathbb{C}} \in p^{-1}(U)$ and $q \in Q$, and therefore also

(1.9b)
$$(Z\tilde{\varphi}_\gamma)(g^{\mathbb{C}}) = -\rho(Z)(\tilde{\varphi}_\gamma(g^{\mathbb{C}}))$$

for $g^{\mathbb{C}} \in p^{-1}(U)$ and $Z \in \mathfrak{q}$, with Z acting as a *real* left-invariant vector field. We recover φ_γ by restricting $\tilde{\varphi}_\gamma$ to $G \cap p^{-1}(U)$. By definition of the complex structure, γ is holomorphic if and only if $\tilde{\varphi}_\gamma$ is holomorphic. We are thus to show that $\tilde{\varphi}_\gamma$ is holomorphic if and only if φ_γ satisfies (1.8).

Suppose $\tilde{\varphi}_\gamma$ is holomorphic. The Cauchy-Riemann equations say that

(1.10)
$$(iZ)\tilde{\varphi}_\gamma = i(Z\tilde{\varphi}_\gamma)$$

for $Z \in \mathfrak{g}$, with Z acting as a real left-invariant vector field. If Z is in $\mathfrak{q}$, write $Z = X + iY$ with X and Y in $\mathfrak{g}_0$. For $g \in G$ we have

$$
\begin{aligned}
-\rho(Z)(\varphi_\gamma(g)) &= -\rho(Z)(\tilde{\varphi}_\gamma(g)) \\
&= (Z\tilde{\varphi}_\gamma)(g) && \text{by (1.9b)} \\
&= (X\tilde{\varphi}_\gamma)(g) + (iY)\tilde{\varphi}_\gamma(g) \\
&= (X\tilde{\varphi}_\gamma)(g) + i(Y\tilde{\varphi}_\gamma)(g) && \text{by (1.10)} \\
&= (X\varphi_\gamma)(g) + i(Y\varphi_\gamma)(g) \\
&= (Z\varphi_\gamma)(g).
\end{aligned}
$$

(1.11)

Thus φ_γ satisfies (1.8).

Conversely suppose φ_γ satisfies (1.8). Unwinding (1.11), we obtain

$$(iY)\tilde{\varphi}_\gamma(g) = i(Y\tilde{\varphi}_\gamma)(g)$$

whenever $Z = X + iY$ is in $\mathfrak{q}$. Replacing Z by iZ, we obtain also

$$(iX)\tilde{\varphi}_\gamma(g) = i(X\tilde{\varphi}_\gamma)(g).$$

Suitable linear combinations of these two equations give

$$(iZ)\tilde{\varphi}_\gamma(g) = i(Z\tilde{\varphi}_\gamma)(g) \qquad \text{and} \qquad (i\overline{Z})\tilde{\varphi}_\gamma(g) = i(\overline{Z}\tilde{\varphi}_\gamma)(g)$$

for Z in $\mathfrak{q}$. Since $\mathfrak{q} + \overline{\mathfrak{q}} = \mathfrak{g}$, (1.10) holds for all $Z \in \mathfrak{g}$ for the special case that g is in $p^{-1}(U) \cap G$. A general member of $p^{-1}(U)$ in $G^{\mathbb{C}}$ is of the form gq with $g \in p^{-1}(U) \cap G$ and $q \in Q$. Taking Z in $\mathfrak{g}$ and letting a dot indicate where a vector field is to act, we have

$$
\begin{aligned}
(iZ)\tilde{\varphi}_\gamma(gq\,\cdot) &= (\mathrm{Ad}(q)(iZ))\tilde{\varphi}_\gamma(g \cdot q) \\
&= \rho(q)^{-1}((\mathrm{Ad}(q)(iZ))\tilde{\varphi}_\gamma(g\,\cdot)) && \text{by (1.9a)} \\
&= \rho(q)^{-1}i(\mathrm{Ad}(q)Z)\tilde{\varphi}_\gamma(g\,\cdot) && \text{by the special case} \\
&= i(\mathrm{Ad}(q)Z)\tilde{\varphi}_\gamma(g \cdot q) && \text{by (1.9a)} \\
&= i(Z\tilde{\varphi}_\gamma)(gq\,\cdot).
\end{aligned}
$$

Thus $\tilde{\varphi}_\gamma$ satisfies (1.10) everywhere on $p^{-1}(U)$ and is holomorphic.

REFERENCE. See Griffiths-Schmid [6, pp. 258-259].

In typical applications to representation theory, ρ in the proposition is given on L and extends holomorphically to $L^{\mathbb{C}}$. The extension to Q is taken to be trivial on the unipotent radical of Q. Equation (1.8) holds for $Z \in \mathfrak{l}_0$ for any C^∞ section, and it extends to $Z \in \mathfrak{l}$ by complex linearity. Thus (1.8) may be replaced in this situation by the condition

(1.12) $$Z\varphi_\gamma = 0 \qquad \text{for all } Z \in \mathfrak{u}.$$

The special case $\rho = 1$ shows how to recognize holomorphic functions on open subsets of G/L.

1.3. Constructions with tangent bundle.

Let M be a complex manifold, and let p be in M. We denote by $T_p(M)$ the tangent space of M (considered as a C^∞ manifold) at p, consisting of derivations of the algebra of smooth germs at p, and we let $T(M)$ be the tangent bundle. Also we denote by $T_{\mathbb{C},p}(M)$ the complex tangent space of M at p, consisting of derivations of the algebra of holomorphic germs at p, and we let $T_{\mathbb{C}}(M)$ be the corresponding complex tangent bundle. There is a canonical $\mathbb{R}$ isomorphism

(1.13a) $$T_p(M) \xrightarrow{\sim} T_{\mathbb{C},P}(M)$$

given by

(1.13b) $$\xi \mapsto \zeta, \quad \text{where } \zeta(u + iv) = \xi(u) + i\xi(v).$$

Let J_p be the member of $GL(T_p(M))$ that corresponds under (1.13) to multiplication by i in $T_{\mathbb{C},p}(M)$. Then $J = \{J_p\}$ is a bundle map from $T(M)$ to itself whose square is -1.

The following proposition allows us to relate these considerations to associated vector bundles.

PROPOSITION 2. *There are canonical bundle isomorphisms*

(1.14a) $$T(G/L) \cong G \times_L (\mathfrak{g}_0/\mathfrak{l}_0)$$

(1.14b) $$T_{\mathbb{C}}(G^{\mathbb{C}}/Q) \cong G^{\mathbb{C}} \times_Q (\mathfrak{g}/\mathfrak{q})$$

with L and Q acting on $\mathfrak{g}_0/\mathfrak{l}_0$ and $\mathfrak{g}/\mathfrak{q}$, respectively, by Ad.

The inclusion $G/L \subseteq G^{\mathbb{C}}/Q$ allows us to regard

(1.15) $$T_{\mathbb{C}}(G/L) \cong GQ \times_Q (\mathfrak{g}/\mathfrak{q}).$$

At any point $p = gL$ of G/L, the left sides of (1.14a) and (1.15), namely $T(G/L)$ and $T_{\mathbb{C}}(G/L)$, are $\mathbb{R}$ isomorphic via (1.13). It is easy to check that the corresponding isomorphism of the right sides of (1.14a) and (1.15) at p is given by

$$(g, X + \mathfrak{l}_0) \mapsto (g, X + \mathfrak{q}) \qquad \text{for } g \in G,\ X \in \mathfrak{g}_0.$$

This result allows us to compute the effect of J.

Complexifying (1.14a), we have

$$T(G/L)^{\mathbb{C}} \cong G \times_L (\mathfrak{g}_0/\mathfrak{l}_0)^{\mathbb{C}},$$

and J acts in the fiber at each point. We let $T(G/L)^{1,0}$ and $T(G/L)^{0,1}$ be the submodules of $T(G/L)^{\mathbb{C}}$ corresponding to the respective eigenvalues i and $-i$ of J, so that

(1.16a) $$T(G/L)^{\mathbb{C}} \cong T(G/L)^{1,0} \oplus T(G/L)^{0,1}.$$

We have

(1.16b) $$(\mathfrak{g}_0/\mathfrak{l}_0)^{\mathbb{C}} \cong \mathfrak{g}/\mathfrak{l} \cong \overline{\mathfrak{u}} \oplus \mathfrak{u}$$

as L modules, and a little calculation shows that (1.16b) gives the decomposition of the fibers under J corresponding to (1.16a). In other words

(1.16c)
$$T(G/L)^{1,0} \cong G \times_L \bar{\mathfrak{u}}$$
$$T(G/L)^{0,1} \cong G \times_L \mathfrak{u}.$$

Taking duals in (1.16a) and forming alternating tensors, we have

(1.17) $$\wedge^{p,q} T^*(G/L)^{\mathbb{C}} \cong G \times_L ((\wedge^p \bar{\mathfrak{u}})^* \otimes (\wedge^q \mathfrak{u})^*).$$

Via (1.17), members of $\mathcal{E}(\wedge^{p,q} T^*(G/L)^{\mathbb{C}})$ correspond to functions from G to $(\wedge^p \bar{\mathfrak{u}})^* \otimes (\wedge^q \mathfrak{u})^*$ transforming on the right under L by $\mathrm{Ad}^* \otimes \mathrm{Ad}^*$.

1.4. $\bar{\partial}$ operator.

The scalar $\bar{\partial}$ operator for a complex manifold M is an operator

$$\bar{\partial} : \mathcal{E}(\wedge^{p,q} T^*(M)^{\mathbb{C}}) \to \mathcal{E}(\wedge^{p,q+1} T^*(M)^{\mathbb{C}}),$$

and it has $\bar{\partial}^2 = 0$. For the case that $M = G/L$, we can interpret $\bar{\partial}$ in terms of (1.17).

Also we can construct a vector-valued version of $\bar{\partial}$. Namely let $G \times_L V$ be a holomorphic vector bundle as above. We introduce $\bar{\partial}_V = \bar{\partial} \otimes 1$ as an operator

$$\bar{\partial}_V : \mathcal{E}(\wedge^{p,q} T^*(G/L)^{\mathbb{C}} \otimes (G \times_L V)) \to \mathcal{E}(\wedge^{p,q+1} T^*(G/L)^{\mathbb{C}} \otimes (G \times_L V));$$

$\bar{\partial}_V$ is well defined because the transition functions for $G \times_L V$ are holomorphic. Also $\bar{\partial}_V^2 = 0$. Using (1.17) and dropping the subscript "V" on $\bar{\partial}_V$, we can interpret $\bar{\partial}_V$ as an operator

$$\bar{\partial} : \mathcal{E}(G \times_L ((\wedge^p \bar{\mathfrak{u}})^* \otimes (\wedge^q \mathfrak{u})^* \otimes V)) \to \mathcal{E}(G \times_L ((\wedge^p \bar{\mathfrak{u}})^* \otimes (\wedge^{q+1} \mathfrak{u})^* \otimes V)).$$

In representation theory one works with the case $p = 0$. We define

$$C^{0,q}(G/L, V) = \mathcal{E}(G \times_L ((\wedge^q \mathfrak{u})^* \otimes V)).$$

As always, this is the representation space for a continuous representation of G. The operator $\bar{\partial}$ is continuous and the kernel is closed. Whether or not the image of $\bar{\partial}$ is closed, we can define the **Dolbeault cohomology space** $H^{0,q}(G/L, V)$ as

(1.18) $$H^{0,q}(G/L, V) = \ker(\bar{\partial}|_{C^{0,q}(G/L,V)}) / \mathrm{image}(\bar{\partial}|_{C^{0,q-1}(G/L,V)}).$$

Since $\bar{\partial}$ commutes with G, the vector space $H^{0,q}(G/L, V)$ carries a representation of G, but possibly the topology is not Hausdorff.

The question whether the image of $\bar{\partial}$ is closed turns out to play a major role in the theory. Sometimes, partly to get around this question, one works with the subspace $\ker \bar{\partial} \cap \ker \bar{\partial}^*$ of $C^{0,q}(G/L, V)$, for a suitably defined "adjoint" $\bar{\partial}^*$, in place of $H^{0,q}(G/L, V)$. In addition to its technical simplicity, this subspace has other advantages that will not be discussed here. Members of $C^{0,q}(G/L, V)$ in $\ker \bar{\partial} \cap \ker \bar{\partial}^*$ are said to be **strongly harmonic**.

2. Bott-Borel-Weil Theorem

The Bott-Borel-Weil Theorem identifies the spaces $H^{0,q}(G/L, V)$ of (1.18) in the case that G is compact. In this situation $\bar{\partial}$ always has closed image, and (1.18) can be computed alternatively as the representation on the strongly harmonic forms in $C^{0,q}(G/L, V)$. (See [19, Chapter V].)

In setting up the complex structure for the case that G is compact, there are two possible ways of proceeding. One is to fix the complex structure (i.e., fix the parabolic subgroup Q) and identify $H^{0,q}(G/L, V)$ for all q and V. The other is to fix V, adapt the complex structure to V, and identify $H^{0,q}(G/L, V)$ for all q. The first way leads to a more general result when G is compact, but the relationship between the two ways is more complicated when G is noncompact.

We begin with the most important special case, where $L = T$. For the first approach, where the complex structure is fixed before V is given, the notation is

$$
\begin{aligned}
&G = \text{compact connected Lie group} \\
&T = \text{maximal torus } (= L) \\
&\Delta = \{\text{roots of } (\mathfrak{g}, \mathfrak{t})\} \\
\text{(2.1)} \quad &\Delta^+ = \text{a positive system for } \Delta \\
&\delta = \tfrac{1}{2} \sum_{\alpha \in \Delta^+} \alpha \\
&W = \text{Weyl group of } \Delta \\
&B = \text{Borel subgroup built from } \textit{negative} \text{ roots } (= Q).
\end{aligned}
$$

The inclusion $G/T \hookrightarrow G^{\mathbb{C}}/B$ is onto since the compactness of G makes the image closed, as well as open. Thus we write $G/T = G^{\mathbb{C}}/B$. If $\lambda \in \mathfrak{t}^*$ is an integral parameter and ξ_λ is the corresponding character of T, we abbreviate the representation of T on $\mathbb{C}$ by ξ_λ as $\mathbb{C}_\lambda$. The role of V is played by $\mathbb{C}_\lambda$.

THEOREM 3. *Let $\lambda \in \mathfrak{t}^*$ be integral.*

(a) *If $\langle \lambda + \delta, \alpha \rangle = 0$ for some $\alpha \in \Delta$, then $H^{0,k}(G/T, \mathbb{C}_\lambda) = 0$ for all k.*

(b) *If $\langle \lambda + \delta, \alpha \rangle \neq 0$ for all $\alpha \in \Delta$, let*

$$
\text{(2.2)} \qquad q = \#\{\alpha \in \Delta^+ \mid \langle \lambda + \delta, \alpha \rangle < 0\}.
$$

Choose $w \in W$ with $w(\lambda + \delta)$ dominant, and put $\mu = w(\lambda + \delta) - \delta$. Then

$$
\text{(2.3)} \qquad H^{0,k}(G/T, \mathbb{C}_\lambda) = \begin{cases} 0 & \text{if } k \neq q \\ F^\mu & \text{if } k = q, \end{cases}
$$

where F^μ is a finite-dimensional irreducible representation of G with highest weight μ.

REFERENCES. For an exposition, see Baston-Eastwood [3, pp. 44-48]. The original paper is Bott [4].

For the second approach, in which V is given and then the complex structure is fixed, we let G, T, and Δ be as in (2.1). Let $\lambda_0 \in \mathfrak{t}^*$ be a given nonsingular parameter (λ_0 corresponds to $\lambda + \delta$ in Theorem 1), and suppose that $\lambda_0 - \delta_0$

is integral for the half sum δ_0 of positive roots in some (or equivalently each) positive system. Take

(2.4)
$$\Delta^+ = \{\alpha \in \Delta \mid \langle \lambda_0, \alpha \rangle > 0\}$$
$$\delta = \tfrac{1}{2} \sum_{\alpha \in \Delta^+} \alpha$$
$$B \text{ built from } \Delta^+ \text{ instead of } -\Delta^+.$$

Again we have $G/T = G^{\mathbb{C}}/B$.

COROLLARY 4. *Let $\lambda_0 \in \mathfrak{t}^*$ be nonsingular with $\lambda_0 - \delta_0$ integral. With Δ^+, δ, and B defined as in (2.4),*

$$H^{0,k}(G/T, \mathbb{C}_{\lambda_0 + \delta}) = \begin{cases} 0 & \text{if } k \neq \dim_{\mathbb{C}}(G/T) \\ F^{\lambda_0 - \delta_0} & \text{if } k = \dim_{\mathbb{C}}(G/T). \end{cases}$$

PROOF. If we put $\lambda = \lambda_0 - \delta$, this becomes a special case of Theorem 3.

The Bott-Borel-Weil Theorem extends to the general G/L with G compact and L the centralizer of a torus. We state the generalization of Theorem 3, omitting the generalization of Corollary 4. The notation is

(2.5)
$$G = \text{compact connected Lie group}$$
$$T = \text{a torus in } G$$
$$L = Z_G(T)$$
$$T \text{ extended to a maximal torus } \tilde{T} \text{ in } L$$
$$\Delta = \{\text{roots of } (\mathfrak{g}, \tilde{\mathfrak{t}})\}$$
$$\Delta(\mathfrak{l}) = \{\text{roots of } (\mathfrak{l}, \tilde{\mathfrak{t}})\} \subseteq \Delta$$
$$\Delta^+ \text{ chosen with } \Delta(\mathfrak{l}) \text{ generated by simple roots}$$
$$\delta = \tfrac{1}{2} \sum_{\alpha \in \Delta^+} \alpha$$
$$W = \text{Weyl group}$$
$$Q = \text{built from } \mathfrak{l} \text{ and } negative \text{ roots.}$$

Then we have $G/L \cong G^{\mathbb{C}}/Q$.

THEOREM 5. *Let V^λ be irreducible for L with highest weight λ.*

(a) *If $\langle \lambda + \delta, \alpha \rangle = 0$ for some $\alpha \in \Delta$, then $H^{0,k}(G/L, V^\lambda) = 0$ for all k.*

(b) *If $\langle \lambda + \delta, \alpha \rangle \neq 0$ for all $\alpha \in \Delta$, define q as in (2.2), choose $w \in W$ so that $w(\lambda + \delta)$ is dominant, and put $\mu = w(\lambda + \delta) - \delta$. Then*

$$H^{0,k}(G/L, V^\lambda) = \begin{cases} 0 & \text{if } k \neq q \\ F^\mu & \text{if } k = q. \end{cases}$$

3. Discrete series

For a unimodular group G, an irreducible unitary representation π is in the **discrete series** if it is a direct summand of the right regular representation on $L^2(G)$, or equivalently if some nonzero matrix coefficient $(\pi(g)v_1, v_2)$ is in $L^2(G)$. (See Godement [5].)

Let G be linear connected reductive, and let K be a maximal compact subgroup. For G compact (so that $K = G$), every irreducible unitary representation is in the discrete series. For G noncompact, the discrete series representations were parametrized by Harish-Chandra. His work can be summarized in the following two theorems.

THEOREM 6. *G has discrete series if and only if rank G =rank K.*

REFERENCES. For an exposition, see [**8**, p. 454]. The original paper is Harish-Chandra [**7**].

The condition on ranks means that a maximal torus T in K is maximal abelian in G. For simple groups that are not complex, the equal rank condition is usually satisfied, but not always. For example, it is satisfied for $SO_e(p,q)$ if p or q is even, for $SU(p,q)$ and $Sp(p,q)$ always, and for $Sp(n,\mathbb{R})$. It is not satisfied for $SL(n,\mathbb{R})$ for $n \geq 3$.

Now let us assume that the equal rank condition is satisfied. Fix a maximal torus T in K. Let Δ_K and Δ be the respective root systems of $\mathfrak{k}$ and $\mathfrak{g}$ with respect to $\mathfrak{t}$, and let W_K and W_G be their Weyl groups. The members of Δ_K are called **compact**, and the other members of Δ are called **noncompact**. Fix a positive system Δ^+ for Δ, and let δ be half the sum of the members of Δ^+.

THEOREM 7. *Assume rank G = rank K. Suppose $\lambda_0 \in \mathfrak{t}^*$ is nonsingular and $\lambda_0 - \delta$ is integral. Then there exists a discrete series representation π_{λ_0} of G such that the global character of π_{λ_0} is given on the conjugates of T by the function*

$$(3.1) \qquad\qquad \pm \frac{\sum_{w \in W_K} (sgn\ w) e^{w\lambda_0}}{\sum_{w \in W_G} (sgn\ w) e^{w\delta}}.$$

Every discrete series is obtained this way, and two such are equivalent if and only if their parameters λ_0 are conjugate under W_K.

REFERENCES. For an exposition, see [**8**, pp. 310, 436, 454]. The original paper is Harish-Chandra [**7**].

REMARKS. For G compact, λ_0 is equal to the sum of the highest weight and δ when λ_0 is dominant; compare with λ_0 in Corollary 4. Also for G compact, (3.1) reduces to the Weyl Character Formula. For any G, if δ is not integral, the numerator and denominator of (3.1) are not separately well defined. But we can replace $e^{w\lambda_0}$ and $e^{w\delta}$ by $e^{w\lambda_0 - \delta}$ and $e^{w\delta - \delta}$, respectively, and then the numerator and denominator are well defined.

Harish-Chandra's proof of Theorem 7 does not give an explicit realization of each discrete series. Instead it produces a discrete series representation for each parameter by finding a subspace of $L^2(G)$ with suitable properties. Soon after Harish-Chandra's work became known, Kostant [**11**] and Langlands [**12**] independently conjectured generalizations of the Bott-Borel-Weil Theorem that would realize all discrete series. Over a period of years beginnning with his thesis, Schmid settled these conjectures. The particular conjecture by Langlands avoids some analytic problems by replacing $H^{0,k}(G/T, \mathbb{C}_\lambda)$ by the space $\mathcal{H}^{0,k}(G/T, \mathbb{C}_\lambda)$ of strongly harmonic square integrable forms. One analytic problem that is avoided in this way is whether the image of $\bar{\partial}$ is closed; another is how to incorporate square integrability into the hypothesis. The notation is

(3.2)

$$\begin{aligned}
&\text{rank } G = \text{rank } K \\
&T = \text{maximal torus in } K, \text{ hence maximal in } G \\
&\Delta = \{\text{roots of } (\mathfrak{g}, \mathfrak{t})\} \\
&\Delta^+ = \text{a positive system for } \Delta \\
&\delta = \tfrac{1}{2}\sum_{\alpha \in \Delta^+} \alpha \\
&B = \text{Borel subgroup built from } \textit{negative} \text{ roots.}
\end{aligned}$$

THEOREM 8. *Let $\lambda \in \mathfrak{t}^*$ be integral.*

(a) *If $\langle \lambda + \delta, \alpha \rangle = 0$ for some $\alpha \in \Delta$, then $\mathcal{H}^{0,k}(G/T, \mathbb{C}_\lambda) = 0$ for all k.*

(b) *If $\langle \lambda + \delta, \alpha \rangle \neq 0$ for all $\alpha \in \Delta$, let*

(3.3)
$$\begin{aligned}
q = \ &\#\{\alpha \in \Delta^+ \mid \alpha \text{ is compact and } \langle \lambda + \delta, \alpha \rangle < 0\} \\
&+ \#\{\alpha \in \Delta^+ \mid \alpha \text{ is noncompact and } \langle \lambda + \delta, \alpha \rangle > 0\}.
\end{aligned}$$

Then

$$\mathcal{H}^{0,k}(G/T, \mathbb{C}_\lambda) = \begin{cases} 0 & \textit{if } k \neq q \\ \pi_{\lambda+\delta} & \textit{if } k = q, \end{cases}$$

where $\pi_{\lambda+\delta}$ is the discrete series representation of G with Harish-Chandra parameter λ.

REFERENCE. Schmid [15].

The particular conjecture by Kostant works with the actual Dolbeault cohomology space $H^{0,k}(G/T, \mathbb{C}_\lambda)$. In this case one has to arrange that the nonzero cohomology appears in the highest possible degree. The result is then a generalization of Corollary 4.

Thus let $\lambda_0 \in \mathfrak{t}^*$ be a given nonsingular parameter, and suppose that $\lambda_0 - \delta_0$ is integral for the half sum of positive roots in some positive system. Take

(3.4)
$$\begin{aligned}
&\Delta^+ = \{\alpha \in \Delta \mid \langle \lambda_0, \alpha \rangle > 0\} \\
&\delta = \tfrac{1}{2}\sum_{\alpha \in \Delta^+} \alpha \\
&B \text{ built from } \Delta^+.
\end{aligned}$$

THEOREM 9. *Let $\lambda_0 \in \mathfrak{t}^*$ be nonsingular with $\lambda_0 - \delta_0$ integral. With Δ^+, δ, and B defined as in (3.4),*

(a) *$H^{0,k}(G/T, \mathbb{C}_{\lambda_0+\delta}) = 0$ if $k \neq \dim_{\mathbb{C}}(K/T)$;*

(b) *$H^{0,k}(G/T, \mathbb{C}_{\lambda_0+\delta})$ is a Frechet space if $k = \dim_{\mathbb{C}}(K/T)$ (i.e., $\bar{\partial}$ has closed image in $C^{0,k}(G/T, \mathbb{C}_{\lambda_0+\delta})$), and the underlying space of K finite vectors of $H^{0,k}(G/T, \mathbb{C}_{\lambda_0+\delta})$ is equivalent with the space of K finite vectors of $\pi_{\lambda_0+\delta}$.*

REFRENCES. Schmid proved this result in [13], except for the identification of $\pi_{\lambda_0+\delta}$, under the additional assumption that λ_0 is very nonsingular. He gave the identification with $\pi_{\lambda_0+\delta}$ in [14], with the same additional assumption. Aguilar-Rodriguez [1] extended the theorem to the form stated here.

4. Schmid's Penrose transform

In his 1967 thesis, Schmid [13] introduced an operator for passing from the top-degree cohomology space in Theorem 9 to the space of sections of a complex vector bundle over G/K, and Wells and Wolf [20] developed the operator further. In fact, the operator readily generalizes to the setting in §1.1, and there it played an important role in [2]. When G/K is complex, the generalized operator reduces to the G equivariant Penrose transform as described in Baston-Eastwood [3].

The setting for this section will be like the one in Corollary 4 or Theorem 9, except that we allow a general L, possibly noncompact, in place of T. That is, we shall in effect adapt the positive system Δ^+ to our given parameter so that the cohomology of interest occurs in the maximum possible degree.

In this exposition we shall assume that rank G = rank K in order to keep matters simple. The notation is

$$
\begin{aligned}
&G = \text{linear connected reductive Lie group} \\
&\text{rank } G = \text{rank } K \\
&T = \text{a torus in } G \\
(4.1) \quad &L = Z_G(T) \\
&T \text{ extended to a maximal torus } \tilde{T} \text{ in } L \\
&\Delta = \{\text{roots of } (\mathfrak{g}, \tilde{\mathfrak{t}})\} \\
&\Delta(\mathfrak{l}) = \{\text{roots of } (\mathfrak{l}, \tilde{\mathfrak{t}})\} \subseteq \Delta
\end{aligned}
$$

Instead of fixing the parameter of a representation of L and then introducing Δ^+, we shall fix Δ^+ and say what parameters are allowed. Thus we use the following additional notation:

$$
\begin{aligned}
&\Delta^+ \text{ chosen with } \Delta(\mathfrak{l}) \text{ generated by simple roots} \\
&Q \text{ built from } \mathfrak{l} \text{ and } positive \text{ roots} \\
&\mathfrak{q} = \mathfrak{l} \oplus \mathfrak{u} \\
(4.2) \quad &\Delta(\mathfrak{u}) = \{\text{roots contributing to } \mathfrak{u}\} \\
&\Delta(\mathfrak{u} \cap \mathfrak{p}) = \{\text{noncompact roots contributing to } \mathfrak{u}\} \\
&\delta(\mathfrak{u}) = \tfrac{1}{2}\sum_{\alpha \in \Delta(\mathfrak{u})} \alpha \\
&\delta(\mathfrak{u} \cap \mathfrak{p}) = \tfrac{1}{2}\sum_{\alpha \in \Delta(\mathfrak{u}\cap\mathfrak{p})} \alpha.
\end{aligned}
$$

We work with an integral parameter $\lambda \in \tilde{\mathfrak{t}}^*$ that satisfies

$$
\begin{aligned}
(4.3) \quad &\langle \lambda, \alpha \rangle \geq 0 \qquad \text{for all } \alpha \in \Delta^+ \\
&\langle \lambda, \alpha \rangle = 0 \qquad \text{for all } \alpha \in \Delta(\mathfrak{l}).
\end{aligned}
$$

Then λ is the (unique) weight of a one-dimensional representation ξ_λ of L, and we write $\mathbb{C}_\lambda$ for the action of ξ_λ on $\mathbb{C}$. For such a parameter λ, the degree of interest for cohomology is

$$s = \dim_{\mathbb{C}}(K/(L \cap K)).$$

The complex manifold $G/L \subseteq G^{\mathbb{C}}/Q$ has s-dimensional compact complex submanifolds, namely $\{g \cdot K/(L \cap K)\}$, and the operator of interest will come from a kind of integration of $(0,s)$ forms over these submanifolds.

First we formulate the operator abstractly. It will help to identify sections γ and functions φ_γ under the isomorphism (1.7). Using the Bott-Borel-Weil Theorem, we have an isomorphism

$$(4.4) \qquad H^{0,s}(K/(K \cap L), \mathbb{C}_{\lambda+2\delta(\mathfrak{u})}) \xrightarrow{\sim} F^{\lambda+2\delta(\mathfrak{u}\cap\mathfrak{p})},$$

the object on the right being a K representation with the indicated highest weight. (This instance of the theorem is a specialization of Theorem 5 in the same way that Corollary 4 is a specialization of Theorem 3.) Noting that the space of top-degree forms $C^{0,s}(K/(K \cap L), \cdot)$ consists only of cocycles, let

$$(4.5) \qquad P : C^{0,s}(K/(K \cap L), \mathbb{C}_{\lambda+2\delta(\mathfrak{u})}) \to F^{\lambda+2\delta(\mathfrak{u}\cap\mathfrak{p})}$$

be the map implementing (4.4). Let $R : K/(L \cap K) \to G/L$ be inclusion, let R^* be the pullback to $(0,s)$ forms, and let

$$(4.6a) \qquad \mathcal{P} : C^{0,s}(G/L, \mathbb{C}_{\lambda+2\delta(\mathfrak{u})}) \to \mathcal{E}(G \times_K F^{\lambda+2\delta(\mathfrak{u}\cap\mathfrak{p})})$$

be given by

$$(4.6b) \qquad \mathcal{P}f(x) = P(R^*(f(x\cdot))).$$

More concretely let $\{\phi_i\}$ be an orthonormal basis of $F^{\lambda+2\delta(\mathfrak{u}\cap\mathfrak{p})}$, and let ϕ be a nonzero highest weight vector. Fix a nonzero $\omega_0 \in \wedge^s(\overline{\mathfrak{u}} \cap \mathfrak{k})^*$, and let

$$\varphi_i = (F^{\lambda+2\delta(\mathfrak{u}\cap\mathfrak{p})}(k)\phi, \phi_i)\omega_0.$$

This is in $C^{s,0}(K/(L \cap K), \mathbb{C}_{-\lambda-2\delta(\mathfrak{u})})$, and its product with a member h of $C^{0,s}(K/(L \cap K), \mathbb{C}_{\lambda+2\delta(\mathfrak{u})})$ is a volume form on $K/(L \cap K)$. For such an h we can therefore define

$$P(h) = \sum_i \left(\int_{K/(L \cap K)} h\varphi_i \right)\phi_i.$$

This version of P coincides with the one in (4.5), and then $\mathcal{P}$ is defined in terms of P by (4.6).

PROPOSITION 10. *The operator $\mathcal{P}$ defined by (4.6) descends to a well defined operator*

$$\mathcal{P} : H^{0,s}(G/L, \mathbb{C}_{\lambda+2\delta(\mathfrak{u})}) \to \mathcal{E}(G \times_K F^{\lambda+2\delta(\mathfrak{u}\cap\mathfrak{p})}).$$

REFERENCES. Schmid [13], Wells-Wolf [20], and Barchini-Knapp-Zierau [2, §10].

5. Zuckerman functors

Zuckerman functors provide an algebraic analog of the analytic construction in §1. They were introduced by Zuckerman [21] in a series of lectures and were developed further by Vogan [18]. Their full theory requires relating them to ring theory, and this step was carried out initially in [10]; for more detail, see [9]. For this section we use the following notation:

(5.1)
$$G = \text{linear connected reductive Lie group}$$
$$K = \text{a maximal compact subgroup}$$
$$T = \text{a torus in } G$$
$$L = Z_G(T)$$
$$Q = \text{parabolic subgroup in } G^{\mathbb{C}} \text{ as in §1}$$
$$\mathfrak{q} = \mathfrak{l} \oplus \mathfrak{u}$$
$$(\sigma, V) = \text{smooth representation of } L.$$

The space V can be infinite-dimensional, but the reader may wish to regard it as finite-dimensional for purposes of motivation. The representation (σ, V) gives us a representation of $\mathfrak{l}$, and we extend this to a representation of $\mathfrak{q}$ by making $\mathfrak{u}$ act as 0. It will be helpful for purposes of motivation to think of the representation of $\mathfrak{q}$ on V as coming from a holomorphic representation of Q on V, but this assumption can be avoided.

In the analytic setting, $\bar{\partial}$ is an operator

(5.2) $$\bar{\partial} : \mathcal{E}(G \times_L ((\wedge^m \mathfrak{u})^* \otimes V)) \to \mathcal{E}(G \times_L ((\wedge^{m+1} \mathfrak{u})^* \otimes V)).$$

Using the isomorphism (1.7), we regard $\bar{\partial}$ as an operator with domain the space of smooth functions φ from G into $(\wedge^m \mathfrak{u})^* \otimes V$ satisfying

(5.3) $$\varphi(gl) = (\text{Ad}(l)^{-1} \otimes \sigma(l)^{-1})\varphi(g) \qquad \text{for } g \in G, l \in L$$

and with range the corresponding space of functions into $(\wedge^{m+1} \mathfrak{u})^* \otimes V$.

In the algebraic analog we try to construct only the K finite vectors of $H^{0,m}$, thus obtaining a $(\mathfrak{g}, K)$ module. (Recall that a $(\mathfrak{g}, K)$ **module** consists of compatible representations of $\mathfrak{g}$ and K on the same vector space with every vector K finite. Let $\mathcal{C}(\mathfrak{g}, K)$ be the category of all $(\mathfrak{g}, K)$ modules.)

The idea is to work with the Taylor coefficients at $g = 1$ of the function φ in (5.3), regarding each coefficient as attached to a left-invariant complex derivative (of some order) of φ at $g = 1$. Thus the idea of passing to Taylor coefficients gives us a linear map

$$\varphi \mapsto \varphi^{\#} \in \text{Hom}_{\mathbb{C}}(U(\mathfrak{g}), (\wedge^m \mathfrak{u})^* \otimes V).$$

The transformation law (5.3) forces

(5.4) $$\varphi^{\#} \in \text{Hom}_{\mathfrak{l}}(U(\mathfrak{g}), (\wedge^m \mathfrak{u})^* \otimes V),$$

where $\mathfrak{l}$ acts on $U(\mathfrak{g})$ on the right. If we assume that φ is K finite, then the action of $L \cap K$ on the left of φ gives an action of $L \cap K$ on $\varphi^{\#}$ by $\mathrm{Hom}(\mathrm{Ad}, \mathrm{Ad}^* \otimes \sigma)$, and $\varphi^{\#}$ will be $L \cap K$ finite. Thus $\varphi^{\#}$ lies in a subspace that we denote

$$(5.5) \qquad \mathrm{Hom}_{\mathfrak{l}}(U(\mathfrak{g}), (\wedge^m \mathfrak{u})^* \otimes V)_{L \cap K}$$

to indicate the $L \cap K$ finiteness. On (5.5) we have a representation of $\mathfrak{g}$ (via the action of $U(\mathfrak{g})$ on the left) and the representation of $L \cap K$, and (5.5) is a $(\mathfrak{g}, L \cap K)$ module.

The passage from the space of φ's as in (5.3) to the space of $\varphi^{\#}$'s in (5.5) loses information because

(a) formal power series do not have to converge and
(b) convergent power series do not have to globalize.

The modification that gets around the difficulties in (a) and (b) is to define away the problem. Let Γ be the functor

$$\Gamma : \mathcal{C}(\mathfrak{g}, L \cap K) \to \mathcal{C}(\mathfrak{g}, K)$$

given by

$$\Gamma(V) = \text{sum of all } \mathfrak{k} \text{ invariant subspaces of } V \text{ for which}$$
$$\text{the action of } \mathfrak{k} \text{ globalizes to } K,$$
$$\Gamma(\psi) = \psi|_{\Gamma(V)} \qquad \text{if } \psi \in \mathrm{Hom}(V, W).$$

The functor Γ is covariant and left exact and is called the **Zuckerman functor**.

IDEA #1. *Impose $\bar{\partial}$ between spaces*

$$(5.6) \qquad \Gamma(\mathrm{Hom}_{\mathfrak{l}}(U(\mathfrak{g}), (\wedge^m \mathfrak{u})^* \otimes V)_{L \cap K}),$$

and take the kernel/image as a $(\mathfrak{g}, K)$ module analog of $H^{0,m}(G/L, V)$.

Now we bring in homological algebra. We *assume temporarily* that $L \subseteq K$. Then we make the following observations:

1) For the case $m = 0$ at least when V is finite-dimensional, the condition that $\varphi^{\#}$ come from a section γ with $\bar{\partial}\gamma = 0$ is that γ be holomorphic, hence that $Z\varphi = 0$ for all $Z \in \mathfrak{u}$, by (1.12). Thus the kernel/image space for $m = 0$ should be regarded as

$$(5.7) \qquad \Gamma(\mathrm{Hom}_{\mathfrak{q}}(U(\mathfrak{g}), V)_{L \cap K}).$$

2) Identification of (5.7) as the space of interest for $m = 0$ suggests looking at the sequence

$$(5.8) \qquad 0 \longrightarrow \mathrm{Hom}_{\mathfrak{q}}(U(\mathfrak{g}), V)_{L \cap K} \longrightarrow \mathrm{Hom}_{\mathfrak{l}}(U(\mathfrak{g}), (\wedge^0 \mathfrak{u})^* \otimes V)_{L \cap K}$$
$$\longrightarrow \mathrm{Hom}_{\mathfrak{l}}(U(\mathfrak{g}), (\wedge^1 \mathfrak{u})^* \otimes V)_{L \cap K} \longrightarrow \cdots$$

in the category $\mathcal{C}(\mathfrak{g}, L \cap K)$. In fact, (5.8) is an injective resolution of $\mathrm{Hom}_{\mathfrak{q}}(U(\mathfrak{g}), V)_{L \cap K}$ in the category $\mathcal{C}(\mathfrak{g}, L \cap K)$. The maps will be made explicit before Theorem 11 below, and a proof that (5.8) is an injective resolution will be given in that theorem.

3) The category $\mathcal{C}(\mathfrak{g}, L \cap K)$ has enough injectives. Combining (2) and Idea #1, we see that the mth space of interest, namely the mth kernel/image of (5.6), is

$$(5.9) \qquad\qquad \Gamma^m(\mathrm{Hom}_{\mathfrak{q}}(U(\mathfrak{g}), V)_{L \cap K}),$$

where Γ^m is the mth right derived functor of Γ. (In fact, (5.9) is defined as the mth cohomology of the complex obtained by applying Γ to (5.8), since (5.8) is an injective resolution.)

4) The space (5.9) gives the underlying $(\mathfrak{g}, K)$ module of K finite vectors of $H^{0,m}(G/L, V)$ for the cases of compact groups and the discrete series. These results are due essentially to Zuckerman [21] and are proved in Vogan [18]. See also [9].

These observations lead us to the second crucial idea.

IDEA #2. *Even when L is not compact*, define *the mth space of interest to be* (5.9).

In short, the Zuckerman construction is to pass from V in $\mathcal{C}(\mathfrak{l}, L \cap K)$ first to $\mathrm{Hom}_{\mathfrak{q}}(U(\mathfrak{g}), V)_{L \cap K}$ in $\mathcal{C}(\mathfrak{g}, L \cap K)$ and then to $\Gamma^m(\mathrm{Hom}_{\mathfrak{q}}(U(\mathfrak{g}), V)_{L \cap K})$ in $\mathcal{C}(\mathfrak{g}, K)$.

Finally let us return to the details of observation (2) above. First we need a formula for the differential in (5.8). From §1.4 we have $\bar{\partial}_V = \bar{\partial} \otimes 1$, and thus it is enough to understand $\bar{\partial}$ for $V = \mathbb{C}$. A formula for $\bar{\partial}$ on a function φ as in (5.3) (but with $V = \mathbb{C}$) is given in Griffiths-Schmid [6, (1.6)]. It works with an expansion

$$\varphi(g) = \sum_{i_1 < \cdots < i_m} f_{i_1 \cdots i_m}(g)\, \omega_{i_1} \wedge \cdots \wedge \omega_{i_m},$$

where the ω_i are left-invariant complex 1-forms obtained from a dual basis to a basis $\{Y_i\}$ of $\mathfrak{u}$. Let us regard the corresponding $\varphi^{\#}$ as in $\mathrm{Hom}_{\mathfrak{l}}(U(\mathfrak{g}) \otimes \wedge^m \mathfrak{u}, \mathbb{C})$. If $Y_{j_1}, \ldots, Y_{j_m}$ have increasing indices, then

$$\varphi^{\#}(u \otimes Y_{j_1} \wedge \cdots \wedge Y_{j_m}) = u f_{j_1 \cdots j_m}(1),$$

where $\mathfrak{u}$ acts by left-invariant complex differentiation. A little computation with the formula for $\bar{\partial}$ in [6] shows that the element $\bar{\partial}\varphi$ corresponds to the element

$(\bar{\partial}\varphi)^{\#}$ given on elements of $U(\mathfrak{g})$ and monomials in $\wedge^{m+1}\mathfrak{u}$ by

$$(\bar{\partial}\varphi)^{\#}(u \otimes X_1 \wedge \cdots \wedge X_{m+1})$$

$$= \sum_{i=1}^{m+1} (-1)^{i+1}\varphi^{\#}(uX_i \otimes X_1 \wedge \cdots \wedge \widehat{X}_i \wedge \cdots \wedge X_{m+1})$$

$$+ \sum_{r<s} (-1)^{r+s}\varphi^{\#}(u \otimes [X_r, X_s] \wedge X_1 \wedge \cdots \wedge \widehat{X}_r \wedge \cdots \wedge \widehat{X}_s \wedge \cdots \wedge X_{m+1}).$$

(5.10)

The formula for general V looks the same; the only difference is that the terms in (5.10) have values in V.

Formula (5.10) defines the maps in (5.8) except for the map going out from $\mathrm{Hom}_{\mathfrak{q}}(U(\mathfrak{g}), V)_{L\cap K}$, which we take to be the obvious inclusion.

THEOREM 11. *Under the assumption that $L \subseteq K$, (5.8) is an injective resolution of $\mathrm{Hom}_{\mathfrak{q}}(U(\mathfrak{g}), V)_{L\cap K}$.*

PROOF. In the notation of [9], we have

(5.11) $$\mathrm{Hom}_{\mathfrak{l}}(U(\mathfrak{g}), (\wedge^m\mathfrak{u})^* \otimes V)_{L\cap K} = I_{\mathfrak{l},L\cap K}^{\mathfrak{g},L\cap K}((\wedge^m\mathfrak{u})^* \otimes V).$$

Since L is compact, $(\wedge^m\mathfrak{u})^* \otimes V$ is injective in $\mathcal{C}(\mathfrak{l}, L \cap K)$. The functor $I_{\mathfrak{l},L\cap K}^{\mathfrak{g},L\cap K}$ carries injectives to injectives, and therefore the members of (5.8) after $\mathrm{Hom}_{\mathfrak{q}}(U(\mathfrak{g}), V)_{L\cap K}$ are injective in $\mathcal{C}(\mathfrak{g}, L \cap K)$.

For exactness we note that exactness at $\mathrm{Hom}_{\mathfrak{q}}(U(\mathfrak{g}), V)_{L\cap K}$ is trivial since the map out is defined as an inclusion. For exactness at

(5.12) $$\mathrm{Hom}_{\mathfrak{l}}(U(\mathfrak{g}), (\wedge^0\mathfrak{u})^* \otimes V)_{L\cap K} \cong \mathrm{Hom}_{\mathfrak{l}}(U(\mathfrak{g}), V)_{L\cap K},$$

we see from (5.10) that the map out has

$$(\bar{\partial}\varphi)^{\#}(u \otimes X) = \varphi^{\#}(uX) \qquad \text{for } u \in U(\mathfrak{g}), X \in \mathfrak{u}.$$

This is 0 for all u and X if and only if $\varphi^{\#}$ respects the $\mathfrak{u}$ action, hence is in $\mathrm{Hom}_{\mathfrak{q}}(U(\mathfrak{g}), V)_{L\cap K}$. Hence we have exactness at (5.12).

For exactness at the other members of (5.8), we rewrite the right side of (5.11) as

$$\cong I_{\mathfrak{q},L\cap K}^{\mathfrak{g},L\cap K}(I_{\mathfrak{l},L\cap K}^{\mathfrak{q},L\cap K}((\wedge^m\mathfrak{u})^* \otimes V))$$

$$\cong I_{\mathfrak{q},L\cap K}^{\mathfrak{g},L\cap K}(\mathrm{Hom}_{\mathfrak{l}}(U(\mathfrak{q}), (\wedge^m\mathfrak{u})^* \otimes V)_{L\cap K}).$$

Now $I_{\mathfrak{q},L\cap K}^{\mathfrak{g},L\cap K}$ is exact, and hence it is enough to prove that

$$\cdots \longrightarrow \mathrm{Hom}_{\mathfrak{l}}(U(\mathfrak{q}), (\wedge^m\mathfrak{u})^* \otimes V)_{L\cap K} \longrightarrow \cdots$$

is exact if the differentials are given as in (5.10). Since passage to $L \cap K$ finite vectors is exact in such a Hom, it is enough to prove that

(5.13) $$\cdots \longrightarrow \mathrm{Hom}_{\mathfrak{l}}(U(\mathfrak{q}), (\wedge^m\mathfrak{u})^* \otimes V) \longrightarrow \cdots$$

is exact. But

$$\begin{aligned}
\operatorname{Hom}_{\mathfrak{l}}(U(\mathfrak{q}), (\wedge^m \mathfrak{u})^* \otimes V) &\cong \operatorname{Hom}_{\mathfrak{l}}(U(\mathfrak{u}) \otimes U(\mathfrak{l}), (\wedge^m \mathfrak{u})^* \otimes V) \\
&\cong \operatorname{Hom}_{\mathfrak{l}}(U(\mathfrak{l}), \operatorname{Hom}_{\mathbb{C}}(U(\mathfrak{u}), (\wedge^m \mathfrak{u})^* \otimes V)) \\
&\cong \operatorname{Hom}_{\mathbb{C}}(U(\mathfrak{u}), (\wedge^m \mathfrak{u})^* \otimes V) \\
&\cong \operatorname{Hom}_{\mathbb{C}}(U(\mathfrak{u}) \otimes \wedge^m \mathfrak{u}, V) \\
&= \operatorname{Hom}_{\mathbb{C}}(X_m, V),
\end{aligned}$$

where X_m is the Koszul (projective) resolution of $\mathbb{C}$ in $\mathcal{C}(\mathfrak{u}, 1)$. (See [9, Theorem 4.6].) It is easy to check that the differentials for (5.13) are the ones induced from the differentials for X_m, and hence (5.13) is exact. This completes the proof.

References

1. R. Aguilar-Rodriguez, *Connections between representations of Lie groups and sheaf cohomology*, Ph.D. dissertation, Harvard University, 1987.
2. L. Barchini, A. W. Knapp, and R. Zierau, *Intertwining operators into Dolbeault cohomology representations*, J. Func. Anal. **107** (1992), 302–341.
3. R. J. Baston and M. G. Eastwood, *The Penrose Transform: Its Interaction with Representation Theory*, Oxford University Press, Oxford, 1989.
4. R. Bott, *Homogeneous vector bundles*, Ann. of Math. **66** (1957), 203–248.
5. R. Godement, *Sur les relations d'orthogonalité de V. Bargmann*, I and II, C. R. Acad. Sci. Paris **225** (1947), 521–523 and 657–659.
6. P. Griffiths and W. Schmid, *Locally homogeneous complex manifolds*, Acta Math. **123** (1969), 253–302.
7. Harish-Chandra, *Discrete series for semisimple Lie groups* II, Acta Math. **116** (1966), 1–111.
8. A. W. Knapp, *Representation Theory of Semisimple Groups: An Overview Based on Examples*, Princeton University Press, Princeton, 1986.
9. ———, *Lie Groups, Lie Algebras, and Cohomology*, Princeton University Press, Princeton, 1988.
10. A. W. Knapp and D. A. Vogan, *Duality theorems in relative Lie algebra cohomology*, duplicated notes, Cornell University and Massachusetts Institute of Technology, 1986.
11. B. Kostant, *Orbits, symplectic structures, and representation theory*, Proceedings of the U.S.-Japan Seminar on Differential Geometry, Kyoto, 1965.
12. R. P. Langlands, *Dimension of spaces of automorphic forms*, Algebraic Groups and Discontinuous Subgroups, Proc. Symp. in Pure Math. 9, American Mathematical Society, Providence, 1966, pp. 253–257.
13. W. Schmid, *Homogeneous complex manifolds and representations of semisimple Lie groups*, Ph.D. dissertation, University of California, Berkeley, 1967, Representation Theory and Harmonic Analysis on Semisimple Lie Groups, Math. Surveys and Monographs, American Mathematical Society, Providence, 1989, pp. 223–286.
14. ———, *On the realization of the discrete series of a semisimple Lie group*, Rice University Studies, Vol. 56, No. 2, 1970, pp. 99–108.
15. ———, *L^2-cohomology and the discrete series*, Ann. of Math. **103** (1976), 375–394.
16. N. Steenrod, *The Topology of Fibre Bundles*, Princeton University Press, Princeton, 1951.
17. J. A. Tirao and J. A. Wolf, *Homogeneous holomorphic vector bundles*, Indiana U. Math. J. **20** (1970), 15–31.
18. D. A. Vogan, *Representations of Real Reductive Lie Groups*, Birkhäuser, Boston, 1981.
19. R. O. Wells, *Differential Analysis on Complex Manifolds*, Springer-Verlag, New York, 1980.
20. R. O. Wells and J. A. Wolf, *Poincaré series and automorphic cohomology on flag domains*, Ann. of Math. **105** (1977), 397–448.

21. G. J. Zuckerman, *Construction of representations via derived functors*, lectures, Institute for Advanced Study, 1978.

DEPARTMENT OF MATHEMATICS, STATE UNIVERSITY OF NEW YORK, STONY BROOK, NEW YORK 11794-3651, USA

E-mail address: aknapp@ccmail.sunysb.edu

Contemporary Mathematics
Volume **154**, 1993

Admissible Representations and Geometry of Flag Manifolds

JOSEPH A. WOLF

ABSTRACT. We describe geometric realizations for various classes of admissible representations of reductive Lie groups. The representations occur on partially holomorphic cohomology spaces corresponding to partially holomorphic homogeneous vector bundles over real group orbits in complex flag manifolds. The representations in question include standard tempered and limits of standard tempered representations, and representations induced from finite dimensional representations of real parabolic subgroups.

Section 1. Introduction.

Harmonic analysis on Lie groups and their homogeneous spaces has been guided and influenced by various geometric constructions of unitary representations. Those unitary representations are the building blocks for the extensions of classical Fourier analysis relevant to the analytic problems in question. Here I'll try to indicate some aspects of the background, concentrating on the interplay between geometry and analysis, I'll indicate some extensions that now seem worth writing down, and I'll mention some interesting open problems.

The best known geometric realization of group representations is the Bott–Borel–Weil Theorem from the 1950's ([**2**], [**19**]). If G is a compact connected Lie group and T is a maximal torus, then a choice $\Phi^+ = \Phi^+(\mathfrak{g}, \mathfrak{t})$ of positive root system defines a G-invariant complex manifold structure on G/T by: $\sum_{\alpha \in \Phi^+} \mathfrak{g}_\alpha$ represents the holomorphic tangent space. Now fix that structure and let $\lambda \in i\mathfrak{t}_0^*$ be integral, that is, e^λ is a well defined character of T. View e^λ as a representation of T on a 1-dimensional vector space E_λ and let $\mathbb{E}_\lambda \to G/T$ denote the associated homogeneous holomorphic hermitian line bundle. We write

1991 *Mathematics Subject Classification.* Primary 22E46; Secondary 14F05, 14F17, 14F25, 22E70, 32A37, 32C36, 32H20.

Research partially supported by NSF Grant DMS 91 00578.

This paper is in final form and no version of it will be submitted for publication elsewhere.

$\mathcal{O}(\mathbb{E}_\lambda) \to G/T$ for the sheaf of germs of holomorphic sections of $\mathbb{E}_\lambda \to G/T$. The group G acts on everything here, including the cohomologies $H^q(G/T; \mathcal{O}(\mathbb{E}_\lambda))$. Let $\rho = \frac{1}{2}\sum_{\alpha \in \Phi^+} \alpha$. The Bott–Borel–Weil Theorem says

THEOREM. *If $\lambda + \rho$ is singular then every $H^q(G/T; \mathcal{O}(\mathbb{E}_\lambda)) = 0$. Now suppose that $\lambda + \rho$ is regular, let w denote the unique Weyl group element such that $\langle w(\lambda + \rho), \alpha \rangle > 0$ for all $\alpha \in \Phi^+$, and let $\ell(w)$ denote its length as a word in the simple root reflections. Then* (i) *$H^q(G/T; \mathcal{O}(\mathbb{E}_\lambda)) = 0$ for $q \neq \ell(w)$, and* (ii) *G acts irreducibly on $H^{\ell(w)}(G/T; \mathcal{O}(\mathbb{E}_\lambda))$ by the representation with highest weight $w(\lambda + \rho) - \rho$.*

In the Bott–Borel–Weil Theorem, $\ell(w)$ can be described as the number of positive roots that w carries to negative roots, the representation of G with highest weight $w(\lambda + \rho) - \rho$ can be described as the discrete series representation with Harish–Chandra parameter $w(\lambda + \rho)$, and, by Kodaira–Hodge Theory, $H^q(G/T; \mathcal{O}(\mathbb{E}_\lambda))$ is naturally G-isomorphic to the space of harmonic differential forms of bidegree $(0, q)$ on G/T with values in $\mathbb{E}_\lambda$.

In 1965 Kostant and Langlands independently conjectured an analog of the Bott–Borel–Weil Theorem for connected noncompact semisimple Lie groups with finite center. The conjecture was proved in the 1970's in two stages by Schmid ([23], [25]), and I extended the result (also in the 70's) to general semisimple Lie groups [31]. The representations in question there are the *discrete series* representations of G. They are the fundamental building blocks for the *tempered representations* of G, which in turn are the representations that enter into the Plancherel formula for G. My structure theory for the geometry of real group orbits $G(z) \subset Z \cong G_{\mathbb{C}}/P$ on complex flag manifolds from the late 1960's [30] also led [31] to corresponding geometric realizations for all standard tempered representations of general semisimple Lie groups G. This followed a line of attack that in retrospect was modelled on the Kostant–Kirillov–Souriau theory of geometric quantization.

In the context of semisimple Lie groups, the theory of geometric quantization seemed to founder on several seemingly intractable technical problems. Typically these involved questions of closed range or of vanishing for cohomology except in a particular degree, especially in regard to representations whose infinitesimal character was singular or even just not very nonsingular.

By the middle 1970's many mathematicians began to look for alternatives to or variations on standard geometric quantization. Methods involving varying polarizations or structure group liftings had specialized success, and the derived functor modules of Vogan and Zuckerman [29] took a central position in the representation theory of semisimple Lie Groups G. Those derived functor modules are simultaneously modules for the Lie algebra $\mathfrak{g}_0$ of G and for the maximal compact subgroup K of G, but are not G–modules. The passage from $(\mathfrak{g}_0, K)$–modules to G–modules, called *globalization*, was understood by Schmid [26] in the middle 1980's in a form that turned out to be suitable for geometric quantization. Schmid and I [27] used exactness of the maximal globalization functor of [26] to make a change of polarization argument, starting with the tempered case, which had become the case of a maximally real polarization. In the setting of real group orbits on the flag manifold $X \cong G_{\mathbb{C}}/B$ of Borel subalgebras of $\mathfrak{g}$, this gave us the connection between hyperfunction quantization and the derived functor modules of Vogan and Zuckerman. That settled the technical problems, mentioned above, for geometric quantization on $X \cong G_{\mathbb{C}}/B$, and at the same time identified the resulting representations.

Of course much of the geometric interest in this requires more general flag manifolds than the flag of Borel subalgebras of $\mathfrak{g}$. There is some work on pushing the hyperfunction quantization method down from the flag $X \cong G_{\mathbb{C}}/B$ to a more general flag manifold $W \cong G_{\mathbb{C}}/P$. This was done in [31] for standard tempered representations and certain well behaved (measurable integrable – defined in §2) G–orbits, in [33] for the realizations of discrete series representations that are "closest" to the realization of holomorphic discrete series as spaces of holomorphic sections of vector bundles, and in [34] for finite rank bundles over measurable open orbits. In this paper we show how those results all fit into a common framework.

I am not going to discuss localization methods here, but rather just indicate some of the work in that area. It starts, of course with the seminal work [1] of Beilinson and Bernstein. There are many unpublished results of Bernstein and Miličić, or at least I have this impression from Miličić. There is some work ([9],[10]) of Hecht, Miličić, Schmid and myself in which we draw the connection between $\mathcal{D}$–module realizations of representations and realizations by Zuckerman derived functor modules, and draw consequences for completeness, vanishing and irreducibility. There are papers of Hecht and Taylor ([11], [12], [13]) where a minimal–globalization form of localization is developed and applied to $\mathfrak{n}$–homology and an elegant geometric character formula. Finally, there is work of Kashiwara, Schmid, Vilonen and many others which would take too much space to catalog.

In Section 2 we specify our class of real Lie groups, and we recall the basic facts [30] concerning real group orbits on complex flag manifolds. We concentrate on the type of orbit that comes into the geometric constructions of representations. Those orbits are the measurable open orbits for the Dolbeault cohomology realization of representations such as those of the discrete series, measurable integrable orbits for the partially holomorphic cohomology realization of representations such as those of the various tempered series.

In Section 3 we recall the solution ([25] for connected semisimple groups of finite center, [31] for more general groups) to the Kostant–Langlands Conjecture. This realizes relative discrete series representations on spaces of square integrable harmonic bundle–valued forms.

In Section 4 we recall the corresponding result [31] for the various series of standard tempered representations. They are realized on spaces of square integrable partially harmonic forms with values in a partially holomorphic vector bundle. Those representations are the ones that enter into the Plancherel formula.

In Section 5 we show how one obtains (partial) Dolbeault coholomogy realizations of standard tempered representations on partially holomorphic negative vector bundles. This material was essentially known ([**22**], [**31**], [**27**]) and we just put it together.

In Section 6 we combine methods of [**56**], [**64**] and [**65**] to describe holomorphic cohomology realizations over measurable open orbits for several classes of representations, not necessarily tempered. In Section 7 we apply the results of §6 to the holomorphic arc components of a measurable integrable orbit, obtaining corresponding partially holomorphic cohomology realizations. Finally, in Section 8, we list some important open questions for this circle of ideas.

Section 2. Real Group Orbits on Complex Flag Manifolds.

We recall some of the main points of [**30**] that we need to describe our geometric constructions of representations, specifically the constructions in [**31**] and some new constructions. If G is a real Lie group then $\mathfrak{g}_0$ denotes its real Lie algebra, $\mathfrak{g}$ is the complexification of $\mathfrak{g}_0$, G^0 is the topological component of the identity of G, and if $A, B \subset G$ are subgroups then $Z_A(B)$ denotes the centralizer of B in A and Z_B denotes the center $Z_B(B)$ of B. We say that G is *reductive* if $\mathfrak{g}_0$ is direct sum of a semisimple ideal and a commutative ideal. If G is reductive then $Int(\mathfrak{g})$ denotes the group of inner automorphisms of $\mathfrak{g}$, group generated by the $\exp(ad(\xi))$ for $\xi \in \mathfrak{g}$. It is the complexification of the adjoint group $Ad(G^0)$.

Here we work with the class of *general semisimple Lie groups*, consisting of all real reductive Lie groups G such that

$$\text{if } g \in G \text{ then } Ad(g) \text{ is an inner automorphism of } \mathfrak{g}$$

(2.1) G has a closed normal abelian subgroup Z such

$$\text{that } Z \subset Z_G(G^0) \text{ and } |G/ZG^0| < \infty.$$

The *Harish–Chandra class* of reductive Lie groups is the case where G/G^0 is finite and $[G^0, G^0]$ has finite center. As in the case of Harish–Chandra class groups, the first part of (2.1) says that irreducible admissible representations of G have well defined infinitesimal characters. The second part says that irreducible admissible representations of G more or less have central characters – that if π is any such representation then the restriction of π to the commutative group ZZ_{G^0} is a sum that involves at most $|G/ZG^0| < \infty$ distinct quasicharacters.

The first condition of (2.1) also says that G has a well defined natural action on all complex flag manifolds for $\mathfrak{g}$. Specifically, if $\mathfrak{q}$ is any parabolic subalgebra of $\mathfrak{g}$ and if $g \in G$ it says that $Ad(g)\mathfrak{q}$ is $Int(\mathfrak{g})$–conjugate to $\mathfrak{q}$. Now G acts on the flag manifold W consisting of all $Int(\mathfrak{g})$–conjugates of $\mathfrak{q}$ by $g : Ad(g_1)\mathfrak{q} \mapsto Ad(gg_1)\mathfrak{q}$. We identify W with the compact complex manifold $Int(\mathfrak{g})/Q$ where Q is the parabolic subgroup of $Int(\mathfrak{g})$ that is the normalizer of $\mathfrak{q}$.

The subgroup $Z_G(G^0) \subset G$ acts trivially on the complex flag manifold W, so for purposes of G–orbit structure we may replace G by the group $\overline{G} = G/Z_G(G^0)$ of Harish–Chandra class. In the remainder of §2 we make that replacement, but we will have to make the distinction in §§3 and 4.

With the replacement just described, $G \subset G_{\mathbb{C}} = Int(\mathfrak{g})$ and $W = G_{\mathbb{C}}/Q$ with G acting as a subgroup of $G_{\mathbb{C}}$. If $w = gQ \in G_{\mathbb{C}}/Q = W$ then we will write Q_w and $\mathfrak{q}_w$ for the isotropy subgroup gQg^{-1} of $G_{\mathbb{C}}$ at w and the isotropy subalgebra of $\mathfrak{g}$ there. We write τ for complex conjugation of $\mathfrak{g}$ over $\mathfrak{g}_0$.

The intersection of any two parabolic subalgebras of $\mathfrak{g}$ contains a Cartan subalgebra. From this,

(2.2) $\qquad \mathfrak{q}_w \cap \tau\mathfrak{q}_w$ contains a τ–stable Cartan subalgebra $\mathfrak{h}$ of $\mathfrak{g}$.

Let $\Phi = \Phi(\mathfrak{g}, \mathfrak{h})$ denote the corresponding root system. There exist a positive root system $\Phi^+ = \Phi^+(\mathfrak{g}, \mathfrak{h})$ and a subset $\Psi \subset \Pi$ of the corresponding system of simple roots such that

$\qquad \mathfrak{q}_w = \mathfrak{q}_\Psi = \mathfrak{q}_\Psi^n + \mathfrak{q}_\Psi^r$, nilradical plus reductive complement, where

(2.3) $\quad \mathfrak{q}_\Psi^n = \sum_{\Psi^n} \mathfrak{g}_\alpha$ and $\mathfrak{q}_\Psi^r = \mathfrak{h} + \sum_{\Psi^r} \mathfrak{g}_\alpha$ with

$\qquad \Psi^r = \{\alpha \in \Phi \mid \alpha \in span(\Psi)\}$ and $\Psi^n = \{\alpha \in \Phi \mid \alpha \in -\Phi^+ \text{ but } \alpha \notin \Psi^r\}$

Since $\mathfrak{g}_0$ has only finitely many G–conjugacy classes of Cartan subalgebras $\mathfrak{h}_0 = \mathfrak{h} \cap \mathfrak{g}_0$, and since $\Phi(\mathfrak{g}, \mathfrak{h})$ admits only finitely many subsystems of positive roots, it follows that there are only finitely many G–orbits on W. In particular there are open orbits, and the union of the open orbits is dense. It also follows that there is just one closed orbit, necessarily the lowest dimensional orbit, and that the closed orbit is in the closure of every orbit. The complexification of the isotropy subalgebra of $\mathfrak{g}_0$ at $w \in W$ is $(\mathfrak{g}_0 \cap \mathfrak{q}_w)_{\mathbb{C}} = \mathfrak{q}_w \cap \tau\mathfrak{q}_w$, sum of

$\qquad$ nilradical: $(\mathfrak{q}_\Psi^n \cap \tau\mathfrak{q}_\Psi^r) + (\mathfrak{q}_\Psi^r \cap \tau\mathfrak{q}_\Psi^n) + (\mathfrak{q}_\Psi^n \cap \tau\mathfrak{q}_\Psi^n) =$

(2.4) $\qquad\qquad \left(\sum_{\Psi^n \cap \tau\Psi^r} + \sum_{\Psi^r \cap \tau\Psi^n} + \sum_{\Psi^n \cap \tau\Psi^n} \right) \mathfrak{g}_\alpha$

$\qquad$ reductive: $(\mathfrak{q}_\Psi^r \cap \tau\mathfrak{q}_\Psi^r) = \mathfrak{h} + \sum_{\Psi^r \cap \tau\Psi^r} \mathfrak{g}_\alpha.$

In particular $|\Psi^n \cap \tau\Psi^n|$ is the real codimension of $G(w)$ in W, and $G(w)$ is open in W just when $\Psi^n \cap \tau\Psi^n$ is empty.

One can be more specific. The real Cartan subalgebra $\mathfrak{h}_0$ has a unique decomposition $\mathfrak{h}_0 = \mathfrak{h}_T + \mathfrak{h}_A$ where the roots are pure imaginary on $\mathfrak{h}_T$ and real on $\mathfrak{h}_A$. Those are the ± 1 eigenspaces on $\mathfrak{h}_0$ for a Cartan involution θ of $\mathfrak{g}_0$ that stabilizes $\mathfrak{h}_0$. Let $\mathfrak{k}_0$ denote the fixed point set of θ on $\mathfrak{g}_0$, Lie algebra of the maximal compact subgroup $K = G^\theta$ of G. These are equivalent: (i) $\mathfrak{h}_T$ is a Cartan subalgebra of $\mathfrak{k}_0$, (ii) $\mathfrak{h}_T$ contains a regular element of $\mathfrak{g}_0$, (iii) there is a system $\Phi^+ = \Phi^+(\mathfrak{g}, \mathfrak{h})$ of positive roots such that $\tau\Phi^+ = -\Phi^+$. The orbit $G(w)$ is open in W precisely when some Cartan subalgebra $\mathfrak{h}_0 \subset \mathfrak{g}_0 \cap \mathfrak{q}_w$, (which necessarily maximizes $\dim \mathfrak{h}_T$), has a positive root system Φ^+ such that (2.3) holds and $\tau\Phi^+ = -\Phi^+$. Since any two maximally compact Cartan subalgebras of $\mathfrak{g}_0$ are G–conjugate, now the open orbits are enumerated by a double coset space $W_K \backslash W_{\mathfrak{g}} / W_{\mathfrak{q}^r}$ of Weyl groups. Note that $W_K = W_G$ catches the action of the various components of G.

The open orbits that seem to enter most strongly into representation theory are the *measurable open orbits*. They are the open orbits $G(w) \subset W$ that carry a G–invariant volume element. If that is the case, then the invariant volume element is the volume element of a G–invariant, possibly indefinite, kaehler metric on the orbit, and the isotropy subgroup $G \cap P_w$ is the centralizer in G of a (compact) torus subgroup of G. In terms of Lie algebras, measurable open orbits are characterized by the following equivalent conditions: (i) $\mathfrak{q}_w \cap \tau\mathfrak{q}_w$ is reductive, i.e. $\mathfrak{q}_w \cap \tau\mathfrak{q}_w = \mathfrak{q}_w^r \cap \tau\mathfrak{q}_w^r$, (ii) $\mathfrak{q}_w \cap \tau\mathfrak{q}_w = \mathfrak{q}_w^r$, (iii) $\tau\Psi^r = \Psi^r$, and $\tau\Psi^n = -\Psi^n$. We know just when this happens: if one open G–orbit on W is measurable, then they all are measurable; and the open G–orbits on W are measurable if and only if $\tau\mathfrak{q}$ is conjugate to the parabolic subalgebra opposite to $\mathfrak{q}$. So in particular the open orbits are measurable in several important situations: the case $\operatorname{rank} K = \operatorname{rank} G$ and the case where $\mathfrak{q}$ is a Borel subalgebra (that is, Ψ is empty) of $\mathfrak{g}$.

There are other useful conditions, mostly automatic for measurable open orbits. For example, an orbit $G(w) \subset W$ is *integrable* if $\mathfrak{q}_w + \tau\mathfrak{q}_w$ is a subalgebra of $\mathfrak{g}$. Let $\mathfrak{u} = \mathfrak{q}_w^n \cap \tau\mathfrak{q}_w^n$ and let $\mathfrak{v}$ denote the normalizer of $\mathfrak{u}$ in $\mathfrak{g}$. Then the following conditions are equivalent to integrality of $G(w)$: (i) $\mathfrak{q}_w + \tau\mathfrak{q}_w = \mathfrak{v}$, (ii) $\mathfrak{q}_w \subset \mathfrak{v}$, (iii) $\mathfrak{u}$ is the nilpotent radical of $\mathfrak{v}$, and (iv) $\mathfrak{q}_w + \tau\mathfrak{q}_w$ is an algebra and $\mathfrak{u}$ is its nilpotent radical. See [**30**, Theorem 7.10] for a complete analysis of integrable orbits.

A *holomorphic arc* in the orbit $G(w) \subset W$ is a holomorphic map from the unit disk in $\mathbb{C}$ to W with image in $G(w)$. A *chain of holomorphic arcs* in $G(w)$ means a sequence $\{f_1, \cdots, f_k\}$ of holomorphic arcs in $G(w)$ such that the image of f_i meets the image of f_{i+1} for $1 \leqq i < k$. The *holomorphic arc components* of $G(w)$ are the equivalence classes of elements of $G(w)$ under the relation: $w_1 \sim w_2$ if there is a chain $\{f_1, \cdots, f_k\}$ of holomorphic arcs in $G(w)$ such that w_1 is in the image of f_1 and w_2 is in the image of f_k. Any connected complex submanifold of W contained in $G(w)$ is contained in a holomorphic arc component. If S is a holomorphic arc component of $G(w)$ and $g \in G$ such that $g(S)$ meets S then $g(S) = S$. It follows that the G–normalizer $N_G(S) = \{g \in G \mid g(S) = S\}$ is a Lie subgroup of G that is transitive on S. In particular S is an embedded C^ω submanifold of W. This notion is nicely set up for combining holomorphic and real induction of group representations, but the catch is that holomorphic arc components might not be complex submanifolds.

Fix an orbit $G(w) \subset W$ and let S_w denote the holomorphic arc component of w. If $g \in G$ then the holomorphic arc component of $g(w)$ is $S_{g(w)} = gS_w$, and $N_G(S_{g(w)}) = gN_G(S_w)g^{-1}$. Write $\mathfrak{n}_G(S_w)_0$ for the real Lie algebra of $N_G(S_w)$, $\mathfrak{n}_G(S_w)$ for its complexification, and $N_G(S_w)_\mathbb{C}$ for the corresponding complex analytic subgroup of $G_{\mathbb{C}}$. Now we need a certain τ–stable subspace $\mathfrak{m}_G(S_w)$ of $\mathfrak{g}$. The linear form $\delta_w = \sum_{\Psi^n \cap \tau \Psi^n} \alpha : \mathfrak{h} \to \mathbb{C}$ defines a τ–stable parabolic subalgebra $\mathfrak{s}_G(S_w) = \mathfrak{s}_G(S_w)^n + \mathfrak{s}_G(S_w)^r$, where $\mathfrak{s}_G(S_w)^n = \sum_{\langle \alpha, \delta_w \rangle > 0} \mathfrak{g}_\alpha$ and $\mathfrak{s}_G(S_w)^r = \mathfrak{h} + \sum_{\alpha \perp \delta_w} \mathfrak{g}_\alpha$. Now $\mathfrak{m}_G(S_w) = \mathfrak{s}_G(S_w) + \sum_{\Gamma} \mathfrak{g}_\alpha$ where $\Gamma = \{\alpha \in \Phi \mid -\alpha \notin \Psi^n \cap \tau \Psi^n, \langle \alpha, \delta_w \rangle < 0, \alpha + \tau \alpha \notin \Phi \}$. Then [30, Theorem 8.9] the following are equivalent: (i) the holomorphic arc components of $G(w)$ are complex submanifolds of W, (ii) $\mathfrak{n}_G(S_w) \subset \mathfrak{q}_w + \tau \mathfrak{q}_w$, (iii) $\mathfrak{n}_G(S_w) = \mathfrak{m}_G(S_w)$, and (iv) $\mathfrak{m}_G(S_w)$ is a subalgebra of $\mathfrak{g}$. When those conditions hold, we say that the orbit $G(w)$ is *partially complex*. In particular, if Γ is empty, then $\mathfrak{n}_G(S_w) = \mathfrak{s}_G(S_w) = \mathfrak{m}_G(S_w)$ and $G(w)$ is partially complex.

We will say that the orbit $G(w) \subset W$ is of *flag type* if $N_G(S_{w'})_{\mathbb{C}}(w')$ is a complex flag manifold for $w' \in G(w)$. We say that $G(w)$ is *measurable* if $S_{w'}$ carries an $N_G(S_{w'})$–invariant positive Radon measure for $w' \in G(w)$. We say that $G(w)$ is *polarized* if the $\mathfrak{q}_{w'}$ have τ–stable reductive parts for $w' \in G(w)$. Set $\mathfrak{t}_w = \sum_{\Psi^n \cap -\tau \Psi^n} \mathfrak{g}_\alpha + \sum_{-\Psi^n \cap \tau \Psi^n} \mathfrak{g}_\alpha$. Then [30, Theorem 9.2] $G(w)$ is measurable if and only if $\mathfrak{n}_G(S_w) = (\mathfrak{q}_w \cap \tau \mathfrak{q}_w) + \mathfrak{t}_w$, in other words just when $\mathfrak{n}_G(S_w)$ has nilpotent radical $(\mathfrak{q}_w \cap \tau \mathfrak{q}_w)^n$ and reductive part $(\mathfrak{q}_w \cap \tau \mathfrak{q}_w)^r + \mathfrak{t}_w$. It follows that if $G(w)$ is measurable then (i) $G(w)$ is partially complex, (ii) $G(w)$ is of flag type, (iii) $G(w)$ is polarized if and only if it is integrable, and (iv) the $N_G(S_w)$–invariant positive Radon measure on S_w is the volume element of an $N_G(S_w)$–invariant, possibly indefinite, kaehler metric. Furthermore [30, Theorem 9.9] if $G(w)$ is polarized then the following are equivalent: (i) $G(w)$ is measurable, (ii) $G(w)$ is integrable, and (iii) $G(w)$ is partially complex and of flag type. Under those conditions, $\mathfrak{n}_G(S_w) = (\mathfrak{q}_w \cap \tau \mathfrak{q}_w) + \mathfrak{t}_w = \mathfrak{s}_G(S_w)$.

Section 3. Harmonic Form Realizations of Relative Discrete Series Representations.

In the setting of the Bott–Borel–Weil Theorem, described in §1 above, the classical theorems of Dolbeault, Hodge and Kodaira tell us that every cohomology class $[c] \in H^q(G/T; \mathcal{O}(\mathbb{E}_\lambda))$ is represented by exactly one harmonic $(0,q)$–form on G/T with values in $\mathbb{E}_\lambda$. In this section we suppose that G is a general semisimple group as in (2.1). We first describe the realization of relative discrete series representations by square integrable harmonic differential forms. Then, using real group orbit results mentioned in §2, we describe the realization of standard tempered representations by square integrable partially harmonic forms.

If $\mathfrak{h}_0$ is a Cartan subalgebra of $\mathfrak{g}_0$, then by definition the corresponding Cartan subgroup of G is given by $H = \{g \in G \mid Ad(g)\xi = \xi \text{ for all } \xi \in \mathfrak{h}_0\}$. Note that $H = Z_G(G^0)(H \cap G^0)$ and that $H \cap G^0$ is the Cartan subgroup of G^0 with Lie algebra $\mathfrak{h}_0$. Let $Z \subset Z_G(G^0)$ as in (2.1). We may (and do) replace Z by ZZ_{G^0}, which still satisfies the requirements of (2.1), but which also satisfies: $Z \cap G^0 = Z_{G^0}$. An irreducible unitary representation $\pi \in \widehat{G}$ belongs to the *relative discrete series* if its coefficients $f_{u,v} : G \to \mathbb{C}$, given by $f_{u,v}(g) = \langle u, \pi(g)v \rangle_{\mathcal{H}_\pi}$, are square integrable on G modulo Z.

Suppose that G has a relatively compact Cartan subgroup, that is, has a Cartan subgroup T such that T/Z is compact. That is the condition ([**4**], [**5**], [**31**]) for the existence of relative discrete series representations of G. Fix a relatively compact Cartan subgroup $T \subset K$ of G. Then $T \cap G^0 = T^0$, in particular the Cartan subgroup $T \cap G^0$ of G^0 is commutative. Let $\Phi = \Phi(\mathfrak{g}, \mathfrak{t})$ be the root system, let $\Phi^+ = \Phi^+(\mathfrak{g}, \mathfrak{t})$ a choice of positive root system, and let $\rho = \frac{1}{2}\sum_{\alpha \in \Phi^+} \alpha$, half the trace of $ad_{\mathfrak{g}}|_{\mathfrak{t}}$ on $\sum_{\alpha \in \Phi^+} \mathfrak{g}_\alpha$.

If π is a relative discrete series representation of G and Θ_π is its distribution character, then the equivalence class of π is determined by the restriction of Θ_π to $T \cap G'$. So we can parameterize the relative discrete series of G by parameterizing those restrictions. Here we follow [**5**], [**6**] and [**31**].

Let $G^\dagger$ denote the finite index subgroup $TG^0 = Z_G(G^0)G^0$ of G. The Weyl group $W^\dagger = W(G^\dagger, T)$ coincides with $W^0 = W(G^0, T^0)$ and is a normal subgroup of $W = W(G, T)$. Let $\chi \in \hat{T}$. It follows from (2.1) that the irreducible unitary representation χ is finite dimensional. Since T^0 is commutative, χ has differential $d\chi(\xi) = \lambda(\xi)I$ where $\lambda \in i\mathfrak{t}_0^*$ and where I is the identity on the representation space of χ. Suppose that $\lambda + \rho$ is regular, i.e., that $\langle \lambda + \rho, \alpha \rangle \neq 0$ for all $\alpha \in \Phi$. Then there are unique relative discrete series representations π_χ^0 of G^0 and $\pi_\chi^\dagger$ of $G^\dagger$ whose distribution characters satisfy

$$(3.1) \qquad \Theta_{\pi_\chi^0}(x) = \pm \frac{\sum_{w \in W^0} sign(w)e^{w(\lambda+\rho)}}{\prod_{\alpha \in \Phi^+}(e^{\alpha/2} - e^{-\alpha/2})} \text{ and } \Theta_{\pi_\chi^\dagger}(zx) = \chi(z)\Theta_{\pi_\chi^0}(x)$$

for $z \in Z_G(G^0)$ and $x \in T^0 \cap G'$. Here note that $\pi_\chi^\dagger = \chi|_{Z_G(G^0)} \otimes \pi_\chi^0$. The same datum χ specifies a relative discrete series representation $\pi_\chi = \text{Ind}_{G^\dagger}^G(\pi_\chi^\dagger)$ of G. π_χ is characterized by the fact that its distribution character is supported in $G^\dagger$, where

$$(3.2) \qquad \Theta_{\pi_\chi} = \sum_{1 \leq i \leq r} \Theta_{\pi_\chi^\dagger} \cdot \gamma_i^{-1}$$

with $\gamma_i = Ad(g_i)|_{G^\dagger}$ where $\{g_1, \ldots, g_r\}$ is any system of coset representatives of G modulo $G^\dagger$. To combine these into a single formula one chooses the g_i so that they normalize T, i.e. chooses the γ_i to be a system of coset representatives of W modulo $W^\dagger$.

Every relative discrete series representation of G is equivalent to a representation π_χ as just described. Relative discrete series representations π_χ and $\pi_{\chi'}$ are equivalent if and only if $\chi' = \chi \cdot w^{-1}$ for some $w \in W$.

A choice $\Phi^+ = \Phi^+(\mathfrak{g}, \mathfrak{t})$ of positive root system defines a G-invariant complex manifold structure on G/T such that $\sum_{\alpha \in \Phi^+} \mathfrak{g}_\alpha$ represents the holomorphic tangent space. In effect, a choice of Φ^+ is a choice of Borel subalgebra $\mathfrak{b} = \mathfrak{t} + \sum_{\alpha \in \Phi^+} \mathfrak{g}_{-\alpha} \subset \mathfrak{g}$. Let X denote the flag variety of Borel subalgebras of $\mathfrak{g}$ and let $x \in X$ stand for the just-described Borel subalgebra $\mathfrak{b}$. Then $gT \mapsto g(x)$ defines a G-equivariant holomorphic diffeomorphism of G/T onto the open real group orbit $G(x) \subset X$.

More generally let W be a complex flag manifold of $\mathfrak{g}$, let $w \in W$, set $Y = G(w)$, and suppose

(3.3) $\quad$ Y is open in W and

$\overline{G} = G/Z_G(G^0)$ has compact isotropy subgroup at w.

Let $\mathfrak{q} \subset \mathfrak{g}$ denote the parabolic subalgebra of $\mathfrak{g}$ corresponding to w, so that W consists of all $Int(\mathfrak{g})$–conjugates of $\mathfrak{q}$. As in §2 we may view $\overline{G} = G/Z_G(G^0)$ inside $\overline{G}_{\mathbb{C}} = Int(\mathfrak{g})$. Let $\overline{L}$ denote the isotropy subgroup of $\overline{G}$ at w and let L denote its inverse image in G under the projection $G \to G/Z_G(G^0) = \overline{G}$ So $\overline{L}$ is compact and L is the isotropy subgroup of G at w. Passing to a conjugate, equivalently moving w within Y, we may suppose $T \subset L$.

Let $\chi \in \widehat{L}$, let E_χ denote the representation space, and let $\mathbb{E}_\chi \to Y \cong G/L$ denote the associated holomorphic homogeneous vector bundle. Using the Mackey machine, Cartan's highest weight theory, and the methods of [31, §2.4, 3.4, 3.5], we see that χ is finite dimensional and is constructed as follows. First, $L \cap G^0 = L^0$ and there is an irreducible representation χ^0 of L^0 with highest weight λ. Second, there is a representation $\psi \in \widehat{Z_G(G^0)}$ that agrees with χ^0 on $Z_{G^0} = Z_G(G^0) \cap L^0$. So we have the irreducible unitary representation $\chi^\dagger = \psi \otimes \chi^0$ of $L^\dagger = Z_G(G^0)L^0$. Third, $\lambda + \rho_{\mathfrak{l}}$ is $\Phi(\mathfrak{l}, \mathfrak{t})$–regular and this implies $\chi = Ind_{L^\dagger}^L(\chi^\dagger)$. We will call λ the *highest weight* of χ.

Note that $\lambda + \rho_{\mathfrak{l}}$ is the Harish–Chandra parameter for the infinitesimal character of χ. This of course is a special case of the relative discrete series picture. We will simply refer to $\lambda + \rho_{\mathfrak{l}}$ as the infinitesimal character of χ.

Since χ is unitary, the bundle $\mathbb{E}_\chi \to Y$ has a G–invariant hermitian metric. Let $\square$ denote the Kodaira–Hodge–Laplace operator $\overline{\partial}\,\overline{\partial}^* + \overline{\partial}^*\overline{\partial}$ on $\mathbb{E}_\chi$. Then we have Hilbert spaces

(3.4) $\quad$ $\mathcal{H}^q(Y; \mathbb{E}_\chi)$: harmonic L_2 $\mathbb{E}_\chi$-valued $(0, q)$-forms on Y

on which G acts naturally, and the natural actions of G on those spaces are unitary representations.

As remarked before for the flag manifold of Borel subalgebras of $\mathfrak{g}$, if G is compact (and in fact if $\overline{G}$ is compact) then the space $\mathcal{H}^q(Y; \mathbb{E}_\chi)$ of L_2 harmonic forms is naturally identified with the sheaf cohomology $H^q(Y; \mathcal{O}(\mathbb{E}_\chi))$.

The root system $\Phi = \Phi(\mathfrak{g}, \mathfrak{t})$ decomposes as the disjoint union of the *compact roots* $\Phi_K = \Phi(\mathfrak{k}, \mathfrak{t}) = \{\alpha \in \Phi : \mathfrak{g}_\alpha \subset \mathfrak{k}\}$ and the *noncompact roots* $\Phi_{G/K} = \Phi \setminus \Phi_K$. Write Φ_K^+ for $\Phi^+ \cap \Phi_K$ and $\Phi_{G/K}^+$ for $\Phi^+ \cap \Phi_{G/K}$.

My proof [31, Theorem 7.2.3] of Theorem 3.5 below only applied to the case where $\lambda + \rho$ is "sufficiently" nonsingular, because it relied on Schmid's proof [23] for connected linear Lie groups. Later [25] Schmid was able to drop the condition of sufficient nonsingularity. With this in mind, the proof of [31, Theorem 7.2.3] now yields

3.5. THEOREM. *Let $\chi \in \hat{L}$ with highest weight λ. Express $\chi = Ind_{L^\dagger}^L(\psi \otimes \chi^0)$. If $\lambda + \rho$ is $\Phi(\mathfrak{g}, \mathfrak{t})$-singular then every $\mathcal{H}^q(Y; \mathbb{E}_\chi) = 0$. Now suppose that $\lambda + \rho$ is $\Phi(\mathfrak{g}, \mathfrak{t})$-regular and define*

$$q(\lambda + \rho) = |\{\alpha \in \Phi_K^+ : \langle \lambda + \rho, \alpha \rangle < 0\}| + |\{\beta \in \Phi_{G/K}^+ : \langle \lambda + \rho, \beta \rangle > 0\}|.$$

Then $\mathcal{H}^q(Y; \mathbb{E}_\chi) = 0$ for $q \neq q(\lambda + \rho)$, and G acts irreducibly on $\mathcal{H}^{q(\lambda+\rho)}(Y; \mathbb{E}_\chi)$ by the relative discrete series representation $\pi_{\psi \otimes e^\lambda}$ of infinitesimal character $\lambda + \rho$.

An interesting variation on this result realizes the relative discrete series on spaces of L_2 bundle-valued harmonic spinors. See [20], [24] and [32].

Section 4. Harmonic Form Realizations of Tempered Representations.

The representations of G that enter into its Plancherel formula are the *tempered representations*. They are constructed from a certain class of real parabolic subgroups of G, the *cuspidal parabolic subgroups*, combining the relative discrete series construction for the reductive part of cuspidal parabolic with unitary induction from the parabolic up to G. We start by recalling the definitions.

Let H be a Cartan subgroup of our general semisimple (2.1) Lie group G. Fix a Cartan involution θ of G such that $\theta(H) = H$. Its fixed point set $K = G^\theta$ is a maximal compactly embedded (compact modulo $Z_G(G^0)$) subgroup of G. We decompose

(4.1)
$$\mathfrak{h}_0 = \mathfrak{t}_0 \oplus \mathfrak{a}_0 \text{ and } H = T \times A$$
$$\text{where } T = H \cap K, \ \theta(\xi) = -\xi \text{ on } \mathfrak{a}, \text{ and } A = \exp_G(\mathfrak{a}_0).$$

Then the centralizer $Z_G(A)$ of A in G has form $M \times A$ where $\theta(M) = M$. Now [31] M is a reductive Lie group in the same class (2.1), and T is a compactly embedded Cartan subgroup of M, so M has relative discrete series representations.

Suppose that the positive root system $\Phi^+ = \Phi^+(\mathfrak{g}, \mathfrak{h})$ is defined by positive root systems $\Phi^+(\mathfrak{m}, \mathfrak{t})$ and $\Phi^+(\mathfrak{g}_0, \mathfrak{a}_0)$. This means that

(4.2a)
$$\Phi^+(\mathfrak{m}, \mathfrak{t}) = \{\alpha|_\mathfrak{t} : \alpha \in \Phi^+(\mathfrak{g}, \mathfrak{h}), \ \alpha|_\mathfrak{a} = 0\}$$

and

(4.2b)
$$\Phi^+(\mathfrak{g}_0, \mathfrak{a}_0) = \{\beta|_{\mathfrak{a}_0} : \beta \in \Phi^+(\mathfrak{g}, \mathfrak{h}), \ \beta|_\mathfrak{a} \neq 0\}.$$

In other words, given $\mathfrak{h}$, the Borel subalgebra

(4.2c)
$$\mathfrak{b} = \mathfrak{b}^n + \mathfrak{b}^r \text{ where } \mathfrak{b}^r = \mathfrak{h} \text{ and } \mathfrak{b}^n = [\mathfrak{b}, \mathfrak{b}] = \sum_{\alpha \in \Phi^+(\mathfrak{g}, \mathfrak{h})} \mathfrak{g}_{-\alpha}$$

is chosen to maximize $\mathfrak{b}^n \cap \overline{\mathfrak{b}^n}$. Note that $\mathfrak{b}^n \cap \overline{\mathfrak{b}^n}$ has real form $(\mathfrak{b}^n \cap \overline{\mathfrak{b}^n})_0 = \mathfrak{b}^n \cap \mathfrak{g}_0$, which is the sum $\sum_{\gamma \in \Phi^+(\mathfrak{g}_0, \mathfrak{a}_0)} (\mathfrak{g}_0)_\gamma$ of the positive restricted root spaces.

A subalgebra $\mathfrak{p}_0 \subset \mathfrak{g}_0$ is a *(real) parabolic subalgebra* if its complexification $\mathfrak{p}$ is a parabolic subalgebra of $\mathfrak{g}$, in other words if $\mathfrak{p}_0 = \mathfrak{p} \cap \mathfrak{g}_0$ for some τ–stable parabolic subalgebra $\mathfrak{p} \subset \mathfrak{g}$. A subgroup $P \subset G$ is a *parabolic subgroup* of G if it is the G–normalizer of a parabolic subalgebra of $\mathfrak{g}_0$. A parabolic subgroup $P \subset G$ is called *cuspidal* if the Levi component (reductive part) has a relatively compact Cartan subgroup. We now have the cuspidal parabolic subgroup $P = MAN$ of G, where M and A are as above, where $MA = M \times A$ is the Levi component of P, and where $N = exp_G((\mathfrak{b}^n \cap \overline{\mathfrak{b}^n})_0)$.

Let $\chi \in \widehat{H}$ and consider the basic datum $(H, \mathfrak{b}, \chi)$. The representation of $\mathfrak{b}$ is determined because χ represents H irreducibly: $\chi(\mathfrak{b}^n) = 0$ and $\chi|_{\mathfrak{h}}$ is the differential of the representation of H. Decompose $\chi = \psi \otimes e^\nu \otimes e^{i\sigma}$, $\psi \in \widehat{Z_G(G^0)}$, $e^\nu \in \widehat{T}$, $\sigma \in \mathfrak{a}_0^*$. Suppose that $\nu + \rho_\mathfrak{m}$ is $\Phi(\mathfrak{m}, \mathfrak{t})$–regular. Then $\psi \otimes e^\nu$ specifies a relative discrete series representation $\eta_{\psi \otimes e^\nu}$ of M. The Levi component $M \times A$ of P acts irreducibly and unitarily on $H_{\eta_{\psi \otimes e^\nu}}$ by $\eta_{\psi \otimes e^\nu} \otimes e^{i\sigma}$. That extends uniquely to a representation (which we still denote $\eta_{\psi \otimes e^\nu} \otimes e^{i\sigma}$) of P on $H_{\eta_{\psi \otimes e^\nu}}$ whose kernel contains N. Now we have the *standard tempered representation*

$$(4.3) \qquad \pi_\chi = \pi_{\psi, \nu, \sigma} = \mathrm{Ind}_{P \uparrow G}(\eta_{\psi \otimes e^\nu} \otimes e^{i\sigma})$$

of G. One can compute the character of π_χ and see that it is independent of the choice of positive root system $\Phi^+(\mathfrak{g}, \mathfrak{h})$ that is defined by choices of $\Phi^+(\mathfrak{m}, \mathfrak{t})$ and $\Phi^+(\mathfrak{g}_0, \mathfrak{a}_0)$. With H fixed up to conjugacy, and as ψ and σ vary, we have the *H-series* of tempered representations of G.

The various tempered series exhaust enough of $\widehat{G}$ for a decomposition of $L_2(G)$ essentially as

$$\sum_{H \in Car(G)} \sum_{\psi \otimes e^\nu \in \widehat{T}} \int_{\widehat{A}} H_{\pi_{\psi, \nu, \sigma}} \otimes H^*_{\pi_{\psi, \nu, \sigma}} m(H : \psi : \nu : \sigma) d\sigma.$$

Here $m(H : \psi : \nu : \sigma) d\sigma$ is the *Plancherel measure* on $\widehat{G}$. This was worked out by Harish–Chandra ([6], [7], [8]) for groups of Harish–Chandra class, and somewhat more generally by Herb and myself ([31]; [14], [15]; [17], [18]). Harish–Chandra's approach is based on an analysis of the structure of the Schwartz space, while Herb and I use explicit character formulae (compare [3], [21], [14], [16]).

Fix a θ-stable Cartan subgroup $H \subset G$ and a positive root system $\Phi^+ = \Phi^+(\mathfrak{g}, \mathfrak{h})$ defined by positive root systems $\Phi^+(\mathfrak{m}, \mathfrak{t})$ and $\Phi^+(\mathfrak{g}_0, \mathfrak{a}_0)$ as in (4.2) above. Then we have the associated cuspidal parabolic subgroup $P = MAN \subset G$.

We now need a complex flag manifold $W \cong \overline{G}_\mathbb{C}/Q$ consisting of the $Int(\mathfrak{g})$–conjugates of a parabolic subalgebra $\mathfrak{q} \subset \mathfrak{g}$, and a real group orbit $Y = G(w) \subset W$, such that

(4.4a) Y is measurable, hence partially complex and of flag type, and

(4.4b) the normalizer $N_G(S_w)$ of the holomorphic arc component S_w has Lie algebra $\mathfrak{p}_0$.

Then S_w will be a topological component of the open M–orbit $M(w)$ in the subflag $\overline{M}_{\mathbb{C}}(w) \subset W$. Thus AN will act trivially on S_w and the isotropy subgroup of G at w will be of the form UAN where $U \subset M$ is of the M–centralizer of a subtorus of T. Suppose in addition that

(4.4c) $U/Z_G(G^0) = \{m \in M \mid m(w) = w\}/Z_G(G^0)$ is compact.

Then $M(w) \cong M/U$ will be a measurable integrable open orbit in $\overline{M}_{\mathbb{C}}(w)$, $U = Z_M(M^0)U^0$ with $U \cap M^0 = U^0$, $UM^0 = M^\dagger$, and $M/M^\dagger$ will enumerate the topological components of $M(w)$.

It is straightforward to construct all pairs (W, w) that satisfy (4.4) and such that $G(w)$ is integrable as well as measurable [**31**, §6.7]. They are given by:

(4.5a) $\mathfrak{u}_0 \subset \mathfrak{m}_0$ is the $\mathfrak{m}_0$–centralizer of a subspace of $\mathfrak{t}$,

(4.5b)   the corresponding analytic subgroup $U \subset M$ has compact
         image in $M/Z_G(G^0)$,

(4.5c) $\mathfrak{r} \subset \mathfrak{m}$ is a parabolic subalgebra with reductive part $\mathfrak{r}^r = \mathfrak{u}$, and

(4.5d) $\mathfrak{q}_w$ is the $\mathfrak{g}$–normalizer of $\mathfrak{r}^n + \mathfrak{n}$.

Then the considerations of [**31**, §6.7] show that

(4.6) $\mathfrak{q}_w^n = \mathfrak{r}^n + \mathfrak{n}$ and $\mathfrak{q}_w^r = \mathfrak{u} + \mathfrak{a}$.

In the case $U = T$ this constructs $|W_{\mathfrak{m}}|$ pairs (W, w) that satisfy (4.4), though of course there will be some identifications under the Weyl group of G. In particular there are many pairs (W, w) that satisfy (4.4).

Now assume that the situation (4.4) is given. Then the holomorphic arc components $gS_w \cong M^\dagger/U$ of $Y = G(w) \cong G/UAN$ are topological components of the fibres of

(4.7a) $Y \to G/P$, G–equivariant fibration with structure group M
           and typical fibre M/U

given by $gUAM \mapsto gMAN$. To say it in a slightly different way, $M(w)$ has finitely many topological components $m_i M^\dagger(w) = m_i S_w$, as $m_i M^\dagger$ ranges over $M/M^\dagger$. The holomorphic arc components of Y are the fibres of

(4.7b) $Y \to G/P^\dagger$, G–equivariant fibration with structure group $M^\dagger$
           and typical fibre $M^\dagger/U$

where $P^\dagger = M^\dagger AN$. The complex structure on the holomorphic arc component S_w, as complex submanifold of W, is the M-invariant complex structure on M/U for which $\sum_{\alpha \in \Phi^+(\mathfrak{m},\mathfrak{t}) \setminus \Phi^+(\mathfrak{m},\mathfrak{u})} \mathfrak{m}_\alpha$ is the holomorphic tangent space.

Consider irreducible unitary representations

(4.8a) $\mu \in \widehat{U}$, with representation space E_μ, and $e^{i\sigma} \in \widehat{A}$ where $\sigma \in \mathfrak{a}_0^*$.

Let $\rho_{\mathfrak{a}} = \frac{1}{2} \sum_{\Phi^+(\mathfrak{g}_0, \mathfrak{a}_0)} (\dim \mathfrak{g}_\phi) \phi$ as usual. This is the quasicharacter on UAN that must be inserted for ordinary induction to become unitary induction from UAN to G. Now UAN acts on E_μ by

$$(4.8b) \qquad \gamma_{\mu,\sigma}(uan) = e^{\rho_{\mathfrak{a}} + i\sigma}(a)\mu(u).$$

That specifies the associated G–homogeneous vector bundle

$$(4.8c) \qquad p : \mathbb{E}_{\mu,\sigma} \to Y \cong G/UAN.$$

This bundle has a natural CR–structure and is holomorphic over every holomorphic arc component of Y. Furthermore K is transitive on Y so the bundle has a natural K–invariant hermitian metric based on the unitary structure of its typical fibre E_μ.

Restrict the holomorphic tangent bundle of W to Y and let $\mathbb{T} \to Y$ denote the sub–bundle whose fibre at $w' \in Y$ is the holomorphic tangent space to $S_{w'}$ at w'. The space of partially smooth (p,q)–forms with values in $\mathbb{E}_{\mu,\sigma}$ is

$$(4.9a) \qquad \begin{aligned} &A^{p,q}(Y; \mathbb{E}_{\mu,\sigma}) : \text{measurable sections of } \mathbb{E}_{\mu,\sigma} \otimes \wedge^p \mathbb{T}^* \otimes \wedge^q \overline{\mathbb{T}}^* \\ &\qquad\qquad \text{that are } C^\infty \text{ on each holomorphic arc component.} \end{aligned}$$

The subspace of square integrable partially smooth forms is defined just as one might guess. Let $\#$ denote the Kodaira–Hodge orthocomplementation mapping $A^{p,q}(Y; \mathbb{E}_{\mu,\sigma}) \to A^{n-p,n-q}(Y; \mathbb{E}_{\mu,\sigma}^*)$ on each holomorphic arc component, and let $\bar{\wedge}$ denote exterior product followed by contraction, so pointwise $\omega \bar{\wedge} \# \omega$ is $\|\omega\|^2$ times the volume element of the holomorphic arc component. Now we have

$$(4.9b) \qquad \begin{aligned} &A_2^{p,q}(Y; \mathbb{E}_{\mu,\sigma}) : \text{all } \omega \in A^{p,q}(Y; \mathbb{E}_{\mu,\sigma}) \text{ such that} \\ &\qquad \int_{S_{kw}} \omega \bar{\wedge} \# \omega < \infty \text{ a.e. } k \in K \text{ and} \\ &\qquad \int_{K/U} \left(\int_{S_{kw}} \omega \bar{\wedge} \# \omega \right) d(kU) < \infty. \end{aligned}$$

The Kodaira–Hodge–Laplace operators on the restrictions of $\mathbb{E}_{\mu,\sigma}$ to the holomorphic arc components fit together to give us essentially self adjoint operators $\square$ on the Hilbert space completions of the $A_2^{p,q}(Y; \mathbb{E}_{\mu,\sigma})$. Their kernels are the spaces

$$(4.9c) \qquad \begin{aligned} &\mathcal{H}^{p,q}(Y; \mathbb{E}_{\mu,\sigma}) : \text{square integrable partially harmonic} \\ &\qquad\qquad \mathbb{E}_{\mu,\sigma}\text{–valued} (p,q)\text{–forms on } Y. \end{aligned}$$

We'll only use the $\mathcal{H}^q(Y; \mathbb{E}_{\mu,\sigma}) = \mathcal{H}^{0,q}(Y; \mathbb{E}_{\mu,\sigma})$.

The natural action of G on $\mathcal{H}^q(Y; \mathbb{E}_{\mu,\sigma})$ is a unitary representation. It is unitarily equivalent to the representation of G on the Hilbert space of L_2 sections of a certain homogeneous vector bundle

$$(4.10) \qquad \begin{aligned} &\mathbb{H}^q(M/U; \mathbb{E}_\mu|_{M/U}) \to G/P, \\ &\qquad \text{fibre } \mathcal{H}^q(M/U; \mathbb{E}_\mu|_{M/U}), \text{ structure group } P = MAN. \end{aligned}$$

Here, as in the discussion of the realization of the relative discrete series, $\mathcal{H}^q(M/U; \mathbb{E}_\mu|_{M/U})$ is the Hilbert space of $L_2(M/U)$ harmonic, $\mathbb{E}_\mu|_{M/U}$–valued, $(0, q)$–forms on M/U. Let η denote the (necessarily unitary) representation of M on $\mathcal{H}^q(M/U; \mathbb{E}_\mu|_{M/U})$. Then MAN acts on $\mathcal{H}^q(M/U; \mathbb{E}_\mu|_{M/U})$ by $\eta \otimes e^{\rho_\mathfrak{a}+i\sigma}(man) = e^{\rho_\mathfrak{a}+i\sigma}(a)\eta(m)$ and $\mathbb{H}^q(M/U; \mathbb{E}_\mu|_{M/U}) \to G/P$ is the associated vector bundle over G/P. Again the $\rho_\mathfrak{a}$ means that the natural action of G on L_2 sections is unitary for $\langle f, f' \rangle = \int_K \langle f(k), f'(k) \rangle dk$.

Theorem 3.5 combines with the considerations above to give the realization of standard tempered representations described in Theorem 4.9 below. Originally I proved Theorem 4.11 only for the case where $\nu + \rho_\mathrm{m}$ is "sufficiently" nonsingular [**31**, Theorem 8.3.4], so that the realization for the corresponding relative discrete series representation of M would be available. As described just before Theorem 3.5, we can now drop that condition.

The representation $\mu \in \widehat{U}$ is of the form $Ind_{U^\dagger}^U(\mu^\dagger)$ where $U^\dagger = Z_M(M^0)U^0$, where $U^0 = U \cap M^0$, and where $\mu^\dagger = \psi \otimes \mu^0$ in such a way that $\psi \in \widehat{Z_M(M^0)}$ agrees with $\mu^0 \in \widehat{U^0}$ on $Z_{M^0} = Z_M(M^0) \cap U^0$. Here μ^0 has some highest weight, say ν, relative to $\Phi^+(\mathfrak{u}, \mathfrak{t})$, and $\nu + \rho_\mathfrak{u}$ is $\Phi(\mathfrak{u}, \mathfrak{t})$–nonsingular. In general we will just refer to ν as the highest weight of μ.

4.11. THEOREM. *Let $\mu \in \widehat{U}$ with highest weight ν. If $\nu + \rho_\mathrm{m}$ is $\Phi(\mathfrak{m}, \mathfrak{t})$–singular then every $\mathcal{H}^q(Y; \mathbb{E}_{\mu,\sigma}) = 0$. Let*

$$q(\nu + \rho_\mathrm{m}) = |\{\alpha \in \Phi^+_{K\cap M} : \langle \nu + \rho_\mathrm{m}, \alpha \rangle < 0\}| + |\{\beta \in \Phi^+_{M/K\cap M} : \langle \nu + \rho_\mathrm{m}, \beta \rangle > 0\}|.$$

Then $\mathcal{H}^q(Y; \mathbb{E}_{\mu,\sigma}) = 0$ for $q \neq q(\nu + \rho_\mathrm{m})$, and G acts on $\mathcal{H}^q(Y; \mathbb{E}_{\mu,\sigma})$ by the standard H–series representation $\pi_{\psi,\nu,\sigma} = Ind_P^G(\eta_{\psi \otimes e^\nu} \otimes e^{i\sigma})$ of G of infinitesimal character $\nu + \rho_\mathrm{m} + i\sigma$.

A variation on this theorem realizes the tempered series on spaces of L_2 bundle-valued partially harmonic spinors. See [**32**].

Whenever $\sigma \in \mathfrak{a}_0^*$ is $\Phi^+(\mathfrak{g}_0, \mathfrak{a}_0)$–regular, the standard H–representation $\pi_{\psi,\nu,\sigma}$ is irreducible. Plancherel measure for G thus is carried by the irreducible representations among the H–series representations realized above, as H varies over the conjugacy classes of Cartan subgroups of G.

Section 5. Sheaf Cohomology Realizations
of Tempered Representations.

An important variation on the Kostant–Langlands Conjecture result — which in fact preceded its solution — is Schmid's Dolbeault cohomology realization [**22**] of discrete series representations of connected semisimple Lie groups with finite center. Suppose that G has a compactly embedded Cartan subgroup T and that we are in the situation of (3.3). Then the open orbit $Y = G(w) \cong G/L$ in W contains $K(w) \cong K/L$ as a maximal compact complex submanifold. We denote

$s = \dim_{\mathbb{C}} K(w)$. Whenever

(5.1) $\lambda + \rho$ is Y–antidominant: $\langle \lambda + \rho, \gamma \rangle < 0$ for all $\beta \in \Phi^+(\mathfrak{g}, \mathfrak{t}) \setminus \Phi^+(\mathfrak{l}, \mathfrak{t})$,

$\langle \lambda, \alpha \rangle \geq 0$ for all $\alpha \in \Phi^+(\mathfrak{l}, \mathfrak{t})$

we have $s = q(\lambda + \rho)$. This is the case where the associated holomorphic vector bundles $\mathbb{E}_\chi \to Y$, $\chi \in \widehat{L}$ with highest weight λ, are negative vector bundles.

If π is a relative discrete series representation of G, we can choose the positive root system $\Phi^+(\mathfrak{g}, \mathfrak{t})$ so that $\pi = \pi_\chi$ where the highest weight λ of χ satisfies (5.1). This is because $\mathfrak{l}$ is the reductive part of a parabolic subalgebra of $\mathfrak{g}$. Thus there is no restriction on $\pi_{\psi \otimes e^\lambda}$ in

5.2. THEOREM. *Let $\chi \in \widehat{L}$ with highest weight λ. Suppose that the infinitesimal character $\lambda + \rho$ is Y–antidominant (5.1). Then $H^q(Y; \mathcal{O}(\mathbb{E}_\chi)) = 0$ for $q \neq s$, $H^s(Y; \mathcal{O}(\mathbb{E}_\chi))$ has a natural structure of infinite dimensional Fréchet space, and the natural action of G on $H^s(Y; \mathcal{O}(\mathbb{E}_\chi))$ is a continuous representation of infinitesimal character $\lambda + \rho$. Express $\chi = Ind_{L^\dagger}^L(\psi \otimes \chi^0)$. The representation of G on $H^s(Y; \mathcal{O}(\mathbb{E}_\chi))$ is infinitesimally equivalent[1] to the relative discrete series representation $\pi_{\psi \otimes e^{\lambda + \rho}}$.*

Theorem 5.2 was proved by Schmid [22] in the case where G is connected with finite center, $L = T$, and λ is sufficiently nonsingular. It follows by now–standard techniques [31, §3] for general semisimple groups G with $L = T$, and λ is sufficiently nonsingular. The analogous statement for Zuckerman's derived functor modules now holds without the requirement of sufficient nonsingularity, because of their analytic continuation properties [29]. In view of [27, §9] the same holds for our cohomology modules. That proves Theorem 5.2 completely for the case $L = T$. The result as stated now is more or less immediate from the Leray spectral sequence for the holomorphic fibration $G/T \to G/L$. Compare [33].

In Theorem 5.2, the infinitesimal equivalence is

(5.3) $$\mathcal{H}^s(Y; \mathbb{E}_\chi)_{(K)} \to H^s(Y; \mathcal{O}(\mathbb{E}_\chi))_{(K)}$$

as a map on spaces of K-finite vectors. That is the map that sends an L_2 harmonic form to (the sheaf cohomology class that corresponds to) its Dolbeault class.

We now look at the tempered case. Fix a Cartan subgroup $H = T \times A$ as in (4.1), a positive root system $\Phi^+(\mathfrak{g}, \mathfrak{h})$ as in (4.2), and a measurable open orbit $Y = G(w) \subset W$ on a complex flag manifold $W \cong \overline{G}_{\mathbb{C}}/Q$ as in (4.4). We are going to replace $\mathcal{H}^q(M/U; \mathbb{E}_\mu|_{M/U})$ by the partial Dolbeault cohomology space that realizes the relative discrete series representations $\eta_{\psi \otimes e^\nu}$ of M and $\eta_{\psi \otimes e^\nu} \otimes e^{i\sigma}$

---

[1]Let π and ϕ be continuous representations of G on complete locally convex topological vector spaces V_π and V_ψ. Let $(V_\pi)_{(K)}$ and $(V_\psi)_{(K)}$ denote the respective subspaces of K-finite vectors. They are modules for the universal enveloping algebra $\mathcal{U}(\mathfrak{g})$. An *infinitesimal equivalence* of V_π with V_ψ means a $\mathcal{U}(\mathfrak{g})$-isomorphism of $(V_\pi)_{(K)}$ onto $(V_\psi)_{(K)}$.

of MA and $P = MAN$. The space $(K \cap M)(w) \cong (K \cap M)/U$ is a maximal compact complex submanifold of $Y \cong G/UAN$. Let $s = \dim_{\mathbb{C}} (K \cap M)/U$. Whenever $\nu + \rho_{\mathfrak{m}}$ is S_w–antidominant (5.1), that is

(5.4)
$$\langle \nu + \rho_{\mathfrak{m}}, \gamma \rangle < 0 \text{ for all } \beta \in \Phi^+(\mathfrak{m}, \mathfrak{t}) \setminus \Phi^+(\mathfrak{u}, \mathfrak{t}) \text{ and}$$
$$\langle \nu, \alpha \rangle \geqq 0 \text{ for all } \alpha \in \Phi^+(\mathfrak{u}, \mathfrak{t}),$$

we have $s = q(\nu + \rho_{\mathfrak{m}})$. As before, this is the case where the bundle $\mathbb{E}_\mu|_{M/U} \to M/U$ is negative.

Let $\mathcal{O}_{\mathfrak{q}}(\mathbb{E}_\mu) \to Y$ denote the sheaf of germs of C^∞ sections of $\mathbb{E}_\mu \to G/Y$ that are holomorphic along the holomorphic arc components of Y, i.e. holomorphic along the fibres of $Y \to G/P$. By irreducibility of μ, these are the sections annihilated by the right action of $\mathfrak{q}^n$. See (5.8) below for the condition when μ may be reducible.

Given a relative discrete series representation η of M, we need a positive root system $\Phi^+(\mathfrak{g}, \mathfrak{h})$ such that:

(1) $\Phi^+(\mathfrak{m}, \mathfrak{t})$ satisfies (5.4),
(2) $\Phi^+(\mathfrak{g}_0, \mathfrak{a}_0)$ is arbitrary, and
(3) $\Phi^+(\mathfrak{g}, \mathfrak{h})$ is defined by $\Phi^+(\mathfrak{m}, \mathfrak{t})$ and $\Phi^+(\mathfrak{g}_0, \mathfrak{a}_0)$.

Given a choice of $\Phi^+(\mathfrak{g}, \mathfrak{h})$ as above, we realize the corresponding standard H–series representations of G on partial Dolbeault cohomology as follows. Given $\mathbb{E}_{\mu,\sigma} \to Y$ as above, we have the sheaf

(5.5a)
$$\mathcal{O}_{\mathfrak{q}}(\mathbb{E}_{\mu,\sigma}) \to G/UA : \text{ germs of } C^\infty \text{ sections } f \text{ of } \mathbb{E}_{\mu,\sigma} \to G/UA$$
$$\text{such that } f(x; \xi) + \chi(\xi) \cdot f(x) = 0 \text{ for all } x \in G \text{ and } \xi \in \mathfrak{q}.$$

$\mathbb{E}_{\mu,\sigma} \to G/UA$ pushes down to a bundle $\mathbb{E}_{\mu,\sigma} \to Y = G(\mathfrak{q}) = G(w) \subset W$, so the sheaf $\mathcal{O}_{\mathfrak{q}}(\mathbb{E}_{\mu,\sigma}) \to G/UA$ pushes down to a sheaf

(5.5b)
$$\mathcal{O}_{\mathfrak{q}}(\mathbb{E}_{\mu,\sigma}) \to Y : \text{ germs of } C^\infty \text{ sections } f \text{ of } \mathbb{E}_{\mu,\sigma} \to Y$$
$$\text{such that } f(x; \xi) + \chi(\xi) \cdot f(x) = 0 \text{ for all } x \in G \text{ and } \xi \in \mathfrak{q}.$$

The germs in $\mathcal{O}_{\mathfrak{q}}(\mathbb{E}_{\mu,\sigma}) \to UA$ are equivariant along fibres of $G/UA \to G/UAN \cong Y$, and the collapse of the Leray spectral sequence of $G/UA \to G/UAN \cong Y$ yields natural G–equivariant isomorphisms

(5.5c)
$$H^q(G/UA; \mathcal{O}_{\mathfrak{q}}(\mathbb{E}_{\mu,\sigma})) \cong H^q(Y; \mathcal{O}_{\mathfrak{q}}(\mathbb{E}_{\mu,\sigma})).$$

When χ is irreducible, (5.5a) and (5.5b) reduce to: $f(x; \xi) = 0$ for all $x \in G$ and $\xi \in \mathfrak{q}^n$. In the more general case considered here, (5.5) is the appropriate condition; see [28]. Now we can state

5.6.   THEOREM. Let $\mu \in \widehat{U}$ and $\sigma \in \mathfrak{a}_0^*$ as in (4.8a), let ν be the highest weight of μ, and suppose that the infinitesimal character $\nu + \rho_{\mathfrak{m}}$ is S_w–antidominant (5.4). So $s = q(\nu + \rho_{\mathfrak{m}})$. If $q \neq s$ then $H^q(Y; \mathcal{O}_{\mathfrak{q}}(\mathbb{E}_{\mu,\sigma})) = 0$. Express $\mu = \mathrm{Ind}_{U^\dagger}^U(\psi \otimes \mu^0)$. Then $H^s(Y; \mathcal{O}_{\mathfrak{q}}(\mathbb{E}_{\mu,\sigma}))$ has a natural structure of infinite dimensional Fréchet space, the natural action of G on $H^s(Y; \mathcal{O}_{\mathfrak{q}}(\mathbb{E}_{\mu,\sigma}))$

is a continuous representation of infinitesimal character $\nu + \rho_{\mathfrak{m}} + i\sigma$, and this representation of G is infinitesimally equivalent to the standard H–series representation $\pi_{\psi,\nu,\sigma} = Ind_P^G(\eta_{\psi \otimes e^\nu} \otimes e^{i\sigma})$.

The Fréchet space structure on the cohomologies $H^q(Y; \mathcal{O}_q(\mathbb{E}_{\mu,\sigma}))$ is described, in a larger context, in §7 below.

On the flag manifold X of Borel subalgebras of $\mathfrak{g}$, our cohomology construction for standard tempered representations can be formulated as follows (compare [27]). Fix a basic datum $(H, \mathfrak{b}, \chi)$: H is a Cartan subgroup of G, $\mathfrak{b}$ is a Borel subalgebra of $\mathfrak{g}$ such that $\mathfrak{h} \subset \mathfrak{b}$, and χ is a finite dimensional representation of $(\mathfrak{b}, H)$. We have the associated homogeneous vector bundles $\mathbb{E}_\chi \to G/H$ and the sheaf $\mathcal{O}_\mathfrak{b}(\mathbb{E}_\chi) \to G/H$ of germs of sections defined by the right action of $\mathfrak{b}$. Note that $Y \cong G/HN$ here and the partial complex structure (CR structure) induced on $G(\mathfrak{b})$ by X is the one for which the holomorphic tangent space of the typical fibre $M(\mathfrak{b}) \cong M/T$ of $G/HN \to G/P$ is $\sum_{\alpha \in \Phi^+(\mathfrak{m},\mathfrak{t})} \mathfrak{m}_\alpha$.

Now the case of Theorem 5.6, where W is the flag manifold X of Borel subalgebras of $\mathfrak{g}$, can be reformulated as

5.7. PROPOSITION. *Every standard tempered series representation $\pi_{\psi,\nu,\sigma}$ of G, $\psi \otimes e^\nu \otimes e^{i\sigma} \in \widehat{H}$ and $H \in Car(G)$, is realized up to infinitesimal equivalence as the natural action of G on a partial Dolbeault cohomology space $H^s(Y, \mathcal{O}_\mathfrak{b}(\mathbb{E}_{\mu,\sigma}))$, $Y = G(\mathfrak{b}) \subset X$, for a basic datum $(H, \mathfrak{b}, \chi)$ as follows. $\mathfrak{b}$ is maximally real subject to the condition $\mathfrak{h} \subset \mathfrak{b}$; $\mu = Ind_{U\dagger}^U(\psi \otimes \mu^o) \in \widehat{U}$ where μ^0 has highest weight ν; χ is given on $H = T \times A$ by $\psi \otimes e^\nu \otimes e^{i\sigma + \rho_\mathfrak{a}}$, on $\mathfrak{h}$ by the differential, and on $\mathfrak{b}^n$ by 0; and $s = \dim_\mathbb{C} (K \cap M)/T$. Here $H^s(Y, \mathcal{O}_\mathfrak{b}(\mathbb{E}_{\mu,\sigma}))$ has a natural Fréchet space structure and the action of G is continuous.*

We may assume $\langle \nu + \rho_\mathfrak{m}, \gamma \rangle < 0$ for all $\gamma \in \Phi^+(\mathfrak{m}, \mathfrak{t})$. With that assumption, if $q \neq s$ then $H^q(Y, \mathcal{O}_\mathfrak{b}(\mathbb{E}_{\mu,\sigma})) = 0$.

Except for the Fréchet space structure on the cohomologies, Proposition 5.7 is the starting point of [27] for the construction of standard admissible representations. The Fréchet space structure itself comes out of some variations (see §§6 and 7 below) on the methods of [27].

In our more general context, a basic datum corresponding the the setup of Theorem 5.6 is of the form $(UA, \mathfrak{q}, \chi)$ where UA is a Levi component (reductive part) of the isotropy subgroup of G at $w \in W$, where $\mathfrak{q}$ is the parabolic subalgebra of $\mathfrak{g}$ represented by w in the complex flag manifold W, and where χ is a finite dimensional representation of $(\mathfrak{q}, UA)$. Then the appropriate sheaf is given by (5.5), and we have natural G–equivariant isomorphisms $H^q(G/UA; \mathcal{O}_\mathfrak{q}(\mathbb{E}_\chi)) \cong H^q(Y; \mathcal{O}_\mathfrak{q}(\mathbb{E}_\chi))$. Then one has the analog for W of Proposition 5.7.

Section 6. Representations for Open Orbits

Fix a complex flag manifold W and a measurable open G–orbit $Y = G(w)$. Let $\mathfrak{q} = \mathfrak{q}_w \subset \mathfrak{g}$ denote the parabolic subalgebra represented by w, and let L

denote the isotropy subgroup of G at w. Then L contains a fundamental Cartan subgroup H of G, $L = Z_G(G^0)L^0$ and $L^0 = L \cap G^0$, by the argument of [31, Lemma 7.1.2]. The image $\overline{L}$ of L in $\overline{G} = G/Z_G(G^0) \subset \overline{G}_\mathbb{C}$ has Lie algebra $\mathfrak{q}_0^r$, real form of a Levi component $\mathfrak{q}^r$ of $\mathfrak{q}$.

We write $s = s_Y$ in for the complex dimension of the maximal compact subvariety $K(w) \cong K/(K \cap L)$ of Y.

For two classes of representations η of $(\mathfrak{q}, L)$ we'll describe a bundle $\mathbb{E}_\eta \to Y$ and a sheaf $\mathcal{O}_\mathfrak{q}(\mathbb{E}_\eta) \to Y$, and we'll discuss the representations of G on to $H^q(Y; \mathcal{O}_\mathfrak{q}(\mathbb{E}_\eta))$.

First, consider the case where η is finite dimensional. Then the main result is Theorem 6.1 below. It was proved in [27] for the case where W is the flag of Borel subalgebras of $\mathfrak{g}$, and in [34] the argument of [27] was reworked to apply to general W. The argument of [34] is for connected linear semisimple groups G, but those restrictions on G can be dropped by the techniques of [31] and [9, Appendix].

6.1. THEOREM. *Suppose χ is a finite dimensional representation of $(\mathfrak{q}, L)$. Then the $\overline{\partial}$ operator for the Dolbeault complex of $\mathbb{E}_\chi \to Y$ has closed range. $H^q(Y; \mathcal{O}_\mathfrak{q}(\mathbb{E}_\chi))$ is an admissible G–module with finite composition series. Its underlying Harish–Chandra module is the Zuckerman derived functor module $A^q(G, L, \mathfrak{q}, \chi)$. If E_χ has infinitesimal character with Harish–Chandra parameter $\lambda + \rho_\mathfrak{l}$ (corresponding to highest weight λ) and if $\chi(\mathfrak{q}^n) = 0$ then $H^q(Y; \mathcal{O}_\mathfrak{q}(\mathbb{E}_\chi))$ has infinitesimal character $\lambda + \rho$; and then if $\lambda + \rho$ is Y–antidominant then $H^q(Y; \mathcal{O}_\mathfrak{q}(\mathbb{E}_\chi)) = 0$ for $q \neq s$.*

Second, we consider the case where η may be infinite dimensional but is constrained to be one of the cohomology space representations described in Theorem 6.1 for an open L–orbit on the flag manifold of Borel subalgebras of $\mathfrak{l} = \mathfrak{q}^r$. As we saw in §5, this includes all fundamental series[2] representations of L, and a moment's thought shows that it includes all standard representations of L whose character has support that meets the elliptic set[3] in G. In particular this class of representations η contains all the representations in the analytic continuation of the fundamental series.

Let X denote the flag manifold of Borel subalgebras of $\mathfrak{g}$ and consider the natural projection

(6.2) $p : X \to W$ defined by $\mathfrak{b} \subset p(\mathfrak{b})$ for all $\mathfrak{b} \in X$.

The typical fibre of $p : X \to W$ is the complex flag manifold $F \cong Q^r/(B \cap Q^r)$ of all Borel subalgebras $'\mathfrak{b} \subset \mathfrak{q}^r$. The real group L acts on F just as G acts on

---

[2]The *fundamental series* for G is the tempered series associated to a fundamental Cartan subgroup. It is the relative discrete series in case that fundamental Cartan subgroup is compact modulo $Z_G(G^0)$.

[3]The *elliptic set* in G is the union of the G–conjugates of $T \cap G'$ where $H = T \times A$ is a fundamental Cartan subgroup, $T = H \cap K$, and G' is the regular set. In other words, the elliptic set in G is the set of all G–conjugates of G–regular elements of K.

X. Let $L(x) \subset F$ be an open orbit. Since L contains the fundamental Cartan subgroup H of G, and all fundamental Cartan subgroups of L are L^0–conjugate, we may assume that H is the isotropy subgroup of L at x. Let $\mathfrak{b} = \mathfrak{b}_x \subset \mathfrak{g}$ denote the Borel subalgebra represented by $x \in X$; then $'\mathfrak{b} = '\mathfrak{b}_x \subset \mathfrak{q}^r$ is the Borel subalgebra represented by $x \in F$. Note that $\mathfrak{b}$ and $'\mathfrak{b}$ determine each other by: $'\mathfrak{b} = \mathfrak{b} \cap \mathfrak{q}^r$ and $\mathfrak{b} = '\mathfrak{b} + \mathfrak{q}^n$.

Let χ be a finite dimensional representation of $(\mathfrak{b}, H)$. Let $'\chi$ be its restriction to a representation of $('\mathfrak{b}, H)$. consider the associated homogeneous holomorphic vector bundles of finite rank

(6.3) $\qquad \mathbb{E}_\chi \to G(x) \cong G/H$ and $'\mathbb{E}_\chi \to L(x) \cong L/H$; so $'\mathbb{E}_\chi = \mathbb{E}_\chi|_{L(x)}$.

They define the sheaves of germs of holomorphic sections

(6.4) $\qquad \mathcal{O}_\mathfrak{b}(\mathbb{E}_\chi) \to G(x)$ and $\mathcal{O}_{'\mathfrak{b}}('\mathbb{E}_\chi) \to L(x)$. so $\mathcal{O}_{'\mathfrak{b}}('\mathbb{E}_\chi) = \mathcal{O}_\mathfrak{b}(\mathbb{E}_\chi)|_{L(x)}$

which conversely define the holomorphic structure of the bundles [**28**].

We now suppose that χ, as a representation of H, is $G(x)$–antidominant with infinitesimal character λ. We also suppose that $\chi(\mathfrak{b}^n) = 0$. Denote dimensions of maximal compact subvarieties by

(6.5) $u = \dim_{\mathbb{C}} K(x)$, $t = \dim_{\mathbb{C}}(K \cap L)(x)$, and $s = \dim_{\mathbb{C}} K(w)$; so $u = t + s$.

Then $H^p(L(x); \mathcal{O}_{'\mathfrak{b}}('\mathbb{E}_\chi)) = 0$ for $p \neq t$ and $H^q(G(x); \mathcal{O}_\mathfrak{b}(\mathbb{E}_\chi)) = 0$ for $q \neq u$ by Theorem 6.1. Thus the Leray spectral sequence of $G(x) \to G(w)$ collapses at E_2,

(6.6) $\quad E_2^{a,b} = H^a(G(w); \mathcal{O}_\mathfrak{q}(\mathbb{H}^b(L(x); \mathcal{O}_{'\mathfrak{b}}('\mathbb{E}_\chi))))$ with $d_2 : E_2^{a,b} \to E_2^{a+2,b-1}$

where $\mathbb{H}^v(L(x); \mathcal{O}_{'\mathfrak{b}}('\mathbb{E}_\chi)) \to G(w)$ is the homogeneous vector bundle whose typical fibre is the $(\mathfrak{q}, L)$–module $H^v(L(x); \mathcal{O}_{'\mathfrak{b}}('\mathbb{E}_\chi))$. Thus

(6.7) $\qquad H^q(G(x); \mathcal{O}_\mathfrak{b}(\mathbb{E}_\chi)) = \sum_{a+b=q} H^a(G(w); \mathcal{O}_\mathfrak{q}(\mathbb{H}^b(L(x); \mathcal{O}_{'\mathfrak{b}}('\mathbb{E}_\chi))))$.

Again by the vanishing of Theorem 6.1, the left side vanishes for $q \neq u$ and the right side vanishes for $b \neq t$, so (6.5) forces

(6.8) $\qquad H^{s+t}(G(x); \mathcal{O}_\mathfrak{b}(\mathbb{E}_\chi)) = H^s(G(w); \mathcal{O}_\mathfrak{q}(\mathbb{H}^t(L(x); \mathcal{O}_{'\mathfrak{b}}('\mathbb{E}_\chi))))$.

Modulo the Fréchet space results described in §7 below, we have proved

6.9. THEOREM. *Let η denote $H^t(L(x); \mathcal{O}_{'\mathfrak{b}}('\mathbb{E}_\chi))$ as a representation of $(\mathfrak{q}, L)$ and let $\mathbb{V}_\eta \to G(w)$ denote the associated homogeneous vector bundle. Then $H^q(G(w); \mathcal{O}_\mathfrak{q}(\mathbb{V}_\eta)) = 0$ for $q \neq s$, $H^s(G(w); \mathcal{O}_\mathfrak{q}(\mathbb{V}_\eta))$ has a natural Fréchet space structure, and the natural action of G on $H^s(G(w); \mathcal{O}_\mathfrak{q}(\mathbb{V}_\eta))$ is a continuous representation. Further, that action of G on $H^s(G(w); \mathcal{O}_\mathfrak{q}(\mathbb{V}_\eta))$ is an admissible representation with finite composition series. It has infinitesimal character $\lambda + \rho$. Its underlying Harish–Chandra module is the Zuckerman derived functor module $A^{s+t}(G, H, \mathfrak{b}, \chi) = A^s(G, L, \mathfrak{q}, \eta)$.*

Section 7. Representations for Measurable Orbits.

Fix a parabolic subgroup $P \subset G$, not necessarily cuspidal. In this Section we study measurable integrable orbits $Y = G(w) \subset W$ in complex flag manifolds. For the appropriate choices of (W, w), which means $\mathfrak{q} = \mathfrak{q}_w$ such that $\mathfrak{p} = \mathfrak{q} + \tau\mathfrak{q}$, we show that the fibres of $Y \to G/P$ are (up to topological components) the holomorphic arc components of Y. With this, we study representations induced from P from the viewpoint of our orbit picture, extending the scope of our geometric construction of the standard tempered representations.

We now look for all complex flag manifolds $W \cong \overline{G}_\mathbb{C}$ consisting of the $Int(\mathfrak{g})$–conjugates of a parabolic subalgebra $\mathfrak{q} \subset \mathfrak{g}$, and all real group orbits $Y = G(w) \subset W$, such that

(7.1a) Y is measurable, thus partially complex and of flag type,

(7.1b) Y is integrable, so $\mathfrak{q}_w + \tau\mathfrak{q}_w$ is an algebra, and

(7.1c)   the normalizer $N_G(S_w)$ of the holomorphic arc
                 component S_w has Lie algebra $\mathfrak{p}_0$.

Recall the Langlands decomposition $P = MAN$. Let H denote a fundamental Cartan subgroup of a Levi component P^r of P, let $\mathfrak{a}_0$ denote the "split component" of the center of $\mathfrak{p}_0^r$,

(7.2)
$$\mathfrak{a}_0 = \{\xi \in \mathfrak{h} \cap [\mathfrak{g}_0, \mathfrak{g}_0] \mid \alpha(\xi) \in \mathbb{R} \text{ for all } \alpha \in \Phi(\mathfrak{g}, \mathfrak{h})$$
$$\text{and } \alpha(\xi) = 0 \text{ for all } \alpha \in \Phi(\mathfrak{p}^r, \mathfrak{h})\}.$$

Then $P^r = M \times A$ and $H = J \times A$ where A is the analytic subgroup of G for $\mathfrak{a}_0$, where $\mathfrak{j}_0 = \mathfrak{a}_0^\perp \cap \mathfrak{h}_0$, and where $\mathfrak{m} = \mathfrak{j} + \sum_{\alpha \in \Phi(\mathfrak{p}^r, \mathfrak{h})} \mathfrak{g}_\alpha$. Also, here, J is a fundamental Cartan subgroup of M and N is the analytic subgroup of G for the real form $\mathfrak{n}_0 = \mathfrak{n} \cap \mathfrak{g}_0$ of $\mathfrak{n} = \mathfrak{p}^n$. The real parabolic P is cuspidal if and only if $J/Z_G(G^0)$ is compact.

Given (7.1), the algebras $\mathfrak{p}$ and $\mathfrak{q}$ are related by

(7.3a) $\mathfrak{u}_0 \subset \mathfrak{m}_0$ is the $\mathfrak{m}_0$–centralizer of a subspace of $\mathfrak{j}$,

(7.3b) $\mathfrak{r} \subset \mathfrak{m}$ is a parabolic subalgebra with reductive part $\mathfrak{r}^r = \mathfrak{u}$, and

(7.3c) $\mathfrak{q} = \mathfrak{q}^r + \mathfrak{q}^n$ where $\mathfrak{q}^r = \mathfrak{u} + \mathfrak{a}$ and $\mathfrak{q}^n = \mathfrak{r}^n + \mathfrak{n}$.

Conversely, given P and its Langlands decomposition, (7.3) gives the construction of all $\mathfrak{q}$ that satisfy (7.1). The proof is essentially the same as the proof [31, §6.7] of the case (4.5) where the analytic group U for $\mathfrak{u}_0$ is compact modulo $Z_G(G^0)$.

Now fix a complex flag manifold W and a measurable integrable orbit $Y = G(w)$ as in (7.1). Then W consists of the $Int(\mathfrak{g})$–conjugates of a parabolic subalgebra $\mathfrak{q} = \mathfrak{q}_w \subset \mathfrak{g}$ that satisfies (7.3). In particular G has isotropy subgroup UAN at w, and M has isotropy subgroup U at w, where $U/Z_G(G^0)$ may be noncompact but the conditions immediately following (4.4c) remain valid. The holomorphic arc components of Y are the $gS_w = S_{gw} \cong M^\dagger/U$, and gS_w is the topological component of w in the fibre of $Y \cong G/UAN \to G/MAN = G/P$ over gP.

S_w is a measurable open M–orbit on the sub–flag $\overline{M}_\mathbb{C}(w) \subset W$ and AN acts trivially on S_w. Now consider representations

(7.4) β : representation of $(\mathfrak{r} + \mathfrak{a}, UA)$, admissible and of finite length on UA.

Let $\rho_\mathfrak{a} = \frac{1}{2}\sum_{\Phi^+(\mathfrak{g}_0,\mathfrak{a}_0)}(\dim \mathfrak{g}_\phi)\phi$ as before. Then as in (4.8), UAN also acts on the representation space E_β by $\gamma_\beta(uan) = e^{\rho_\mathfrak{a}}(a)\beta(ua)$, and that specifies a homogeneous vector bundle $\mathbb{E}_{\gamma_\beta} \to Y \cong G/UAN$. Then γ_β is also a representation of $\mathfrak{q} = \mathfrak{r} + \mathfrak{n}$ because β is defined on $\mathfrak{r}$ and we defined γ_β to annihilate $\mathfrak{n}$. Now γ_β is a representation of $(\mathfrak{q}, UAN)$. Thus we have

(7.5a) $\pi_{\beta,\mathfrak{q}}$: representation of G on $H^q(Y; \mathcal{O}_\mathfrak{q}(\mathbb{E}_{\gamma_\beta}))$ and

(7.5b) $\eta_{\beta,\mathfrak{q}}$: representation of MAN on $H^q(M/U; \mathcal{O}_\mathfrak{r}(\mathbb{E}_{\gamma_\beta}|_{M/U}))$

such that $\pi_{\beta,\mathfrak{q}} = Ind_P^G(\eta_{\beta,\mathfrak{q}})$.

First consider the case where β is finite dimensional. Apply Theorem 6.1 to $M(w) \subset \overline{M}_\mathbb{C}(w)$. The $\overline{\partial}$ operator for the Dolbeault complex of $\mathbb{E}_{\gamma_\beta} \to M(w) \cong MA/UA$ has closed range. The cohomologies $H^q(M(w); \mathcal{O}_\mathfrak{q}(\mathbb{E}_{\gamma_\beta}))$ are admissible Fréchet MA–modules with finite composition series and underlying Harish–Chandra modules $A^s(MA, UA, \mathfrak{r} + \mathfrak{a}, \gamma_\beta)$. The representations here are the $\eta_{\beta,\mathfrak{q}}$ of (7.5b).

Given $gP \in G/P$ we have the $\overline{\partial}$ operator for the Dolbeault complex of $\mathbb{E}_{\gamma_\beta}|_{gM(w)} \to gM(w)$. The base space there is the fibre of
$$Y \cong G/UAN \to G/MAN = G/P \text{ over } gP.$$
These Dolbeault operators $\overline{\partial}_{gP}$ for the fibre–restrictions of $\mathbb{E}_{\gamma_\beta}$ fit together to form

(7.6)
$$\overline{\partial}_Y : \text{Cauchy-Riemann operator for}$$
$$\text{the partial Dolbeault complex of } \mathbb{E}_{\gamma_\beta} \to Y.$$

The partial Dolbeault complex here consists of the spaces $C^{-\omega}(Y; \mathbb{E}_{\gamma_\beta} \otimes \wedge^\bullet \mathbb{N}_Y^*)$ with the operators $\overline{\partial}_Y$. Here $C^{-\omega}$ denotes hyperfunction sections. $\mathbb{N}_Y \to Y$ is the antiholomorphic tangent bundle, intersection of the complexified tangent bundle of Y with the antiholomorphic tangent bundle of W, so $\mathbb{E}_{\gamma_\beta} \otimes \wedge^q \mathbb{N}_Y^* \to Y$ consists of the $\mathbb{E}_{\gamma_\beta}$–valued $(0, q)$–forms on Y.

The cohomology of that partial Dolbeault complex is computed from a subcomplex with hyperfunction coefficients that are C^∞ along the fibres of $Y \to G/P$. More precisely the inclusion

(7.7) $\{C_{G/P}^{-\omega}(Y; \mathbb{E}_{\gamma_\beta} \otimes \wedge^\bullet \mathbb{N}_Y^*), \overline{\partial}_Y\} \hookrightarrow \{C^{-\omega}(Y; \mathbb{E}_{\gamma_\beta} \otimes \wedge^\bullet \mathbb{N}_Y^*), \overline{\partial}_Y\}$

of that subcomplex in the partial Dolbeault complex of $\mathbb{E}_{\gamma_\beta}$ induces isomorphisms of cohomology. The point is that the subcomplex has a natural Fréchet topology adapted to the fibration $Y \to G/P$, and in that topology the operator $\overline{\partial}_Y$ is continuous. This is the content of [**27**, §7].

7.8. LEMMA. *Each* $\overline{\partial}_Y : C_{G/P}^{-\omega}(Y; \mathbb{E}_{\gamma_\beta} \otimes \wedge^q \mathbb{N}_Y^*) \to C_{G/P}^{-\omega}(Y; \mathbb{E}_{\gamma_\beta} \otimes \wedge^{q+1} \mathbb{N}_Y^*)$ *has closed range.*

PROOF. For each fibre $gM(w)$ of $Y \to G/P$ and each integer $q \geq 0$ we write C_{gP}^q for the space $C^\infty(gM(w); (\mathbb{E}_{\gamma_\beta} \otimes \wedge^q \mathbb{N}_Y^*)|_{gM(w)})$ of C^∞ bundle valued forms over that fibre, we write Z_{gP}^q for the kernel of $\overline{\partial}_{gP} : C_{gP}^q \to C_{gP}^{q+1}$, and we write B_{gP}^q for the image of $\overline{\partial}_{gP} : C_{gP}^{q-1} \to C_{gP}^q$. We know from Theorem 6.1 that Z_{gP}^q and B_{gP}^q are closed subspaces of the Fréchet space C_{gP}^q. In particular $\overline{\partial}_{gP}$ induces a Fréchet space isomorphism of C_{gP}^q/Z_{gP}^q onto B_{gP}^{q+1}.

Now write $\mathbb{C}^q \to G/P$, $\mathbb{Z}^q \to G/P$ and $\mathbb{B}^q \to G/P$ for the G–homogeneous Fréchet bundles over G/P with fibre over gP given by C_{gP}^q, Z_{gP}^q and B_{gP}^q, respectively. $C_{G/P}^{-\omega}(Y; \mathbb{E}_{\gamma_\beta} \otimes \wedge^q \mathbb{N}_Y^*)$ is the space of $C^{-\omega}$ sections of $\mathbb{C}^q \to G/P$, and there the kernel and image of $\overline{\partial}_Y$ are the respective spaces of $C^{-\omega}$ sections of $\mathbb{Z}^q \to G/P$ and $\mathbb{B}^q \to G/P$.

If $\phi \in C_{G/P}^{-\omega}(Y; \mathbb{E}_{\gamma_\beta} \otimes \wedge^q \mathbb{N}_Y^*)$ then $\overline{\partial}_Y(\phi) = 0 \iff$ each $\overline{\partial}_{gP}(\phi|_{gM(w)}) = 0$. It follows that the subbundle $\mathbb{Z}^q \to G/P$ is the kernel of $\overline{\partial}_Y : \mathbb{C}^q \to \mathbb{C}^{q+1}$. Thus $\overline{\partial}_Y$ induces an injective bundle map $\gamma^q : \mathbb{C}^q/\mathbb{Z}^q \to \mathbb{B}^{q+1}$. But γ^q is invertible on each fibre, so it must be invertible. Now γ^q is surjective, so $\overline{\partial}_Y$ maps $\mathbb{C}^q$ onto $\mathbb{B}^{q+1}$. In other words, the range of $\overline{\partial}_Y : C_{G/P}^{-\omega}(Y; \mathbb{E}_{\gamma_\beta} \otimes \wedge^q \mathbb{N}_Y^*) \to C_{G/P}^{-\omega}(Y; \mathbb{E}_{\gamma_\beta} \otimes \wedge^{q+1} \mathbb{N}_Y^*)$ is the space of $C^{-\omega}$ sections of the Fréchet subbundle $\mathbb{B}^{q+1} \to G/P$ of $\mathbb{C}^{q+1} \to G/P$. Thus $\overline{\partial}_Y$ has closed range as asserted. $\square$

Now we have a serious extension of Theorem 6.1:

7.9. THEOREM. *Let β be a finite dimensional representation of $(\mathfrak{r} + \mathfrak{a}, UA)$. Then the cohomologies $H^q(Y; \mathcal{O}_q(\mathbb{E}_{\gamma_\beta}))$ are admissible Fréchet G–modules with finite composition series. Their underlying Harish–Chandra modules are the Zuckerman derived functor modules $A^q(G, UA, \mathfrak{r}+\mathfrak{a}, \beta)$. If E_β has highest weight λ, then the representation $\pi_{\beta,q}$ of G on $H^q(Y; \mathcal{O}_q(\mathbb{E}_{\gamma_\beta}))$ has infinitesimal character $\lambda + \rho_{\mathfrak{m}}$. If $\lambda + \rho_{\mathfrak{m}}$ is S_w–antidominant (5.4), then $H^q(Y; \mathcal{O}_q(\mathbb{E}_{\gamma_\beta})) = 0$ for $q \neq s_{S_w}$.*

Finally we consider the case where β may be infinite dimensional but is constrained to be one of the cohomology representations of Theorem 7.9 for an open M–orbit on the flag manifold of all Borel subalgebras of $\mathfrak{m}$.

Start with the projection $p : X \to W$ from the flag manifold of all Borel subalgebras of $\mathfrak{g}$, as in (6.2). Fix a Borel subalgebra $\mathfrak{b} = \mathfrak{h} + \sum_{\alpha \in \Phi^+(\mathfrak{g}, \mathfrak{h})} \mathfrak{g}_{-\alpha} \subset \mathfrak{q}$ where $\mathfrak{h} = \mathfrak{j} + \mathfrak{a}$ corresponds to a fundamental Cartan subgroup of $MA = P^r$ and where $\tau \Phi^+(\mathfrak{m}, \mathfrak{j}) = -\Phi^+(\mathfrak{m}, \mathfrak{j})$. Now $p(x) = w$ where $\mathfrak{b} = \mathfrak{b}_x$ and $M(x)$ is open in the flag manifold $\overline{M}_{\mathbb{C}}(x)$ of Borel subalgebras of $\mathfrak{m}$.

Let χ be a finite dimensional representation of $(\mathfrak{b}, H)$. Then we have the bundles and sheaves of (6.3) and (6.4), except that those over $G(x)$ are holomorphic only over the holomorphic arc components. Now suppose that χ, as a representation of H, is $G(x)$–antidominant with infinitesimal character λ. Suppose also that $\chi(\mathfrak{b}^n) = 0$. Essentially as in (6.5) write

(7.10) $\qquad u = \dim_{\mathbb{C}}(K \cap M)(x), \ t = \dim_{\mathbb{C}}(K \cap U)(x), \ \text{and}$
$\qquad\qquad s = \dim_{\mathbb{C}}(K \cap M)(w); \ \text{so } u = t + s \ .$

Then $H^p(U(x); \mathcal{O}'_{\mathfrak{b}}('\mathbb{E}_{\chi \otimes e^{\rho_{\mathfrak{a}}}})) = 0$ for $p \neq t$ and $H^q(G(x); \mathcal{O}_{\mathfrak{b}}(\mathbb{E}_{\chi \otimes e^{\rho_{\mathfrak{a}}}})) = 0$ for $q \neq u$ by Theorem 7.9. As in (6.6) and (6.7), the Leray spectral sequence of $G(x) \to G(w)$ collapses at E_2 and

$$(7.11) \ \ H^q(G(x); \mathcal{O}_{\mathfrak{b}}(\mathbb{E}_{\chi \otimes e^{\rho_{\mathfrak{a}}}})) = \sum_{a+b=q} H^a(G(w); \mathcal{O}_{\mathfrak{q}}(\mathbb{H}^b(U(x); \mathcal{O}'_{\mathfrak{b}}('\mathbb{E}_{\chi \otimes e^{\rho_{\mathfrak{a}}}})))).$$

Here we use the fact that the isotropy subgroup UAN of G at w has orbit $UAN(x) = U(x)$ because $\mathfrak{a} + \mathfrak{n} \subset \mathfrak{b}$. The vanishing in Theorem 7.9 now tells us that

$$(7.12) \quad H^{s+t}(G(x); \mathcal{O}_{\mathfrak{b}}(\mathbb{E}_{\chi \otimes e^{\rho_{\mathfrak{a}}}})) = H^s(G(w); \mathcal{O}_{\mathfrak{q}}(\mathbb{H}^t(U(x); \mathcal{O}'_{\mathfrak{b}}('\mathbb{E}_{\chi \otimes e^{\rho_{\mathfrak{a}}}})))).$$

It also follows from Theorem 7.9 that the space (7.12) has a natural Fréchet space structure for which the action of G is a continuous representation. Now we have the extension of Theorem 6.9 from measurable open orbits to measurable integrable orbits:

7.13. THEOREM. *Let η denote $H^t(U(x); \mathcal{O}'_{\mathfrak{b}}('\mathbb{E}_{\chi \otimes e^{\rho_{\mathfrak{a}}}}))$ as a representation of $(\mathfrak{q}, UAN)$ and let $\mathbb{V}_\eta \to G(w)$ denote the associated homogeneous vector bundle. Then $H^q(G(w); \mathcal{O}_{\mathfrak{q}}(\mathbb{V}_\eta)) = 0$ for $q \neq s$, $H^s(G(w); \mathcal{O}_{\mathfrak{q}}(\mathbb{V}_\eta))$ has a natural Fréchet space structure and the natural action of G on $H^s(G(w); \mathcal{O}_{\mathfrak{q}}(\mathbb{V}_\eta))$ is a continuous representation. Further, that action of G on $H^s(G(w); \mathcal{O}_{\mathfrak{q}}(\mathbb{V}_\eta))$ is an admissible representation with finite composition series. It has infinitesimal character $\lambda + \rho_{\mathrm{m}}$. Its underlying Harish–Chandra module is the Zuckerman derived functor module $A^{s+t}(G, H, \mathfrak{b}, \chi) = A^s(G, UA, \mathfrak{q}, \eta)$.*

Section 8. Open Problems.

The first obvious open problem is to remove the requirement of finite dimensionality from the representations χ of Theorem 6.1. This is done, but only in special cases, in Theorem 6.9. The problem divides into several parts: a clean functorial definition of the topology of the Dolbeault complex, the closed range problem for the $\overline{\partial}$–operator, keeping track of the infinitesimal character, and a vanishing theorem for the antidominant case. It would be especially interesting here to understand whether the infinitesimal character and the vanishing theorem really need $\chi(\mathfrak{q}^n) = 0$, even in the finite dimensional case. Most of this was done by H.-W. Wong [34] for finite dimensional χ, but it is not at all clear how to proceed in the infinite dimensional case.

The second obvious open problem is to remove the requirement of finite dimensionality from the representations β of Theorem 7.9. This is done, again only in special cases, in Theorem 7.13, respectively. The problems include those of the measurable open orbit case, but here one must first find a good subcomplex of the partial Dolbeault complex that computes the cohomologies, and then one must have the solution to the first problem for the holomorphic arc components.

Third, it would be good to use this geometric setting to obtain character formulas. This is done in another setting by Hecht and Taylor [13].

Fourth, one needs a better connection between representations constructed as in this paper from the G–orbit structure of a complex flag manifold W and those constructed by localization methods from the $\overline{K}_{\mathbb{C}}$–orbit structure. This is fine up on the flag of Borel subalgebras of $\mathfrak{g}$ ([9], [10]), but not yet satisfactory in general.

Finally, of course, one can ask how much of this goes over to the quantum group setting.

References

1. A. Beilinson and J. Bernstein, *Localisation des $\mathfrak{g}$-modules*, C. R. Acad. Sci. Paris **292** (1981), 15–18.
2. R. Bott, *Homogeneous vector bundles*, Annals of Math. **66** (1957), 203–248.
3. Harish–Chandra, *The Plancherel formula for complex semisimple Lie groups*, Trans. Amer. Math. Soc. **76** (1954), 485–528.
4. _____, *Representations of semisimple Lie groups*, VI, Amer. J. Math. **78** (1956), 564–628.
5. _____, *Discrete series for semisimple Lie groups*, II, Acta Math. **116** (1966), 1–111.
6. _____, *Harmonic analysis on real reductive groups*, I, J. Functional Analysis **19** (1975), 104–204.
7. _____, *Harmonic analysis on real reductive groups*, II, Inventiones Math. **36** (1976), 1–55.
8. _____, *Harmonic analysis on real reductive groups*, III, Annals of Math. **104** (1976), 117–201.
9. H. Hecht, D. Miličić, W. Schmid and J. A. Wolf, *Localization and standard modules for semisimple Lie Groups, I: The duality theorem*, Invent. Math. **90** (1987), 297–332.
10. _____, *Localization and standard modules for semisimple Lie Groups, II: Vanishing and irreducibility theorems*, to appear.
11. H. Hecht and J. Taylor, *Analytic localization of group representations*, Advances in Math. **79** (1990), 139–212..
12. _____, *A comparison theorem for n-homology*, Compositio Math., to appear.
13. _____, *A geometric formula for characters of semisimple Lie groups*, to appear.
14. R. A. Herb, *Fourier inversion and the Plancherel formula for semisimple Lie groups*, Amer. J. Math. **104** (1982), 9–58.
15. _____, *Discrete series characters and Fourier inversion on semisimple real Lie groups*, Trans. Amer. Math. Soc. **277** (1983), 241–261.
16. _____, *The Schwartz space of a general semisimple group, II: Wave packets associated to Schwartz functions*, Trans. Amer. Math. Soc. **327** (1991), 1–69.
17. R. A. Herb and J. A. Wolf, *The Plancherel theorem for general semisimple Lie groups*, Compositio Math. **57** (1986), 271–355.
18. _____, *Rapidly decreasing functions on general semisimple Lie groups*, Compositio Math. **58** (1986), 73–110.
19. B. Kostant, *Lie algebra cohomology and the generalized Borel–Weil theorem*, Annals of Math. **74** (1961), 329–387.
20. R. Parthasarathy, *Dirac operator and the discrete series*, Annals of Math. **96** (1972), 1–30.

21. P. Sally and G. Warner, *The Fourier transform on semisimple Lie groups of real rank one*, Acta Math. **131** (1973), 11–26.

22. W. Schmid, *Homogeneous complex manifolds and representations of semisimple Lie groups*, Ph.D. thesis, University of California at Berkeley, 1967, Representation Theory and Harmonic Analysis on Semisimple Lie Groups, Mathematical Surveys and Monographs **31** (eds. P. Sally and D. Vogan), 223–286, Amer. Math. Soc., 1988.

23. _____, *On a conjecture of Langlands*, Annals of Math. **93** (1971), 1–42.

24. _____, *Some properties of square integrable representations of semisimple Lie groups*, Annals of Math. **102** (1975), 535–564.

25. _____, *L²-cohomology and the discrete series*, Annals of Math. **103** (1976), 375–394.

26. _____, *Boundary value problems for group invariant differential equations*, Proceedings, Cartan Symposium, Lyon, 1984, Astérisque (1985), 311–322.

27. W. Schmid and J. A. Wolf, *Geometric quantization and derived functor modules for semisimple Lie groups*, J. Functional Analysis **90** (1990), 48–112.

28. J. A. Tirao and J. A. Wolf, *Homogeneous holomorphic vector bundles*, Indiana Univ. Math. J. **20** (1970), 15–31.

29. D. Vogan, *Representations of real reductive Lie groups*, Progress in Math. **15**, Birkhaüser (1981).

30. J. A. Wolf, *The action of a real semisimple Lie group on a complex manifold, I: Orbit structure and holomorphic arc components*, Bull. Amer. Math. Soc. **75** (1969), 1121–1237.

31. _____, *The action of a real semisimple Lie group on a complex manifold, II: Unitary representations on partially holomorphic cohomology spaces*, Memoirs Amer. Math. Soc. **138** (1974).

32. _____, *Partially harmonic spinors and representations of reductive Lie groups*, J. Functional Analysis **15** (1974), 117–154.

33. _____, *Geometric realizations of discrete series representations in a nonconvex holomorphic setting*, Bull. Soc. Math. Belgique **42** (1990), 797–812.

34. H.-W. Wong, *Dolbeault cohomologies and Zuckerman modules associated with finite rank representations*, Ph.D. thesis, Harvard University, 1991.

DEPARTMENT OF MATHEMATICS, UNIVERSITY OF CALIFORNIA, BERKELEY, CALIFORNIA 94720, USA

E-mail address: jawolf@math.berkeley.edu

Contemporary Mathematics
Volume **154**, 1993

Unipotent Representations and Cohomological Induction

DAVID A. VOGAN, JR.

1. Introduction.

Suppose G is a real reductive Lie group in Harish-Chandra's class. (The definition and basic properties of this class may be found in [**5**], section 3.) This paper is concerned with an algebraic construction, called *cohomological induction*, of unitary representations of G related to certain complex homogeneous spaces. Finding a geometric construction parallel to the algebraic one is a topic of current research. We will outline some aspects of the algebraic construction for which it seems possible to give "best" formulations. Our hope is that the details of these formulations may ultimately help to suggest proofs for their geometric analogues.

We begin by describing the class of homogeneous spaces under consideration. Write $\mathfrak{g}_0 = \mathrm{Lie}(G)$, and $\mathfrak{g} = \mathfrak{g}_0 \otimes_{\mathbb{R}} \mathbb{C}$ for its complexification. Write bar for the complex conjugation on $\mathfrak{g}$ corresponding to the real form $\mathfrak{g}_0$. Analogous notation will be used for other real Lie groups. A parabolic subalgebra $\mathfrak{q}$ of $\mathfrak{g}$ is called *purely complex* if $\bar{\mathfrak{q}}$ is opposite to $\mathfrak{q}$; that is, if the intersection $\mathfrak{l} = \mathfrak{q} \cap \bar{\mathfrak{q}}$ is a Levi subalgebra of $\mathfrak{q}$. In this case $\mathfrak{l}$ is the complexification of the real Lie algebra

$$(1.1)(a) \qquad \mathfrak{l}_0 = \mathfrak{l} \cap \mathfrak{g}_0 = \mathfrak{q} \cap \mathfrak{g}_0 = \bar{\mathfrak{q}} \cap \mathfrak{g}_0.$$

This Lie algebra is the Lie algebra of the subgroup

$$(1.1)(b) \qquad L = \{g \in G | \mathrm{Ad}(g)(\mathfrak{q}) = \mathfrak{q}\} = \{g \in G | \mathrm{Ad}(g)(\bar{\mathfrak{q}}) = \bar{\mathfrak{q}}\},$$

which we call the *Levi subgroup of* $\mathfrak{q}$. Write $\mathfrak{u}$ for the nil radical of $\mathfrak{q}$; then $\bar{\mathfrak{u}}$ is the nil radical of $\bar{\mathfrak{q}}$. The triangular decomposition of $\mathfrak{g}$ with respect to $\mathfrak{q}$ (and

---

1991 *Mathematics Subject Classification*. Primary 22E47; Secondary 17B10.

Supported in part by NSF Grant DMS 90 11483.

This paper is in final form and no version of it will be submitted for publication elsewhere.

the Levi subalgebra $\mathfrak{l}$) is

$$(1.1)(c) \qquad\qquad \mathfrak{g} = \mathfrak{u} \oplus \mathfrak{l} \oplus \overline{\mathfrak{u}}.$$

Now the tangent space at eL to the homogeneous space G/L may be identified with $\mathfrak{g}_0/\mathfrak{l}_0$. Its complexification is $\mathfrak{g}/\mathfrak{l}$, which according to $(1.1)(c)$ has a natural splitting

$$(1.1)(d) \qquad\qquad T_{eL}(G/L)_{\mathbb{C}} \simeq \mathfrak{g}/\mathfrak{l} \simeq \mathfrak{u} \oplus \overline{\mathfrak{u}}.$$

Complex conjugation interchanges these summands, so $(1.1)(d)$ defines a complex structure on $T_{eL}(G/L)$, with $\mathfrak{u}$ the antiholomorphic subspace. According to $(1.1)(b)$, this structure is preserved by $\mathrm{Ad}(L)$; so it gives rise to a G-invariant almost complex structure on G/L. The structure is integrable because $\mathfrak{q}$ is a Lie subalgebra of $\mathfrak{g}$; so $\mathfrak{q}$ defines a G-invariant complex structure on G/L.

If G has a complexification $G_{\mathbb{C}}$ (a complex connected reductive Lie group with Lie algebra $\mathfrak{g}$, containing G as a subgroup with Lie algebra $\mathfrak{g}_0$), then this construction can be simplified. There is a parabolic subgroup $Q_{\mathbb{C}}$ of $G_{\mathbb{C}}$ with Lie algebra $\mathfrak{q}$, and $L = Q_{\mathbb{C}} \cap G$. This provides an embedding

$$(1.1)(e) \qquad\qquad G/L \hookrightarrow G_{\mathbb{C}}/Q_{\mathbb{C}}.$$

It is an open embedding as a consequence of $(1.1)(c)$. The (compact) homogeneous space $G_{\mathbb{C}}/Q_{\mathbb{C}}$ carries a natural invariant complex structure since $Q_{\mathbb{C}}$ is a complex subgroup of the complex group $G_{\mathbb{C}}$; so the open subset G/L inherits a complex structure.

Example 1.2. Suppose $G = GL(2n, \mathbb{R})$, the group of invertible linear transformations of $\mathbb{R}^{2n}$. Write X for the space of complex structures on $\mathbb{R}^{2n}$. One way to think of such a complex structure is as a $2n$ by $2n$ real matrix J with the property that $J^2 = -I$. The minimal polynomial of any such matrix is obviously $x^2 + 1$, which is irreducible over $\mathbb{R}$. It follows from the theory of rational canonical forms that any two such matrices are similar; that is, that G acts transitively on X. We can fix a base point J_0 in X by fixing an identification of $\mathbb{R}^{2n}$ with $\mathbb{C}^n$. Then the isotropy group of the action of G at J_0 will consist of all invertible real-linear transformations respecting the complex structure; that is, of all invertible complex-linear transformations. Consequently

$$(1.2)(a) \qquad\qquad X \simeq GL(2n, \mathbb{R})/GL(n, \mathbb{C}).$$

To see X as a complex manifold, it is helpful to look at complex structures in a slightly different way. The complexification of $\mathbb{R}^{2n}$ is $\mathbb{C}^{2n}$. A complex structure on $\mathbb{R}^{2n}$ may be regarded as a complex subspace V of $\mathbb{C}^{2n}$ with the property that

$$(*) \qquad\qquad \mathbb{C}^{2n} = V \oplus \overline{V}.$$

Equivalently, V should be of complex dimension n and satisfy

$$(**) \qquad\qquad V \cap \overline{V} = 0.$$

Now let $Gr(n, 2n)$ be the complex Grassmann variety of n-dimensional complex subspaces of $\mathbb{C}^{2n}$. This is a complex manifold (even a complex projective algebraic variety) on which $GL(2n, \mathbb{C})$ acts holomorphically. The subgroup $G = GL(2n, \mathbb{R})$ therefore acts holomorphically on Z as well. By $(**)$,

$$(1.2)(b) \qquad X \hookrightarrow Gr(n, 2n), \qquad X \simeq \{V \in Gr(n, 2n) | V \cap \overline{V} = 0\}.$$

This condition on V is open, so X is an open subset of $Gr(n, 2n)$. It therefore inherits a complex structure.

Why are these particular complex homogeneous spaces of special interest? One answer comes from the "method of coadjoint orbits" of Kirillov and Kostant (see for example [4] or [11], Chapter 10). The spaces G/L as in (1.1) are exactly the coadjoint orbits of G admitting totally complex polarizations, and these polarizations are exactly the complex structures of (1.1). As explained in [12] or [11], the orbit method suggests that most interesting unitary representations of reductive groups should arise in three steps. The first step (which is very poorly understood) is attaching representations to nilpotent coadjoint orbits. The second (which is the subject of this paper) is cohomological induction in the setting of (1.1), going from representations of L to representations of G. The third (going back to work of Gelfand, Harish-Chandra, and Mackey more than forty years ago) is parabolic induction from Levi subgroups of real parabolic subgroups.

Perhaps surprisingly, this very abstract philosophy offers some concrete guidance about how to do representation theory in the setting (1.1). Parabolic induction from a real parabolic subgroup can be applied to any unitary representation of the Levi subgroup to yield a unitary representation of G. Cohomological parabolic induction is not so well behaved: one can get unitary representations of G only from certain representations of L. If we think of cohomological induction as just a step in the method of coadjoint orbits, as in the preceding paragraph, we see why this is so: we should apply it only to representations of L attached to nilpotent coadjoint orbits. These are the "unipotent representations" of the title. The simplest examples are the one-dimensional characters of L, and geometric work on cohomological induction has concentrated on this case. Geometrically the next obvious case to consider might be finite-dimensional representations of L. But finite-dimensional representations are usually not unipotent, and the orbit method suggests (correctly) that the unitarity of the corresponding cohomologically induced representations is difficult to understand. A better generalization is to infinite-dimensional unipotent representations of L. There is a serious difficulty with this program, however: unipotent representations have not yet been defined. To circumvent this problem, we recall from [10] the notion of "weakly unipotent representation" (Definition 4.6 below). Whatever unipotent representations turn out to be, the evidence suggests that they will at least be weakly unipotent.

This outline omits a second subtlety, also illuminated by the method of coadjoint orbits. When the orbit method presents a unipotent representation of L to which cohomological induction should be applied, it also provides a particular choice of complex structure on G/L. If we think of fixing G/L with a complex structure given by $\mathfrak{q}$ as in (1.1), then (in order to stay in the context of the orbit method) we should consider just representations of L whose central character is in some sense "positive" in the direction of $\mathfrak{q}$.

To summarize, the orbit method suggests that in the setting (1.1), each "positive unipotent" unitary representation of L should give rise to a unitary representation of G. Our goal is to give a precise formulation of this statement (Theorem B_2 at the end of section 4). Along the way, we will formulate two other results that suggest the rôle of cohomological induction in the classification of all unitary representations (Theorems A_1 and B_1 after Definition 3.7).

All of these theorems can be extended in various ways. The hypotheses are "best" only in the sense of being very natural. An analogous "best" theorem in the setting of induction from real parabolic subgroups is that unitary representations go to unitary representations. There is much more to say (the theory of complementary series), but to say it requires much harder and less natural statements and proofs, often peculiar to a particular family of groups or representations. What is presented here is supposed to be the easiest (and perhaps most important) part of the theory.

Here are some general remarks about proofs and references. I have included outlines of only a few of the easiest proofs, mostly to shed light on definitions, or to indicate how to translate a result in a reference to what is given here. As compensation, I have tried to provide fairly complete references. For a few results there is no published reference, mostly because the basic definitions have been improved somewhat since the main sources were written. Tony Knapp and I are writing a book that is intended to include complete proofs of almost everything here, together with additional material relevant to the classification of all unitary representations. His work on that project has done much to shape the perspective adopted in this paper, and I am grateful for his insights.

2. Algebraic setting.

Suppose θ is an involutive automorphism of G. Write K for the group of fixed points of θ and $\mathfrak{s}_0$ for the -1 eigenspace of the differential of θ on $\mathfrak{g}_0$. We say that θ is a *Cartan involution* of G if K is compact, and the natural map

$$(2.1) \qquad\qquad K \times \mathfrak{s}_0 \to G, \qquad (k, X) \mapsto k \exp(X)$$

is a diffeomorphism. Such involutions always exist; this is a consequence of the requirement that G be in Harish-Chandra's class (see [5]). The decomposition

(2.1) is called a *Cartan decomposition*; it implies easily that K is a maximal compact subgroup of G.

Lemma 2.2 ([**16**], 1.2 and 1.3). *Suppose H is a Cartan subgroup of G. Then there is a Cartan involution of G preserving H.*

Another way to phrase essentially the same result is that if θ is a Cartan involution of G, then every conjugacy class of Cartan subgroups of G has a θ-stable representative.

Suppose G/L is a complex homogeneous space as in (1.1).

Proposition 2.3. *There is a Cartan involution θ of G preserving L. Such an involution automatically preserves the parabolic subalgebra $\mathfrak{q}$, and therefore $\bar{\mathfrak{q}}$, $\mathfrak{u}$, and so on. Its restriction to L is a Cartan involution of L.*

Proof. Choose a Cartan subgroup H of L. Because L is a Levi subgroup of G, H must be a Cartan subgroup of G. By Lemma 2.2, there is a Cartan involution θ of G preserving H. Then θ and complex conjugation both act on the roots of $\mathfrak{h}$ in $\mathfrak{g}$, and these actions are related by

$$(2.4)(a) \qquad\qquad \theta(\alpha) = -\bar{\alpha}$$

(see for example [**9**], Lemma 0.2.2). Whenever $\mathfrak{v}$ is an ad($\mathfrak{h}$)-stable subspace of $\mathfrak{g}$, we write $\Delta(\mathfrak{v}, \mathfrak{h})$ for the corresponding set of roots. Because of the triangular decomposition (1.1)(c), we have

$$(2.4)(b) \qquad\qquad \Delta(\mathfrak{u}, \mathfrak{h}) = -\Delta(\bar{\mathfrak{u}}, \mathfrak{h}) = -\overline{\Delta(\mathfrak{u}, \mathfrak{h})}$$

By (2.4)(a), it follows that

$$(2.4)(c) \qquad\qquad \Delta(\mathfrak{u}, \mathfrak{h}) = \theta\Delta(\mathfrak{u}, \mathfrak{h}),$$

and therefore that $\theta\mathfrak{u} = \mathfrak{u}$. For similar reasons θ preserves $\mathfrak{l}$, so it preserves $\mathfrak{q} = \mathfrak{l} + \mathfrak{u}$. Because of (1.1)(b), it follows that θ preserves L. It remains only to check that θ provides a Cartan decomposition of L, and this is an easy consequence of the Cartan decomposition for G. Q.E.D.

For the balance of the paper, we fix a Cartan involution θ as in Proposition 2.3, with fixed point group K; then $L \cap K$ is a maximal compact subgroup of L. If we write $\mathfrak{s}$ for the -1 eigenspace of θ on $\mathfrak{g}$, then we have decompositions

$$(2.5)(a) \qquad L = (L \cap K)\exp(\mathfrak{l}_0 \cap \mathfrak{s}), \qquad \mathfrak{k} = (\mathfrak{u} \cap \mathfrak{k}) \oplus (\mathfrak{l} \cap \mathfrak{k}) \oplus (\bar{\mathfrak{u}} \cap \mathfrak{k}),$$

and so on. In particular, $\mathfrak{q} \cap \mathfrak{k}$ defines an invariant complex structure on $K/L \cap K$, making it a compact complex submanifold of G/L. Put

$$(2.5)(b) \qquad\qquad s = \dim_{\mathbb{C}} K/L \cap K = \dim \mathfrak{u} \cap \mathfrak{k}.$$

We will make use of Harish-Chandra's results relating representations of G to $(\mathfrak{g}, K)$-modules, and representations of L to $(\mathfrak{l}, L \cap K)$-modules (see for example

[9], 0.3). Recall that an *invariant Hermitian form* on a $(\mathfrak{g}, K)$-module V is a Hermitian form $\langle , \rangle$ on V with the property that

$$(2.6) \qquad \langle k \cdot v, k \cdot v' \rangle = \langle v, v' \rangle, \qquad \langle X \cdot v, v' \rangle = \langle v, -\overline{X} \cdot v' \rangle$$

for $k \in K$, $X \in \mathfrak{g}$, and v and v' in V. A $(\mathfrak{g}, K)$-module is called *unitarizable* if it admits a positive definite invariant Hermitian form.

Theorem 2.7 (Harish-Chandra). *Passage to K-finite vectors defines a bijection from the set of equivalence classes of irreducible unitary representations of G onto the set of equivalence classes of irreducible unitarizable $(\mathfrak{g}, K)$-modules.*

We want cohomological induction to provide unitary representations of G from unitary representations of L. Because of Theorem 2.7, it is enough to give a construction of unitarizable $(\mathfrak{g}, K)$-modules from unitarizable $(\mathfrak{l}, L \cap K)$-modules. The basic construction has three steps. Suppose Z is an $(\mathfrak{l}, L \cap K)$-module. The first step is to use the quotient mapping $\mathfrak{q} \to \mathfrak{l}$ to regard Z as a $(\mathfrak{q}, L \cap K)$-module (on which $\mathfrak{u}$ acts by zero). The second step is to apply "algebraic production" to form a $(\mathfrak{g}, L \cap K)$-module

$$(2.8)(a) \qquad W_{\mathfrak{q}} = \mathrm{pro}_{\mathfrak{q}, L \cap K}^{\mathfrak{g}, L \cap K}(Z)$$

([9], Definition 6.1.21). By definition,

$$(2.8)(b) \qquad W_{\mathfrak{q}} = \mathrm{Hom}_{\mathfrak{q}}(U(\mathfrak{g}), Z)_{L \cap K - \text{finite}} \simeq \mathrm{Hom}_{\mathbb{C}}(U(\overline{\mathfrak{u}}), Z)_{L \cap K - \text{finite}}.$$

The third step is to apply a Zuckerman functor $\Gamma^i = (\Gamma_{\mathfrak{g}, L \cap K}^{\mathfrak{g}, K})^i$ ([9], Definition 6.2.11) to get a $(\mathfrak{g}, K)$-module:

$$(2.8)(c) \qquad \mathcal{R}^i Z = \Gamma^i W_{\mathfrak{q}}.$$

When $K/L \cap K$ is connected, $\Gamma^0 W$ is just the largest subspace of W on which the Lie algebra action of $\mathfrak{k}_0$ exponentiates to a representation of K_0 that agrees with the given action of $L \cap K_0$ on W. The higher Γ^i are the right derived functors of Γ^0.

Let us try to understand these steps by examining their geometric analogues. Suppose Z has finite length. It is then the Harish-Chandra module of a representation of L on a nice topological vector space V. From V we can form a G-equivariant bundle

$$(2.9) \qquad \mathcal{V} = G \times_L V \to G/L$$

on G/L. The first step in the preceding paragraph (regarding Z as a representation of $\mathfrak{q}$) amounts to regarding $\mathcal{V}$ as a holomorphic bundle (compare [11], Proposition 1.21). (H. Hecht and T. Bratten have pointed out a serious technical difficulty overlooked in [11]: if V is infinite-dimensional, then it is not clear that $\mathcal{V}$ has a local holomorphic trivialization; so it should perhaps not be called a holomorphic vector bundle. Nevertheless there is no dificulty in defining its holomorphic sections, Dolbeault cohomology, and so on, and this is all we will need here.) The second step is equally natural.

Proposition 2.10 ([**11**], Proposition 6.17). *In the setting of* (2.9), *the* $(\mathfrak{g}, L \cap K)$-*module* $W_{\mathfrak{q}}$ *of* (2.8)(b) *may be identified with the space of* $L \cap K$-*finite formal power series holomorphic sections of the holomorphic bundle* $\mathcal{V}$.

From the holomorphic bundle $\mathcal{V}$ we can form Dolbeault cohomology groups

$$(2.11) \qquad\qquad H^i(G/L, \mathcal{V}),$$

the first of which is the space of holomorphic sections of $\mathcal{V}$. Formally these are representations of G, but for $i > 0$ there is a difficulty: to put a Hausdorff topology on a cohomology space, we need to know that the coboundary operator has closed range. Now a K-finite holomorphic section of $\mathcal{V}$ certainly gives rise to an $L \cap K$-finite formal power series section with a locally finite action of $\mathfrak{k}$. Examining the definition of the Zuckerman functor Γ^0, we find

Corollary 2.12. *There is a natural injection from K-finite holomorphic sections of $\mathcal{V}$ into* $\Gamma^0 W_{\mathfrak{q}}$. *That is,*

$$H^0(G/L, \mathcal{V})_K \hookrightarrow \mathcal{R}^0 Z.$$

The map of Corollary 2.12 ought to be an isomorphism. Here is why. Assume for simplicity that $K/L \cap K$ is connected. An element of $\mathcal{R}^0 Z$ is then a K-finite formal power series section of $\mathcal{V}$. Such a formal power series defines a (K-finite) holomorphic section of $\mathcal{V}$ over a formal neighborhood of $K/L \cap K$ in G/L. Very roughly speaking, the complex manifold G/L is Stein in the directions away from $K/L \cap K$. (We refer to [**8**] for a precise formulation.) For this reason, one can hope that a nice section of $\mathcal{V}$ near $K/L \cap K$ should extend to a section over all of G/L; that is, that the map in the corollary should be surjective. In fact, one can make

Conjecture 2.13. Suppose Z is an $(\mathfrak{l}, L \cap K)$-module of finite length, V is a corresponding representation of L, and $\mathcal{V}$ is the associated holomorphic bundle on G/L. Then the coboundary operators of Dolbeault cohomology all have closed range, and the corresponding Harish-Chandra modules for G are

$$H^i(G/L, \mathcal{V})_K \simeq \mathcal{R}^i Z.$$

When Z is finite-dimensional, this has been proved by H. Wong [**17**] (following the treatment of a number of important special cases by Schmid and others). In general, the conjecture is probably easier to prove if we take for V Schmid's maximal globalization of Z; for then $H^i(G/L, \mathcal{V})$ should be the maximal globalization of $\mathcal{R}^i Z$.

At any rate, we see that the functors $\mathcal{R}^i(\cdot)$ provide a reasonable algebraic analogue of the Dolbeault cohomology functors $H^i(G/L, G \times_L \cdot)$. In order to discuss unitarity, we need to see how these functors affect Hermitian forms. Here we find an unpleasant surprise: $\mathcal{R}^i$ cannot naturally be applied to Hermitian forms. Roughly speaking, the problem is that $W_{\mathfrak{q}} \simeq \mathrm{Hom}_{\mathbb{C}}(U(\bar{\mathfrak{u}}), Z)_{L \cap K-\text{finite}}$ is simply too large to carry such a form. (Actually the imposed $L \cap K$-finiteness allows one to find invariant forms on $W_{\mathfrak{q}}$, but not in any natural way.) The only consolation is that the Dolbeault cohomology has a parallel problem: $H^i(G/L, \mathcal{V})$ is known to carry an invariant Hermitian form only when it is finite-dimensional. (The forms arising in Schmid's proof of the Kostant-Langlands conjecture, for example, are defined only on certain dense subspaces of the cohomology.)

So we start over. Suppose again that Z is an $(\mathfrak{l}, L \cap K)$-module. The first step this time is to use the quotient mapping $\bar{\mathfrak{q}} \to \mathfrak{l}$ to regard Z as a $(\bar{\mathfrak{q}}, L \cap K)$-module (on which $\bar{\mathfrak{u}}$ acts by zero). The second step is to apply "algebraic induction" to form a $(\mathfrak{g}, L \cap K)$-module

$$(2.14)(a) \qquad\qquad W_{\bar{\mathfrak{q}}} = \mathrm{ind}_{\bar{\mathfrak{q}}, L \cap K}^{\mathfrak{g}, L \cap K}(Z)$$

([9], Definition 6.1.21). By definition,

$$(2.14)(b) \qquad\qquad W_{\bar{\mathfrak{q}}} = U(\mathfrak{g}) \otimes_{\bar{\mathfrak{q}}} Z \simeq U(\mathfrak{u}) \otimes_{\mathbb{C}} Z.$$

The third step is to apply a projective Zuckerman functor $\Pi_j = (\Pi_{\mathfrak{g}, L \cap K}^{\mathfrak{g}, K})_j$ ([7], page 431) to get a $(\mathfrak{g}, K)$-module:

$$(2.14)(c) \qquad\qquad \mathcal{L}_j Z = \Pi_j W_{\bar{\mathfrak{q}}}.$$

The projective Zuckerman functor, introduced by Bernstein, is related to "largest K-finite quotients" just as Γ is related to "largest K-finite subspaces." Its definition requires a little more care, and a complete treatment is not yet available in the literature. For the moment one can take Theorem 2.16(a) below as a definition, although this is certainly not the most natural approach.

It is not at all clear how to make analogous constructions in the geometric setting, as we did in (2.9)–(2.13) for $\mathcal{R}^i$. At least we can relate this construction to the one in (2.8), however. Write a superscript h for the Hermitian dual functor in the category of $(\mathfrak{l}, L \cap K)$-modules or $(\mathfrak{g}, L \cap K)$-modules or $(\mathfrak{g}, K)$-modules.

Proposition 2.15. *Suppose we are in the setting of* (2.8) *and* (2.14).
(a) *There is a natural* $(\mathfrak{g}, L \cap K)$-*module map*

$$W_{\bar{\mathfrak{q}}} = \mathrm{ind}_{\bar{\mathfrak{q}}, L \cap K}^{\mathfrak{g}, L \cap K}(Z) \to \mathrm{pro}_{\mathfrak{q}, L \cap K}^{\mathfrak{g}, L \cap K}(Z) = W_{\mathfrak{q}}.$$

(b) *Suppose* Z *has finite length. Then* $W_{\mathfrak{q}}$ *and* $W_{\bar{\mathfrak{q}}}$ *have finite length, and they have the same irreducible composition factors and multiplicities.*

(c) *There is a natural isomorphism*

$$[\mathrm{ind}_{\bar{\mathfrak{q}}, L \cap K}^{\mathfrak{g}, L \cap K}(Z)]^h \simeq \mathrm{pro}_{\mathfrak{q}, L \cap K}^{\mathfrak{g}, L \cap K}(Z^h).$$

(d) *Any invariant Hermitian form* $\langle, \rangle_L$ *on* Z *induces an invariant Hermitian form* $\langle, \rangle_{\mathfrak{g}}$ *on* $W_{\bar{\mathfrak{q}}}$.

This is elementary. Part (a) is [10], Lemma 5.15; (c) is [10], Lemma 5.13; and (d) is [10], Lemma 5.19. Part (b) is proved for "positive" Z in [10], Lemma 5.17; the general case may be reduced to this by a tensor product argument. We omit the details.

To continue, we need to know how the Zuckerman functors affect Hermitian forms.

Theorem 2.16 ([3]; see [10], Theorem 5.7). *Suppose we are in the setting of* (2.8) *and* (2.14).

(a) *There is a natural equivalence of functors*

$$(\Pi_{\mathfrak{g},L\cap K}^{\mathfrak{g},K})_j \simeq (\Gamma_{\mathfrak{g},L\cap K}^{\mathfrak{g},K})^{2s-j}.$$

(b) *If W is any $(\mathfrak{g}, L \cap K)$-module, there is a natural isomorphism*

$$[\Pi_j(W)]^h \simeq \Gamma^j(W^h).$$

(c) *Any invariant Hermitian form $\langle,\rangle_{\mathfrak{g}}$ on a $(\mathfrak{g}, L \cap K)$-module W induces an invariant Hermitian form $\langle,\rangle_G$ on $\Pi_s(W)$.*

In order to deduce this from [10] (which does not use the projective Zuckerman functors Π_j), one can take (a) as a definition; part (b) is then the Enright-Wallach theorem. If Π_j is defined as in [7], then (b) is elementary, and (a) is the difficult result. In either case (c) is essentially a formal consequence.

We can now formulate our first two main theorems.

Theorem A$_0$. *In the setting of* (2.8) *and* (2.14), *suppose Z has finite length. Then all the $(\mathfrak{g}, K)$-modules $\mathcal{R}^i Z$ and $\mathcal{L}_j Z$ have finite length, and*

$$\sum_i (-1)^i \mathcal{R}^i Z = \sum_j (-1)^j \mathcal{L}_j Z$$

in the Grothendieck group of finite length $(\mathfrak{g}, K)$-modules.

Proof. In the case of $\mathcal{R}$, the finiteness is [9], Lemma 6.3.33. The result for $\mathcal{L}$ and the equality follow from Theorem 2.16(a) and Proposition 2.15(b), using the long exact sequences for the Zuckerman functors. Q.E.D.

Theorem B$_0$. *In the setting of* (2.8) *and* (2.14), *an invariant Hermitian form $\langle,\rangle_L$ on Z induces an invariant Hermitian form $\langle,\rangle_G$ on $\mathcal{L}_s Z$.*

Proof. This is immediate from Proposition 2.15(d) and Theorem 2.16(c). Q.E.D.

In order to construct interesting unitary representations using cohomological induction, these theorems suggest that we should consider two problems:

Problem A. When is $\mathcal{L}_s Z$ irreducible, and $\mathcal{L}_j Z = 0$ for $j \neq s$?

Problem B. When is $\mathcal{L}_s Z$ unitary?

In the following sections, we will offer various partial solutions of these two problems.

3. Infinitesimal characters, positivity and the good range.

As mentioned in the introduction, cohomological induction is very well behaved only under appropriate "positivity" hypotheses on the representation Z of L. In this section we will begin to study such hypotheses. At first they will be formulated in terms of infinitesimal characters, so we begin by recalling some standard structure theory for the center of the enveloping algebra.

As always we work with a parabolic subalgebra $\mathfrak{q} = \mathfrak{l} + \mathfrak{u}$ as in (1.1), and a Cartan involution θ as in Proposition 2.3. Define

$$(3.1)(a) \qquad \mathcal{Z}(\mathfrak{g}) = \text{center of } U(\mathfrak{g}), \qquad \mathcal{Z}(\mathfrak{l}) = \text{center of } U(\mathfrak{l}).$$

These algebras are related by a Harish-Chandra homomorphism, defined as follows. The triangular decomposition (1.1)(c) provides a decomposition of the enveloping algebra

$$(3.1)(b) \qquad U(\mathfrak{g}) = U(\mathfrak{l}) \oplus (\mathfrak{u}U(\mathfrak{g}) + U(\mathfrak{g})\overline{\mathfrak{u}}).$$

Write $\tilde{\xi}$ (or $\tilde{\xi}^{\mathfrak{l}}_{\mathfrak{g}}$) for the corresponding projection on the first summand, a linear map from $U(\mathfrak{g})$ to $U(\mathfrak{l})$. The restriction of $\tilde{\xi}$ to $\mathfrak{z}(\mathfrak{g})$ is an algebra homomorphism

$$(3.1)(c) \qquad \tilde{\xi} : \mathfrak{z}(\mathfrak{g}) \to \mathfrak{z}(\mathfrak{l}),$$

the *unnormalized Harish-Chandra homomorphism*. This map is related to the constructions of section 2 by the following elementary result.

Lemma 3.2. *In the setting of* (2.8) *and* (2.14), *suppose that* $\mathfrak{z}(\mathfrak{l})$ *acts on* Z *by scalars, through an algebra homomorphism*

$$\chi_L : \mathfrak{z}(\mathfrak{l}) \to \mathbb{C}.$$

Then $\mathfrak{z}(\mathfrak{g})$ *acts on* $W_{\mathfrak{q}}$ *and on* $W_{\overline{\mathfrak{q}}}$ *by scalars, through the algebra homomorphism*

$$\chi_G = \chi_L \circ \tilde{\xi} : \mathfrak{z}(\mathfrak{g}) \to \mathbb{C}.$$

Consequently $\mathfrak{z}(\mathfrak{g})$ *acts through* χ_G *on all the representations* $\mathcal{R}^i Z$ *and* $\mathcal{L}_j Z$ *as well.*

The first assertions are proved in exactly the same way as the familiar special cases for Verma modules (see for example [**6**]). The last include an assertion about the Zuckerman functors that may be found (at least for Γ) in [**9**], Lemma 6.2.13.

Fix now a Cartan subalgebra $\mathfrak{h} \subset \mathfrak{l}$, and a Borel subalgebra $\mathfrak{b}_{\mathfrak{l}} = \mathfrak{h} + \mathfrak{n}_{\mathfrak{l}}$. Then

$$(3.3)(a) \qquad \mathfrak{b} = \mathfrak{b}_{\mathfrak{l}} + \mathfrak{u} = \mathfrak{h} + (\mathfrak{n}_{\mathfrak{l}} + \mathfrak{u}) = \mathfrak{h} + \mathfrak{n}$$

is a Borel subalgebra of $\mathfrak{g}$. Using these Borel subalgebras instead of $\mathfrak{q}$ in (3.1), we can define Harish-Chandra homomorphisms from $\mathfrak{z}(\mathfrak{g})$ and $\mathfrak{z}(\mathfrak{l})$ to $S(\mathfrak{h})$. This leads to the commutative diagram

$$(3.3)(b) \qquad \begin{array}{ccc} \mathfrak{Z}(\mathfrak{g}) & \xrightarrow{\tilde{\xi}^{\mathfrak{l}}_{\mathfrak{g}}} & \mathfrak{Z}(\mathfrak{l}) \\ {\scriptstyle \tilde{\xi}^{\mathfrak{h}}_{\mathfrak{g}}} \searrow & & \swarrow {\scriptstyle \tilde{\xi}^{\mathfrak{h}}_{\mathfrak{l}}} \\ & S(\mathfrak{h}) & \end{array}$$

The difficulty is that the Harish-Chandra homomorphisms to $S(\mathfrak{h})$ are (for a wealth of excellent reasons) usually normalized by composition with certain "ρ-shifts" on $S(\mathfrak{h})$. The shifts are different for $\mathfrak{g}$ and for $\mathfrak{l}$; so in order to preserve the commutativity of (3.3)(b) after normalization, we need to normalize the top homomorphism as well. To do that, notice that the adjoint action of the group L provides a one-dimensional character of L on the top exterior power of $\mathfrak{u}$. We write

$$\chi_{2\rho(\mathfrak{u})} : L \to \mathbb{C}^{\times}$$

for this character, which may also be defined as the determinant of the adjoint action of L on $\mathfrak{u}$. The differential of $\chi_{2\rho(\mathfrak{u})}$ is a Lie algebra homomorphism

$$2\rho(\mathfrak{u}) : \mathfrak{l} \to \mathbb{C}$$

(the trace of the adjoint action of $\mathfrak{l}$ on $\mathfrak{u}$). We define

$$(3.3)(c) \qquad\qquad \rho(\mathfrak{u}) : \mathfrak{l} \to \mathbb{C}$$

to be one half of $2\rho(\mathfrak{u})$; this need not be the differential of a character of L, but it is still a perfectly good Lie algebra homomorphism. In terms of this homomorphism, we define a map

$$T_{-\rho(\mathfrak{u})} : \mathfrak{l} \to U(\mathfrak{l}), \qquad T_{-\rho(\mathfrak{u})}(X) = X + \rho(\mathfrak{u})(X).$$

Because $\rho(\mathfrak{u})$ is a Lie algebra homomorphism, one checks immediately that $T_{-\rho(\mathfrak{u})}$ is as well (using the commutator Lie algebra structure on $U(\mathfrak{l})$). By the universality property of $U(\mathfrak{l})$, $T_{-\rho(\mathfrak{u})}$ extends to a homomorphism of associative algebras

$$(3.3)(d) \qquad\qquad T_{-\rho(\mathfrak{u})} : U(\mathfrak{l}) \to U(\mathfrak{l}).$$

The inverse of $T_{-\rho(\mathfrak{u})}$ is $T_{\rho(\mathfrak{u})}$, so these maps are actually automorphisms of $U(\mathfrak{l})$. We define the *normalized Harish-Chandra homomorphism* $\xi = \xi^{\mathfrak{l}}_{\mathfrak{g}}$ as

$$(3.3)(e) \qquad\qquad \xi : \mathfrak{Z}(\mathfrak{g}) \to \mathfrak{Z}(\mathfrak{l}), \qquad \xi = T_{-\rho(\mathfrak{u})} \circ \tilde{\xi}.$$

The normalized versions of $\xi^{\mathfrak{h}}_{\mathfrak{g}}$ and $\xi^{\mathfrak{h}}_{\mathfrak{l}}$ are defined analogously, using the actions of $\mathfrak{h}$ on $\mathfrak{n}$ and $\mathfrak{n}_{\mathfrak{l}}$ respectively (see [6], page 130.) Because of (3.3)(a), we have

$$\rho(\mathfrak{n}) = \rho(\mathfrak{n}_{\mathfrak{l}}) + \rho(\mathfrak{u})$$

as characters of $\mathfrak{h}$. From this it is not difficult to deduce that the normalized version of (3.3)(b) is also a commutative diagram:

$$(3.3)(f) \qquad \begin{array}{ccc} \mathfrak{Z}(\mathfrak{g}) & \xrightarrow{\;\xi_{\mathfrak{g}}^{\mathfrak{l}}\;} & \mathfrak{Z}(\mathfrak{l}) \\[2pt] {\scriptstyle \xi_{\mathfrak{g}}^{\mathfrak{h}}}\searrow & & \nearrow{\scriptstyle \xi_{\mathfrak{l}}^{\mathfrak{h}}} \\[2pt] & S(\mathfrak{h}) & \end{array}$$

Here is the fundamental theorem of Harish-Chandra that justifies the normalizations.

Theorem 3.4 (see [6], page 130). *In the setting of* (3.3), *the normalized Harish-Chandra homomorphisms are isomorphisms onto the invariants of the Weyl group:*

$$\xi_{\mathfrak{g}}^{\mathfrak{h}} : \mathfrak{Z}(\mathfrak{g}) \to S(\mathfrak{h})^{W(\mathfrak{g},\mathfrak{h})}, \qquad \xi_{\mathfrak{l}}^{\mathfrak{h}} : \mathfrak{Z}(\mathfrak{l}) \to S(\mathfrak{h})^{W(\mathfrak{l},\mathfrak{h})}.$$

Suppose $\lambda \in \mathfrak{h}^*$. We may regard λ as an algebra homomorphism from $S(\mathfrak{h})$ to $\mathbb{C}$, and we define

$$(3.5)(a) \qquad\qquad \xi_\lambda : \mathfrak{Z}(\mathfrak{g}) \to \mathbb{C}, \qquad \xi_\lambda = \lambda \circ \xi_{\mathfrak{g}}^{\mathfrak{h}}.$$

If $\mathfrak{Z}(\mathfrak{g})$ acts by scalars on a representation X through the homomorphism ξ_λ, then we say that X has *infinitesimal character* λ. Of course we can apply the same terminology to representations of $\mathfrak{l}$. We will suppose from now on that

$(3.5)(b)$ the representation Z of $\mathfrak{l}$ has infinitesimal character $\lambda_{\mathfrak{l}} \in \mathfrak{h}^*$.

Because everything else will be phrased in terms of this weight, it is worthwhile to pause for a moment to review exactly what this assumption means in various special cases. If $\mathfrak{l}$ is abelian (so that $\mathfrak{l} = \mathfrak{h}$) and Z is a one-dimensional character of the group L, then $\lambda_{\mathfrak{l}}$ is just the differential of Z. If Z is finite-dimensional, then it has a highest weight space: the subspace annihilated by the nilpotent Lie algebra $\mathfrak{n}_{\mathfrak{l}}$ of (3.3)(a). Since Z has an infinitesimal character, $\mathfrak{h}$ acts by scalars on this highest weight space, say by a weight $\mu_{\mathfrak{l}} \in \mathfrak{h}^*$. Then the weight of (3.5)(b) is given by $\lambda_{\mathfrak{l}} = \mu_{\mathfrak{l}} + \rho(\mathfrak{n}_{\mathfrak{l}})$ (the shift being half the sum of the roots of $\mathfrak{h}$ in $\mathfrak{n}_{\mathfrak{l}}$). Finally, if Z is an irreducible $(\mathfrak{l}, L \cap K)$-module, then it has a "distribution character" Θ_Z, which is a function on the regular set in L. If $\mathfrak{h}$ is the complexified Lie algebra of a Cartan subgroup H of L, then the restriction of Θ_Z to H will locally be a finite linear combination of exponential functions divided by the Weyl denominator. If $\lambda_{\mathfrak{l}} \in \mathfrak{h}^*$ is the differential of one of those exponential functions, then (3.5)(b) is satisfied.

Define now

$$(3.5)(c) \qquad\qquad \lambda_{\mathfrak{g}} = \lambda_{\mathfrak{l}} - \rho(\mathfrak{u}).$$

Lemma 3.6. *In the setting of* (2.8) *and* (2.14), *suppose that Z has infinitesimal character $\lambda_{\mathfrak{l}}$ (cf.* (3.5)(b)*). Then $W_{\mathfrak{q}}$, $W_{\bar{\mathfrak{q}}}$, and all the $\mathcal{R}^i Z$ and $\mathcal{L}_j Z$ have infinitesimal character $\lambda_{\mathfrak{g}} = \lambda_{\mathfrak{l}} - \rho(\mathfrak{u})$ (cf.* (3.5)(c)*).*

This is an immediate consequence of Lemma 3.2 and the definitions in (3.3).

In order to formulate positivity conditions, it is convenient to fix an $\mathrm{Ad}(G)$-invariant non-degenerate symmetric bilinear form $\langle,\rangle$ on $\mathfrak{g}_0$. This form may be chosen so that its restriction to $\mathfrak{k}_0$ is negative definite; its restriction to $\mathfrak{s}_0$ is positive definite (cf. (2.1)); and so that $\mathfrak{k}_0$ and $\mathfrak{s}_0$ are orthogonal. The form extends by complexification to all of $\mathfrak{g}$, by restriction to non-degenerate forms on Levi subalgebras (including Cartan subalgebras), and by dualization to forms on the duals of these algebras. In particular, we get a Weyl group invariant form (still written $\langle,\rangle$) on the dual $\mathfrak{h}^*$ of a Cartan subalgebra. This form is positive definite on the real span of the roots.

Definition 3.7. Suppose we are in the setting of (3.5). Fix an invariant bilinear form on $\mathfrak{g}_0$ as in the preceding paragraph. We say that Z is in the *good range*, or that Z is *good* (for $\mathfrak{q}$ and $\mathfrak{g}$) if for every root $\alpha \in \Delta(\mathfrak{u}, \mathfrak{h})$ we have

$$\mathrm{Re}\langle \alpha, \lambda_\mathfrak{g} \rangle > 0.$$

We say that Z is *weakly good* if for every $\alpha \in \Delta(\mathfrak{u}, \mathfrak{h})$ we have

$$\mathrm{Re}\langle \alpha, \lambda_\mathfrak{g} \rangle \geq 0.$$

It is not difficult to see that these conditions are independent of the choice of invariant form on $\mathfrak{g}_0$.

Here are some results about Problems A and B from the end of section 2.

Theorem A_1 ([10], section 4). *In the setting of (2.8) and (2.14), suppose that Z is good (Definition 3.7).*

(a) *The $(\mathfrak{g}, K)$-modules $\mathcal{R}^i Z$ and $\mathcal{L}_j Z$ are all zero for i and j not equal to s (cf. (2.5)(b)).*

(b) *There is an isomorphism $\mathcal{R}^s Z \simeq \mathcal{L}_s Z$.*

(c) *The $(\mathfrak{g}, K)$-module $\mathcal{L}_s Z$ is irreducible if and only if Z is irreducible.*

If we assume only that Z is weakly good, then (a) and (b) still hold, but (c) must be replaced by

(c′) *If Z is irreducible, then $\mathcal{L}_s Z$ is irreducible or zero.*

There is a parallel result about unitarity.

Theorem B_1 ([10], Theorem 6.3). *In the setting of (2.14), suppose that Z is good (Definition 3.7). Then every invariant Hermitian form $\langle,\rangle_G$ on $\mathcal{L}_s Z$ is induced by a unique invariant Hermitian form $\langle,\rangle_L$ on Z (Theorem B_0 of section 2). This correspondence relates non-degenerate forms to non-degenerate forms, and positive forms to positive forms. In particular, $\mathcal{L}_s Z$ is unitarizable if and only if Z is unitarizable.*

If we assume only that Z is weakly good, then the correspondence of Theorem B_0 still sends non-degenerate forms to non-degenerate forms, and positive forms to positive forms. In particular, if Z is unitarizable, then so is $\mathcal{L}_s Z$.

Example 3.8. Suppose $L = H = T$ is a Cartan subgroup of G contained in K. (Of course this can happen if and only if the rank of K is equal to the rank of G.) For simplicity assume that G is connected; then T is as well. We take for Z a one-dimensional character of T of weight $\lambda_{\mathfrak{l}} \in \mathfrak{t}^*$. Because T is compact, Z is automatically unitary. If Z is in the good range, then it turns out that $\mathcal{L}_s Z$ is the discrete series representation of G with Harish-Chandra parameter $\lambda_{\mathfrak{g}}$; all discrete series representations of G arise in this way. If Z is weakly good, then $\mathcal{L}_s Z$ is a "limit of discrete series" representation attached to the parameter $\lambda_{\mathfrak{g}}$ and the positive root system defined by $\mathfrak{u}$. This is a tempered unitary representation. It turns out to be zero if and only if there is a compact simple root α of $\mathfrak{t}$ in $\mathfrak{u}$ with the property that $\langle \alpha, \lambda_{\mathfrak{g}} \rangle = 0$.

Example 3.9. Suppose G is a complex connected group, and that $L = H = TA$ is a Cartan subgroup (with T a maximal torus in K, and A the vector group corresponding to the -1 eigenspace of θ). Again we take Z to be a one-dimensional character of H, of differential $\lambda_{\mathfrak{l}}$. Write

$$\lambda_{\mathfrak{g}} = (\mu, \nu) \in \mathfrak{t}^* + \mathfrak{a}^*.$$

The character is unitary if and only if ν takes purely imaginary values on $\mathfrak{a}_0$.

Assume from now on that μ is (weakly) dominant as a weight for K; that is, that $\langle \beta, \mu \rangle \geq 0$ for every root β of $\mathfrak{t}$ in $\mathfrak{u} \cap \mathfrak{k}$. (This is a necessary but not sufficient condition for Z to be weakly good; weakly good means in addition that $\mathrm{Re}(\nu)$ is sufficiently small with respect to μ. Precisely, the condition is

$$|\mathrm{Re}\langle \alpha, \nu \rangle| \leq \langle \alpha, \mu \rangle$$

for every root α of $\mathfrak{h}$ in $\mathfrak{u}$.) By [9], Proposition 8.2.15, we get the vanishing conclusion of Theorem $A_1(a)$ above, and $\mathcal{L}_s Z$ is a standard principal series representation $I(\mu, \nu)$ (having a unique Langlands quotient).

For regular μ, it is known that $I(\mu, \nu)$ is unitary if and only if ν is unitary. Now the regularity of μ is a necessary but not sufficient condition for Z to be good; good means in addition that $\mathrm{Re}(\nu)$ is sufficiently small with respect to μ. This is therefore consistent with Theorem B_1. For singular μ, $I(\mu, \nu)$ is also unitary for certain small real values of ν (the complementary series). These unitary representations are not predicted by Theorem B_1, although of course their existence does not contradict it.

Example 3.10. Suppose $G = Sp(4, \mathbb{C})$ (a group of real rank two), and $L = GL(1, \mathbb{C}) \times Sp(2, \mathbb{C})$. (A detailed discussion of the unitary representations of G may be found in [2].) We may choose a Cartan subgroup $H = GL(1, \mathbb{C}) \times GL(1, \mathbb{C})$. Coordinates may be chosen so that $\mathfrak{h}^*$ is identified with the product of two copies of $\mathbb{C}^2$, interchanged by θ. (This corresponds to the identification of the complexified real Lie algebra with a product of two copies of $\mathfrak{sp}(4, \mathbb{C})$.) The roots of $\mathfrak{h}$ in $\mathfrak{u}$ are then $[(2,0),(0,0)]$, $[(1,1),(0,0)]$, $[(1,-1),(0,0)]$, and the same weights in the second factor. Consequently

$$\rho(\mathfrak{u}) = [(2,0),(2,0)].$$

The compact part of H is $T = S^1 \times S^1$, and the vector part is a product of two copies of the multiplicative group of the positive real numbers. The dual Lie algebras $\mathfrak{t}^*$ and $\mathfrak{a}^*$ may each be identified with $\mathbb{C}^2$; then the restriction of a weight $[\lambda_1, \lambda_2]$ in $\mathfrak{h}^*$ to $\mathfrak{t}$ is $\lambda_1 + \lambda_2$, and the restriction to $\mathfrak{a}$ is $\lambda_1 - \lambda_2$.

We will assume that the representation Z of L is given by the character $(\det /|\det|)^5$ on the $GL(1, \mathbb{C})$ factor, tensored with an irreducible spherical representation Z_0 on the $Sp(2, \mathbb{C})$ factor. This implies that the infinitesimal character $\lambda_{\mathfrak{l}}$ is given by

$$\lambda_{\mathfrak{l}} = [(5/2, \nu/2), (5/2, -\nu/2)].$$

Using the formula for $\rho(\mathfrak{u})$, we get

$$\lambda_{\mathfrak{g}} = [(1/2, \nu/2), (1/2, -\nu/2)].$$

The unitarity of Z is determined by Z_0, and is therefore a question about $SL(2, \mathbb{C})$. It turns out that Z_0 is unitary if and only if ν is purely imaginary (the unitary spherical principal series), or $\nu \in (-2, 2)$ (the complementary series), or $\nu = \pm 2$ (the trivial representation).

Now L is also the Levi factor of a real parabolic subgroup of G. It turns out that $\mathcal{L}_s Z$ is more or less the representation induced by $(\det /|\det|) \otimes Z_0$ from this real parabolic. Therefore $\mathcal{L}_s Z$ is always unitary when Z is, which is consistent with Theorem B_1. More precisely, Z is weakly good if and only if $\mathrm{Re}(\nu) \leq 1$; this allows all possible unitary Z_0 except half the complementary series and the trivial representation. In this way we see that Theorem B_1 fails to find certain apparently natural unitary representations, as for example the degenerate series representation unitarily induced from $(\det /|\det|) \otimes Z_0$. This representation is isomorphic to $\mathcal{L}_s((\det /|\det|)^5 \otimes \mathbb{C})$.

On the other hand, we cannot weaken the positivity condition too much. For $\nu = 3$, Z_0 is an irreducible non-unitary spherical representation. Nevertheless $\mathcal{L}_s Z$ and the representation induced by $(\det /|\det|) \otimes Z_0$ (which are both reducible) in this case share a unitary composition factor, namely the "odd" half of the metaplectic representation of $Sp(4, \mathbb{C})$. The perfect correspondence between unitary representations for G and for L provided by Theorem B_1 in the good range therefore breaks down.

The relation between cohomological and ordinary induction exploited in Examples 3.9 and 3.10 can be extended to all complex groups. This is one reason for believing that the functor $\mathcal{L}_s$ should preserve unitarity under much weaker hypotheses than "weakly good." Another reason comes from the work of Oshima and Matsuki on discrete series for semisimple symmetric spaces. They show that these representations (which are unitary by their definition) are all of the form $\mathcal{L}_s Z$, with a Z one-dimensional unitary character, for appropriate $\mathfrak{q}$ and L (see

[**13**]). Most but not all of these characters lie in the good range. One might certainly expect that a "best" unitarizability theorem should predict the unitarity of all these discrete series representations (as was the case with Harish-Chandra's discrete series in Example 3.8). In the next section we will see how to extend Theorems A_1 and B_1 to achieve that.

4. The fair range and unipotent representations.

Definition 4.1. Suppose $\mathfrak{q} = \mathfrak{l} + \mathfrak{u}$ is a purely complex parabolic as in (1.1), θ is a Cartan involution as in Proposition 2.3, and Z is an $(\mathfrak{l}, L \cap K)$-module. Fix an invariant bilinear form on $\mathfrak{g}_0$ as in Definition 3.7. Write

$$(4.1)(a) \qquad\qquad \mathfrak{z} = \mathfrak{z}(\mathfrak{l})$$

for the center of the Lie algebra $\mathfrak{l}$. Assume that $\mathfrak{z}$ acts by scalars in Z, according to a linear functional

$$(4.1)(b) \qquad\qquad \lambda_{\mathfrak{l}}^{\mathfrak{z}} \in \mathfrak{z}^*.$$

Define

$$(4.1)(c) \qquad\qquad \lambda_{\mathfrak{g}}^{\mathfrak{z}} = \lambda_{\mathfrak{l}}^{\mathfrak{z}} - \rho(\mathfrak{u})|_{\mathfrak{z}} \in \mathfrak{z}^*.$$

We say that Z is in the *fair range*, or that Z is *fair* if for every weight β of $\mathfrak{z}$ in $\mathfrak{u}$, we have

$$(4.1)(d) \qquad\qquad \operatorname{Re}\langle \beta, \lambda_{\mathfrak{g}}^{\mathfrak{z}} \rangle > 0.$$

We say that Z is *weakly fair* if for every such β we have

$$(4.1)(e) \qquad\qquad \operatorname{Re}\langle \beta, \lambda_{\mathfrak{g}}^{\mathfrak{z}} \rangle \geq 0.$$

Suppose in addition that we are in the setting of (3.5); that is, that we fix a Cartan subalgebra $\mathfrak{h}$ of $\mathfrak{l}$, and that Z has infinitesimal character $\lambda_{\mathfrak{l}}$. Then $\mathfrak{z} \subset \mathfrak{h}$, and

$$(4.1)(f) \qquad\qquad \lambda_{\mathfrak{l}}^{\mathfrak{z}} = \lambda_{\mathfrak{l}}|_{\mathfrak{z}}, \qquad \lambda_{\mathfrak{g}}^{\mathfrak{z}} = \lambda_{\mathfrak{g}}|_{\mathfrak{z}}.$$

We also have orthogonal decompositions

$$(4.1)(g) \qquad\qquad \mathfrak{l} = \mathfrak{z} \oplus [\mathfrak{l}, \mathfrak{l}], \qquad \mathfrak{h} = \mathfrak{z} \oplus ([\mathfrak{l}, \mathfrak{l}] \cap \mathfrak{h}).$$

Using this decomposition, we regard $\mathfrak{z}^*$ as a subspace of $\mathfrak{h}^*$. The condition (4.1)(d) may then be phrased in terms of $\lambda_{\mathfrak{g}}$, as follows: for every root $\alpha \in \Delta(\mathfrak{u}, \mathfrak{h})$, the restriction $\alpha^{\mathfrak{z}}$ of α to $\mathfrak{z}$ satisfies

$$(4.1)(d') \qquad\qquad \operatorname{Re}\langle \alpha^{\mathfrak{z}}, \lambda_{\mathfrak{g}} \rangle > 0.$$

(Implicitly we are extending $\alpha^{\mathfrak{z}}$ to a linear functional on all of $\mathfrak{h}$, making it zero on $[\mathfrak{l}, \mathfrak{l}] \cap \mathfrak{h}$.) Similarly, weakly fair is equivalent to

$$(4.1)(e') \qquad\qquad \operatorname{Re}\langle \alpha^{\mathfrak{z}}, \lambda_{\mathfrak{g}} \rangle \geq 0.$$

for every $\alpha \in \Delta(\mathfrak{u}, \mathfrak{h})$.

If L is a Cartan subgroup, then $\mathfrak{l} = \mathfrak{h} = \mathfrak{z}$, and the notions of good and fair coincide. In general, we will show that the notion of fair is actually weaker than that of good.

Lemma 4.2. *In the setting of Definition 4.1, fix a root α of $\mathfrak{h}$ in $\mathfrak{u}$, and write $\alpha^{\mathfrak{z}}$ for its restriction to $\mathfrak{z}$, extended by zero to all of $\mathfrak{h}$. Then*

$$\alpha^{\mathfrak{z}} = 1/|W(\mathfrak{l}, \mathfrak{h})| \sum_{w \in W(\mathfrak{l}, \mathfrak{h})} w \cdot \alpha.$$

Proof. Because $W(\mathfrak{l}, \mathfrak{h})$ acts trivially on $\mathfrak{z}$, the right side has the same restriction to $\mathfrak{z}$ as α. On the other hand, the right side is clearly fixed by $W(\mathfrak{l}, \mathfrak{h})$. Since the Weyl group of a semisimple Lie algebra fixes only zero, the right side vanishes on $[\mathfrak{l}, \mathfrak{l}] \cap \mathfrak{h}$. Q.E.D.

Essentially the same argument shows that the weight $\rho(\mathfrak{u}) \in \mathfrak{h}^*$ actually belongs to the subspace $\mathfrak{z}^*$.

Proposition 4.3. *in the setting of Definition 4.1, if Z is in the good range, then Z is in the fair range. If Z is weakly good, then Z is weakly fair.*

Proof. The Weyl group $W(\mathfrak{l}, \mathfrak{h})$ acts on $\Delta(\mathfrak{u}, \mathfrak{h})$. Fix $\alpha \in \Delta(\mathfrak{u}, \mathfrak{h})$. In the terminology of Definition 4.1, Lemma 4.2 says that

$$\operatorname{Re}\langle \alpha^{\mathfrak{z}}, \lambda_{\mathfrak{g}} \rangle = 1/|W(\mathfrak{l}, \mathfrak{h})| \sum_{w \in W(\mathfrak{l}, \mathfrak{h})} \operatorname{Re}\langle w \cdot \alpha, \lambda_{\mathfrak{g}} \rangle.$$

If Z is in the good range, then every term on the right is positive; so the left side must be positive as well, and Z is in the fair range. Exactly the same argument applies in the weakly good range. Q.E.D.

Example 4.4. Suppose $\mathfrak{g} = \mathfrak{sp}(2n)$, and that $\mathfrak{l} = \mathfrak{gl}(1)^{n-p} \times \mathfrak{sp}(2p)$. We can take $\mathfrak{h} \simeq \mathbb{C}^n$; these coordinates may be chosen so that $\mathfrak{z} \simeq \mathbb{C}^{n-p}$ (the first $n - p$ coordinates). For an appropriate choice of $\mathfrak{q}$, we have

$$\rho(\mathfrak{u}) = (n, n - 1, \dots, p + 1, 0, \dots, 0).$$

Assume that Z is one-dimensional. Then the restriction of $\lambda_{\mathfrak{l}}$ or $\lambda_{\mathfrak{g}}$ to $[\mathfrak{l}, \mathfrak{l}] \cap \mathfrak{h}$ is half the sum of a set of positive roots for $\mathfrak{l}$. We may therefore choose

$$\lambda_{\mathfrak{g}} = (\lambda_1, \lambda_2, \dots, \lambda_{n-p}, p, p - 1, \dots, 2, 1).$$

Then

$$\lambda_{\mathfrak{g}}^{\mathfrak{z}} = (\lambda_1, \lambda_2, \dots, \lambda_{n-p}, 0, \dots, 0).$$

In terms of these coordinates, here are the four positivity conditions we have considered.

$$
\begin{aligned}
\text{good:} \quad & \lambda_1 > \cdots \lambda_{n-p} > p \\
\text{weakly good:} \quad & \lambda_1 \geq \cdots \lambda_{n-p} \geq p \\
\text{fair:} \quad & \lambda_1 > \cdots \lambda_{n-p} > 0 \\
\text{weakly fair:} \quad & \lambda_1 \geq \cdots \lambda_{n-p} \geq 0
\end{aligned}
$$

(It would be a little more straightforward to put $\mathrm{Re}(\lambda_i)$ everywhere in these formulas. But it turns out that the assumption that $\mathfrak{q}$ is purely complex forces the center of L to be compact in this example, so that $\lambda_{\mathfrak{l}}^{\mathfrak{i}}$ must take purely imaginary values. Because of our choice of coordinates, this means that the λ_i are automatically real.) Notice that fair is strictly weaker than good unless $p = 0$; that is, unless $\mathfrak{l}$ is abelian.

Unfortunately, the fair hypothesis does not imply analogues of Theorems A_1 and B_1 from section 3 in general. But here is a first hint that there are positive results to be found.

Theorem A'. *In the setting of (2.8) and (2.14), suppose that Z is weakly fair (Definition 4.1) and one-dimensional.*

    (a) *The $(\mathfrak{g}, K)$-modules $\mathcal{R}^i Z$ and $\mathcal{L}_j Z$ are all zero for i and j not equal to s (cf. (2.5)(b)).*

    (b) *There is an isomorphism $\mathcal{R}^s Z \simeq \mathcal{L}_s Z$.*

    (c) *The action of $U(\mathfrak{g})$ on $\mathcal{L}_s Z$ extends naturally to the algebra $\mathcal{D}(G_{\mathbb{C}}/Q_{\mathbb{C}})_{\lambda_{\mathfrak{g}}^{\mathfrak{s}}}$ of twisted differential operators. As a $(\mathcal{D}_{\lambda_{\mathfrak{g}}^{\mathfrak{s}}}, K)$-module, $\mathcal{L}_s Z$ is irreducible or zero.*

Parts (a) and (b) may be found in **[10]**, Theorem 7.1 and Proposition 8.5. Part (c) is a consequence of Theorem A_1 and **[14]**, Corollary 7.14. Unfortunately these results cannot be strengthened to closer analogues of Theorem A_1 by strengthening the hypothesis from weakly fair to fair; for example, one cannot guarantee that $\mathcal{L}_s(Z)$ is non-zero, or that it is irreducible or zero as a $(\mathfrak{g}, K)$-module.

There is again a parallel result for unitarity.

Theorem B'. *In the setting of (2.14), suppose that Z is weakly fair (Definition 4.1), one-dimensional, and unitary. Then the invariant Hermitian form $\langle , \rangle_G$ given by Theorem B_0 of section 2 is positive definite, so $\mathcal{L}_s Z$ is unitary.*

In the context of the orbit method, this theorem is sufficient (together with real parabolic induction) to attach a unitary representation to each admissible semisimple coadjoint orbit. (In fact the "fair" version of the theorem is all that is needed, but this would not simplify the algebraic proof.)

The result of Oshima and Matsuki mentioned at the end of section 3 says that any discrete series for a semisimple symmetric space is of the form $\mathcal{L}_s Z$, with Z a one-dimensional unitary character in the fair range (for an appropriate family of complex parabolic subalgebras $\mathfrak{q}$, depending on the symmetric space). Not every such representation occurs as a discrete series; the character Z must also be trivial on a certain subgroup of L.

The examples of the discrete series for G (Example 3.8) and more generally for symmetric spaces suggest a possible interpretation of the fair range. If a geometric proof of Theorem B$'$ can be found, then the fair condition may be needed for the convergence of some integrals defining Hermitian forms. Unitarity in the weakly fair range will perhaps appear by a more subtle limiting argument.

Example 4.5. Speh's representations. Suppose $G = GL(2n, \mathbb{R})$, and $\mathfrak{q}$ is as in Example 1.2; in particular, $L = GL(n, \mathbb{C})$. We fix a Cartan subgroup $H = GL(1, \mathbb{C})^n$. Then $H \cap K = T \simeq (S^1)^n$ is a maximal torus in $K \simeq O(2n)$. It is convenient to put coordinates on $\mathfrak{h} \simeq \mathbb{C}^{2n}$ in such a way that the roots are $\{e_i - e_j | 1 \leq i, j \leq 2n\}$, and the roots in L are those with i and j either both between 1 and n or both between $n + 1$ and $2n$. The Cartan involution acts on $\mathfrak{h}^*$ by

$$\theta(x, y) = (-y, -x)$$

for x and y in $\mathbb{C}^n$. The restriction of (x, y) to $\mathfrak{t} \simeq \mathbb{C}^n$ is $x - y$. (This choice of coordinates is not consistent with Example 3.10.) For appropriate choices, we get

$$\Delta(\mathfrak{u}, \mathfrak{h}) = \{e_i - e_j | 1 \leq i \leq n, \quad n + 1 \leq j \leq 2n\}.$$

Therefore

$$\rho(\mathfrak{u}) = (n/2, \ldots, n/2, -n/2, \ldots, -n/2).$$

For a standard choice of positive roots in $\mathfrak{l}$, we have

$$\rho(\mathfrak{n}_\mathfrak{l}) = ((n-1)/2, (n-3)/2, \ldots, -(n-1)/2, (n-1)/2, \ldots, -(n-1)/2).$$

We take for Z a one-dimensional character of L. Necessarily it is of the form

$$l \mapsto (\det l/|\det l|)^{p+n}|\det l|^\nu,$$

with $p \in \mathbb{Z}$ and $\nu \in \mathbb{C}$. (The shift by n in the parameter will simplify later formulas.) In these coordinates, its differential on $\mathfrak{h}$ is

$$((\nu + p + n)/2, \ldots, (\nu + p + n)/2, (\nu - p - n)/2, \ldots, (\nu - p - n)/2).$$

We therefore calculate

$$\lambda_\mathfrak{l} = ((\nu + p + 2n - 1)/2, \ldots, (\nu + p + 1)/2, (\nu - p - 1)/2, \ldots, (\nu - p - 2n + 1)/2)$$
$$\lambda_\mathfrak{g} = ((\nu + p + n - 1)/2, \ldots, (\nu + p - n + 1)/2, (\nu - p + n - 1)/2, \ldots, (\nu - p - n + 1)/2)$$
$$\lambda_\mathfrak{g}^s = ((\nu + p)/2), \ldots, (\nu + p)/2), (\nu - p)/2, \ldots, (\nu - p)/2).$$

Consequently

$$\begin{array}{ccc}
Z \text{ is good} & \Leftrightarrow & p > n - 1 \\
Z \text{ is weakly good} & \Leftrightarrow & p \geq n - 1 \\
Z \text{ is fair} & \Leftrightarrow & p > 0 \\
Z \text{ is weakly fair} & \Leftrightarrow & p \geq 0.
\end{array}$$

Write $I(p, \nu) = \mathcal{L}_s Z$. The character Z is unitary if and only if ν is pure imaginary. By Theorem B$'$, it follows that $I(p, \nu)$ is unitary whenever ν is imaginary and $p \geq 0$. Now $I(p, \nu) = I(p, 0) \otimes |\det|^{\nu/2}$, so the dependence on the parameter ν is not very interesting. The dependence on p, on the other hand, is quite subtle. Recall that we have fixed a maximal torus $T = (S^1)^n$ of K, and identified $\mathfrak{t}^*$ with $\mathbb{C}^n$. For every $p \geq 0$ there is a unique representation μ_{p+1} of $K = O(2n)$ having extremal weight $(p + 1, p + 1, \ldots, p + 1)$. It turns out that μ_{p+1} is the unique lowest K-type of $I(p, \nu)$ whenever $p \geq 0$; that is, in the weakly fair range.

For $p = n$ (that is, at the last point of the good range) μ_{n+1} is the representation of K generated inside the exterior algebra of $\mathfrak{s}$ by the top exterior power of $\mathfrak{u} \cap \mathfrak{s}$. We have in fact

$$\operatorname{Hom}_K(\textstyle\bigwedge \mathfrak{s}, I(n, \nu)) = \operatorname{Hom}_K(\textstyle\bigwedge \mathfrak{s}, \mu_{n+1}) \simeq \operatorname{Hom}_{L \cap K}(\textstyle\bigwedge(\mathfrak{l} \cap \mathfrak{s}), \mathbb{C})$$

$$\simeq \text{deRham cohomology of } U(n)$$

(see [15]). This is the beginning of the determination of the relative Lie algebra cohomology of the representation $I(n, 0)$, which turns out to be isomorphic to the cohomology of $U(n)$ except for a degree shift. (Notice that the formula for the infinitesimal character $\lambda_\mathfrak{g}$ of $I(p, \nu)$ is nice when $p = n$: the terms decrease steadily by one, without a "jump" in the middle.)

For $p = 0$, μ_1 is the representation of $O(2n)$ on $\bigwedge^n(\mathbb{C}^{2n})$. The representation $I(0, \nu)$ has a direct elementary construction, which we now explain. Let $P = MN$ be the parabolic subgroup of G consisting of block-upper triangular matrices, with $M = GL(n, \mathbb{R}) \times GL(n, \mathbb{R})$. Let $Z_M(\nu)$ be the one-dimensional character of M given by

$$(m_1, m_2) \mapsto (\operatorname{sgn}(\det(m_1)))|\det(m_1 m_2)|^{\nu/2}.$$

This is unitary exactly when ν is purely imaginary. It turns out that

$$I(0, \nu) \simeq \operatorname{Ind}_P^G(Z_M(\nu)).$$

Of course one can make a similar construction without the sign character; the resulting unitary representation is unrelated to the series $I(p, \nu)$. It is possible to realize $I(p, \nu)$ as a subquotient of a representation induced from a non-unitary one-dimensional character of P. This was one of Speh's original constructions of the representations.

In order to formulate our last theorems, we need to understand what it is about one-dimensional representations that makes Theorems A$'$ and B$'$ work. The only reasonable way to do that is to look at the proofs in [10], notably that of Proposition 8.5. This exercise leads to

Definition 4.6 ([10], Definition 8.16). Suppose we are in the setting (3.5); that is, that the representation Z of $\mathfrak{l}$ has infinitesimal character $\lambda_\mathfrak{l} \in \mathfrak{h}^*$. Define

$\mathfrak{r} = ([\mathfrak{l}, \mathfrak{l}] \cap \mathfrak{h})$, so that $\mathfrak{h} = \mathfrak{z} \oplus \mathfrak{r}$ (cf. (4.1)(g)). We say that Z is *weakly unipotent* if the following two conditions are satisfied.

(i) The weight $\lambda_{\mathfrak{l}}|_{\mathfrak{r}}$ belongs to the real span of the roots.

(ii) Suppose F is a finite-dimensional representation of $\mathfrak{l}$, and that $\lambda'_{\mathfrak{l}}$ is the infinitesimal character of a subquotient of $Z \otimes F$. Then

$$\langle \lambda'_{\mathfrak{l}}|_{\mathfrak{r}}, \lambda'_{\mathfrak{l}}|_{\mathfrak{r}} \rangle \geq \langle \lambda_{\mathfrak{l}}|_{\mathfrak{r}}, \lambda_{\mathfrak{l}}|_{\mathfrak{r}} \rangle.$$

Notice that being weakly unipotent is a property of $Z|_{[\mathfrak{l}, \mathfrak{l}]}$, and indeed only of the annihilator of Z in $U([\mathfrak{l}, \mathfrak{l}])$.

Examples 4.7. Although Definition 4.6 is meant to abstract some property of the trivial representation, it actually admits an even simpler example. If $\lambda_{\mathfrak{l}}$ vanishes on $\mathfrak{r}$, then Z is weakly unipotent: condition (i) is trivial, and (ii) is a consequence of the positivity of our invariant form on the real span of the roots. The corresponding representations are all tempered (modulo the center of L). A typical example is (for a quasisplit group L) the spherical principal series representation with A parameter 0; that is, the representation unitarily induced from the trivial character of a minimal parabolic subgroup.

The next natural example is a one-dimensional representation Z. In that case the infinitesimal character $\lambda_{\mathfrak{l}}$ may be taken (on the semisimple part $\mathfrak{r}$) to be the half sum $\rho(\mathfrak{n}_{\mathfrak{l}})$ of a set of positive roots. Condition (i) of Definition 4.6 is therefore satisfied. For (ii), we may suppose F is irreducible with highest weight μ. Then $F \otimes Z$ is just F (at least on $[\mathfrak{l}, \mathfrak{l}]$), so $\lambda'_{\mathfrak{l}}$ may be taken equal to $\mu + \rho(\mathfrak{n}_{\mathfrak{l}})$ (cf. (3.5)). The inequality we must verify in Definition 4.6(ii) is therefore

$$\langle \mu|_{\mathfrak{r}} + \rho(\mathfrak{n}_{\mathfrak{l}}), \mu|_{\mathfrak{r}} + \rho(\mathfrak{n}_{\mathfrak{l}}) \rangle \geq \langle \rho(\mathfrak{n}_{\mathfrak{l}}), \rho(\mathfrak{n}_{\mathfrak{l}}) \rangle.$$

This is equivalent to

$$\langle \mu|_{\mathfrak{r}} + 2\rho(\mathfrak{n}_{\mathfrak{l}}), \mu|_{\mathfrak{r}} \rangle \geq 0.$$

Because of the positivity of the form on the real span of the roots (to which the integral weight $\mu|_{\mathfrak{r}}$ must belong), it suffices to prove

$$\langle 2\rho(\mathfrak{n}_{\mathfrak{l}}), \mu|_{\mathfrak{r}} \rangle \geq 0.$$

The right side is the sum over positive roots α for $\mathfrak{l}$ of $\langle \alpha, \mu \rangle$. Since μ is dominant, each such term is non-negative. This completes the verification that a one-dimensional representation is weakly unipotent.

A third class of weakly unipotent representations (actually including the first two) is Arthur's "special unipotent" representations. These are defined in [1] in terms of their annihilators in the enveloping algebra. (The definition is given only for complex G, but the real case is identical.) The conditions of Definition 4.6 are verified in Proposition 5.18 of [1].

Finally, we mention a fourth example of a weakly unipotent representation: the metaplectic representation of a (real or complex) metaplectic group. The

verification of condition (ii) in Definition 4.6 requires a little basic structure theory for primitive ideals (essentially the notion of the τ-invariant), and we will
not develop it here. The point of the example is that the metaplectic represe-
entation is an interesting unitary representation that is in some sense attached
to a nilpotent coadjoint orbit (so that one would like to be able to apply cohomological induction to it as part of a general orbit correspondence). Although
it is excluded by Arthur's notion of special unipotent representation, it is still
allowed by the "weakly unipotent" condition of Definition 4.6.

Here are our last main theorems. To formulate the first, it is helpful to use the
notion of a Dixmier algebra for G. This is explained carefully in [**14**], Definition
2.1. For the moment it is enough to know that a Dixmier algebra is an associative
algebra A_G equipped with an algebra homomorphism

$$(4.8) \qquad\qquad \phi_G : U(\mathfrak{g}) \to A_G$$

having a very large image. An excellent example is an algebra $\mathcal{D}(G_\mathbb{C}/Q_\mathbb{C})_\lambda$ of
twisted differential operators on a flag variety, with ϕ_G the map coming from the
action of G. For most λ, ϕ_G is surjective. When we say that a module M for
$U(\mathfrak{g})$ "extends" to A_G, we mean first that the kernel of ϕ_G acts trivially on M,
and second that the action of the subalgebra $\phi_G(U(\mathfrak{g})) \subset A_G$ on M extends to
an action of all of A_G. The reason for introducing Dixmier algebras in irreducibility theorems is that quotients of $U(\mathfrak{g})$ are sometimes a little smaller than they
"ought" to be, as for example when not all twisted differential operators on a flag
variety come from $U(\mathfrak{g})$. For that reason, $U(\mathfrak{g})$ may fail to act irreducibly when
we would like it to. In this case we can sometimes add the missing elements to
make a Dixmier algebra that *does* act irreducibly.

Theorem A$_2$. *In the setting of (2.8) and (2.14), suppose that Z is weakly
fair (Definition 4.1) and weakly unipotent (Definition 4.6).*

(a) *The $(\mathfrak{g}, K)$-modules $\mathcal{R}^i Z$ and $\mathcal{L}_j Z$ are all zero for i and j not equal to s
(cf. (2.5)(b)).*

(b) *There is an isomorphism $\mathcal{R}^s Z \simeq \mathcal{L}_s Z$.*

(c) *Suppose that Z is a module for a weakly unipotent Dixmier algebra A_L
for $\mathfrak{l}$. Define*

$$A_G = \mathrm{Ind}_{Dix}(\mathfrak{q} \uparrow \mathfrak{g})(A_L)$$

*to be the induced Dixmier algebra for G ([**14**], Definition 4.7). Then
the action of $U(\mathfrak{g})$ on $\mathcal{L}_s Z$ extends naturally to the algebra A_G. If Z is
irreducible as an $(A_L, L \cap K)$-module, then $\mathcal{L}_s Z$ is irreducible or zero as
an (A_G, K)-module.*

Parts (a) and (b) may be found in [**10**], Theorem 7.1 and Proposition 8.17.
There is no proof of (c) in the literature. The first part (getting A_G to act) is
standard; one can for example imitate some of the ideas in [**3**]. The irreducibility
is a little harder. The proof has two steps. First, one assumes that Z is in the

good range. Then one can take advantage of the very close relationship between reducibilty for L and for G suggested by Theorem $A_1(c)$. (This is the main step.) The reduction of the general case to the good range uses [14], Corollary 7.14.

Here is the result for unitarity.

Theorem B_2. *In the setting of (2.14), suppose that Z is weakly fair (Definition 4.1), weakly unipotent (Definition 4.6), and unitary. Then the invariant Hermitian form $\langle , \rangle_G$ given by Theorem B_0 of section 2 is positive definite, so $\mathcal{L}_s Z$ is unitary.*

The first hypothesis on Z is a positivity condition on its central character. In the language of the orbit method, it may be taken as a positivity hypothesis on the "polarization" $\mathfrak{q}$. The second refers only to the restriction of Z to the commutator subgroup; it says that this restriction should have very small infinitesimal character. In light of the relation between infinitesimal characters and coadjoint orbits, the second hypothesis may be taken as a weak version of the assumption that Z is associated to a nilpotent coadjoint orbit. The examples in 4.7 are intended to support this view.

REFERENCES

1. D. Barbasch and D. Vogan, *Unipotent representations of complex semisimple Lie groups*, Ann. of Math. **121** (1985), 41–110.
2. M. Duflo, *Représentations unitaires irréductibles des groupes semi-simples complexes de rang deux*, Bull. Soc. Math. France **107** (1979), 55–96.
3. T. J. Enright and N. R. Wallach, *Notes on homological algebra and representations of Lie algebras*, Duke Math. J. **47** (1980), 1–15.
4. V. Guillemin and S. Sternberg, *Geometric Asymptotics*, Math. Surveys vol. 14, American Mathematical Society, 1978.
5. Harish-Chandra, *Harmonic analysis on reductive groups I, The theory of the constant term*, J. Func. Anal. **19** (1975), 104–204.
6. J. E. Humphreys, *Introduction to Lie Algebras and Representation Theory*, Graduate Texts in Math. vol. 9, Springer, 1972.
7. A. Knapp, *Lie Groups, Lie Algebras, and Cohomology*, Math. Notes vol. 34, Princeton University Press, 1988.
8. W. Schmid and J. Wolf, *A vanishing theorem for open orbits on complex flag manifolds*, Proc. Amer. Math. Soc. **92** (1984), 461–464.
9. D. Vogan, *Representations of Real Reductive Lie Groups*, Birkhäuser, 1981.
10. D. Vogan, *Unitarizability of certain series of representations*, Ann. of Math. **120** (1984), 141–187.
11. D. Vogan, *Unitary Representations of Reductive Lie Groups*, Annals of Mathematics Studies, Princeton University Press, 1987.
12. D. Vogan, *Representations of reductive Lie groups*, Proceedings of the International Congress of Mathematicians 1986, volume I, American Mathematical Society, 1987, pp. 245–266.
13. D. Vogan, *Irreducibility of discrete series representations for semisimple symmetric spaces*, Representations of Lie groups, Kyoto, Hiroshima, 1986, (K. Okamoto and T. Oshima, eds.), Advanced Studies in Pure Mathematics, volume 14. Kinokuniya 1988, pp. 191–221.
14. D. Vogan, *Dixmier algebras, sheets, and representation theory*, Operator Algebras, Unitary Representations, Enveloping Algebras, and Invariant Theory, (A. Connes, M. Duflo, A. Joseph, and R. Rentschler, eds.), Birkhäuser, 1990, pp. 333–395.

15. D. Vogan and G. Zuckerman, *Unitary representations with non-zero cohomology*, Compositio Math. **53** (1984), 51–90.

16. G. Warner, *Harmonic Analysis on Semisimple Lie Groups*, vol. I, Springer, 1972.

17. H. Wong, *Dolbeault cohomologies and Zuckerman modules associated with finite rank representations*, Ph.D. Dissertation, Harvard University, 1991.

DEPARTMENT OF MATHEMATICS, MASSACHUSETTS INSTITUTE OF TECHNOLOGY, CAMBRIDGE, MASSACHUSETTS 02139, USA

E-mail address: dav@math.mit.edu

Contemporary Mathematics
Volume **154**, 1993

Introduction to Penrose Transform

MICHAEL EASTWOOD

If R is a simply connected region in the complex plane, then any real-valued harmonic function on R may be written as the real part of a holomorphic function. It follows that any complex-valued harmonic function $\phi(x, y)$ may be written

$$\phi(x, y) = f_+(x + iy) - f_-(x - iy)$$

where $f_\pm$ are holomorphic functions. If we set

$$f(p, \zeta) = \begin{cases} f_+(p) & \text{if } \zeta = 1 \\ f_-(p) & \text{if } \zeta = -1 \end{cases}$$

then we can rewrite this formula as

$$\phi(x, y) = f(x + iy, 1) - f(x - iy, -1) = \int_{S^0} f(x + iy\zeta, \zeta) \, d\zeta \, .$$

The function f is determined by ϕ up to an additive constant $f \mapsto f + C$.

In 1904, Bateman proposed an analogue of this formula for harmonic functions of four variables [7]. If $f(p, q, \zeta)$ is a holomorphic function of three complex variables, then

$$\phi(w, x, y, z) = \int_{S^1} f((w + ix) + (iy + z)\zeta, (iy - z) + (w - ix)\zeta, \zeta) \, d\zeta$$

is a harmonic function as is easily verified by differentiating under the integral sign. The indeterminacy in f, however, is quite subtle. For one thing, the contour of integration S^1 may be deformed and, by Cauchy's theorem, ϕ is unchanged. Moreover, if f is holomorphic inside this contour, then Cauchy's theorem implies that ϕ vanishes. The geometric interpretation of this formula is also not immediately apparent though it clearly resembles the Radon transform [20]. Finally, it is not clear that all harmonic ϕ may be obtained in this way.

1991 *Mathematics Subject Classification.* Primary 32L25.

The author was supported by the Australian Research Council.

This paper is in final form and no version of it will be submitted for publication elsewhere.

In 1966, Penrose rediscovered this formula [17]. In fact, he found a generalization which generates solutions of the so-called *massless field equations* (e.g. Maxwell's equations). Since his discovery was geometrically motivated, the geometric significance of the formula was clear. In 1976, he realized that the indeterminacy in f is precisely accounted for by interpreting f as a Čech representative for an analytic cohomology class [19]. With this interpretation it is then possible to show that this *Penrose transform* is an isomorphism between cohomology and massless fields [11]. In particular, all harmonic ϕ may be obtained in this way. In the past few years, the transform has undergone considerable generalization and refinement (e.g. [3], [9]). It has also become apparent that various versions have existed for some time in representation theory—for example in [23], Schmid identifies the cohomology of a homogeneous holomorphic line bundle over G/T in an appropriate degree with solutions of a first order elliptic differential equation on G/K (notation as in Knapp's talk [14] with rank G = rank K) and another version appears in the recent work of Barchini, Knapp, and Zierau [5] (see Barchini's talk [6] and Wong's talk [26]).

Here is a geometric interpretation of Bateman's formula. Define

$$\pi : \mathbb{C}^3 \longrightarrow \mathbb{R}^4$$

$$\uplus \qquad\qquad \uplus$$

$$(p, q, \zeta) \qquad\qquad (w, x, y, z)$$

by solving the equations

$$
\begin{aligned}
p &= (w + ix) + (iy + z)\zeta \\
q &= (iy - z) + (w - ix)\zeta
\end{aligned}
$$

as in the integrand of Bateman's formula. We find that

$$w + ix = \frac{p + \bar{q}\zeta}{1 + \zeta\bar{\zeta}} \quad \text{and} \quad iy + z = \frac{p\bar{\zeta} - \bar{q}}{1 + \zeta\bar{\zeta}}.$$

If we include $\mathbb{C}^3$ in $\mathbb{CP}_3$ with homogeneous coördinates $[\xi_1, \xi_2, \xi_3, \xi_4]$ in the usual way by $(p, q, \zeta) \mapsto [p, q, \zeta, 1]$, then π extends to

$$\pi : \mathbb{CP}_3 \setminus \{\xi_3 = \xi_4 = 0\} \longrightarrow \mathbb{R}^4$$

defined by

$$w + ix = \frac{\xi_1\bar{\xi}_4 + \xi_3\bar{\xi}_2}{\xi_3\bar{\xi}_3 + \xi_4\bar{\xi}_4} \quad \text{and} \quad iy + z = \frac{\xi_1\bar{\xi}_3 - \xi_4\bar{\xi}_2}{\xi_3\bar{\xi}_3 + \xi_4\bar{\xi}_4}.$$

Finally, if we include $\mathbb{R}^4 \hookrightarrow S^4$ by stereographic projection, then we can extend π to all of $\mathbb{CP}_3$ by sending $\{\xi_3 = \xi_4 = 0\}$ to the projection point in S^4. The resulting mapping

$$\pi : \mathbb{CP}_3 \longrightarrow S^4$$

given explicitly by

$$[\xi_1, \xi_2, \xi_3, \xi_4] \longmapsto \begin{pmatrix} \dfrac{\xi_1\bar{\xi}_4+\xi_4\bar{\xi}_1+\xi_2\bar{\xi}_3+\xi_3\bar{\xi}_2}{\xi_1\bar{\xi}_1+\xi_2\bar{\xi}_2+\xi_3\bar{\xi}_3+\xi_4\bar{\xi}_4} \\[2ex] \dfrac{i\xi_1\bar{\xi}_4-i\xi_4\bar{\xi}_1-i\xi_2\bar{\xi}_3+i\xi_3\bar{\xi}_2}{\xi_1\bar{\xi}_1+\xi_2\bar{\xi}_2+\xi_3\bar{\xi}_3+\xi_4\bar{\xi}_4} \\[2ex] \dfrac{i\xi_1\bar{\xi}_3-i\xi_3\bar{\xi}_1+i\xi_2\bar{\xi}_4-i\xi_4\bar{\xi}_2}{\xi_1\bar{\xi}_1+\xi_2\bar{\xi}_2+\xi_3\bar{\xi}_3+\xi_4\bar{\xi}_4} \\[2ex] \dfrac{\xi_1\bar{\xi}_3+\xi_3\bar{\xi}_1-\xi_2\bar{\xi}_4-\xi_4\bar{\xi}_2}{\xi_1\bar{\xi}_1+\xi_2\bar{\xi}_2+\xi_3\bar{\xi}_3+\xi_4\bar{\xi}_4} \\[2ex] \dfrac{\xi_1\bar{\xi}_1+\xi_2\bar{\xi}_2-\xi_3\bar{\xi}_3-\xi_4\bar{\xi}_4}{\xi_1\bar{\xi}_1+\xi_2\bar{\xi}_2+\xi_3\bar{\xi}_3+\xi_4\bar{\xi}_4} \end{pmatrix}$$

has holomorphic fibres isomorphic to $\mathbb{CP}_1$. This is the Riemannian version of Penrose's original *twistor construction* [16] first applied in [2]. It is easily interpreted as a mapping of homogeneous spaces for $O_o(5,1)$. The action on $\mathbb{CP}_3$ is a projective spin representation and on S^4 the action is obtained by regarding S^4 as the null quadric in $\mathbb{RP}_5$. This realizes $O_o(5,1)$ as the group of conformal motions of S^4. As a mapping between appropriate homogeneous line bundles, the Laplacian is invariant under such motions. Bateman's formula is now subsumed by the following:

THEOREM. *For any open subset $R \subset S^4$, there is a natural isomorphism*

$$\mathcal{P} : H^1(\pi^{-1}(R), \mathcal{O}(-2)) \xrightarrow{\simeq} \{\text{Harmonic } \phi \text{ on } R\}.$$

The notation $\mathcal{O}(k)$ for the homogeneous holomorphic line bundles on $\mathbb{CP}_3$ is standard [12]. The cohomology $H^1(\mathbb{CP}_1, \mathcal{O}(-2))$ is one-dimensional. Thus, as $\mathbf{x} \in S^4$ varies, $H^1(\pi^{-1}(\mathbf{x}), \mathcal{O}(-2))$ defines a homogeneous line bundles on S^4. It is on this homogeneous line bundle that the conformally invariant Laplacian acts. The Penrose transform $\mathcal{P}$ is simply obtained by restricting a cohomology class to $\pi^{-1}(\mathbf{x})$ and allowing $\mathbf{x} \in R$ to vary.

As a simple variation on this theme, $H^1(\pi^{-1}(R), \mathcal{O}(-n-2))$ for $n > 0$ is interpreted as solutions of a first order elliptic differential equation on R (known as the massless field equation of helicity $n/2$). The cohomology $H^1(\mathbb{CP}_1, \mathcal{O}(-n-2))$ has dimension $n+1$ so a homogeneous vector bundle is obtained. For example, $H^1(\pi^{-1}(R), \mathcal{O}(-4))$ is isomorphic to the space of closed self-dual two-forms on R. The Penrose transform also interprets $H^1(\pi^{-1}(R), \mathcal{O}(-n-2))$ for $n < 0$ on R though its implementation is less direct. In this case the transform yields an isomorphism with the cohomology of an elliptic complex on R. For example, $H^1(\pi^{-1}(R), \mathcal{O})$ is isomorphic to the cohomology of the complex

$$\{\text{functions on } R\} \longrightarrow \{\text{1-forms on } R\} \longrightarrow \{\text{self-dual two-forms on } R\}$$

with differentials induced by the exterior derivative.

There are two ways of proving these isomorphims. One way is to interpret the Dolbeault complex on $\pi^{-1}(R)$ directly in terms of operators on R. This is explained in [22]. The other method, explained for example in [3] or [11], regards the twistor projection $\pi : \mathbb{CP}_3 \to S^4$ as a real form of the double fibration

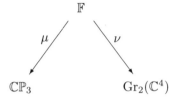

where $\mathbb{F}$ is the flag manifold of lines inside planes in $\mathbb{C}^4$ and the projections μ and ν are the obvious forgetful ones. The Klein correspondence realizes $\mathrm{Gr}_2(\mathbb{C}^4)$ as a non-singular quadric in $\mathbb{CP}_5$ and hence a complexification of S^4. When μ is restricted to $\nu^{-1}(S^4)$, it provides an isomorphism with $\mathbb{CP}_3$ and so π is recovered. The Penrose transform in this complex environment is well understood—it is the subject of [9] where general double fibrations of complex homogeneous spaces are considered. In this classical case the group is $\mathrm{SO}(6, \mathbb{C})$ or, equivalently, $\mathrm{SL}(4, \mathbb{C})$. The differential equations and complexes which arise on $\mathrm{Gr}_2(\mathbb{C}^4)$ are consequences of constancy along the fibres of μ. To deduce the real Penrose transform notice that the Laplacian is an elliptic operator so its kernel consists of real-analytic functions. Therefore these functions always extend to a sufficiently small neighbourhood of R in $\mathrm{Gr}_2(\mathbb{C}^4)$. For other real forms, the double fibration persists. This is true for the case studied in [11] where the spaces involved are open orbits of $\mathrm{O}_\circ(4, 2)$.

In [5], the Penrose transform is used to map cohomology of G/L to fields on G/K (as explained in §4 of Knapp's talk [14]). A simple example is when $G = \mathrm{SL}(2, \mathbb{C})$ in which case G/K is hyperbolic 3-space. The space G/L is known as hyperbolic *minitwistor space* [1]. A related construction (which is, in a sense, a limiting case) is the *minitwistor correspondence* for $\mathbb{R}^3$ and, in particular, Whittaker's formula [24] from 1902

$$\phi(x, y, z) = \int_0^{2\pi} f(x \cos t + y \sin t + iz, t) \, dt$$

giving harmonic functions of three variables (and very much related to Weierstraß's 1866 description of minimal surfaces in $\mathbb{R}^3$ (see [13])). A footnote in Whittaker and Watson [25, pp. 390–391] laments the indeterminacy in f when ϕ is given!

Finally, it should be mentioned that the Penrose transform is not restricted to homogeneous spaces. Generally, some special differential geometric structure is required. The classical case is in four dimensions where self-duality is the appropriate requirement [2], [18], [22]. There are useful generalizations to *quaternionic* manifolds [4], [10], [21] (see also LeBrun's talk [15]) and *almost Hermitian symmetric* manifolds [8].

References

1. M. F. Atiyah, *Magnetic monopoles in hyperbolic space*, Vector Bundles on Algebraic Varieties, Oxford University Press, 1987, pp. 1–34.

2. M. F. Atiyah, N. J. Hitchin, and I. M. Singer, *Self-duality in four-dimensional Riemannian geometry*, Proc. Roy. Soc. Lond. **A362** (1978), 425–461.

3. T. N. Bailey and M. A. Singer, *Twistors, massless fields and the Penrose transform*, Twistors in Mathematics and Physics, Lond. Math. Soc. Lecture Note Series, vol. 156, Cambridge University Press, 1990, pp. 299–338.

4. T. N. Bailey and M. G. Eastwood, *Complex paraconformal manifolds—their differential geometry and twistor theory*, Forum Math. **3** (1991), 61–103.

5. L. Barchini, A. W. Knapp, and R. Zierau, *Intertwining operators into Dolbeault cohomology*, Jour. Funct. Anal. **107** (1992), 302–341.

6. L. Barchini, *Strongly harmonic differential forms on elliptic orbits*, in this volume, pp. 77–88.

7. H. Bateman, *The solution of partial differential equations by means of definite integrals*, Proc. Lond. Math. Soc. **1(2)** (1904), 451–458.

8. R. J. Baston, *Almost Hermitian symmetric manifolds*, I and II, Duke Math. Jour. **63** (1991), 81–112 and 113–138.

9. R. J. Baston and M. G. Eastwood, *The Penrose Transform; its Interaction with Representation Theory*, Oxford University Press, 1989.

10. L. Bérard Bergery and T. Ochiai, *On some generalization of the construction of twistor spaces*, Global Riemannian Geometry, Ellis Horwood, 1984, pp. 52–59.

11. M. G. Eastwood, R. Penrose, and R. O. Wells, Jr., *Cohomology and massless fields*, Commun. Math. Phys. **78** (1981), 305–351.

12. P. A. Griffiths and J. Harris, *Principles of Algebraic Geometry*, Wiley, 1978.

13. N. J. Hitchin, *Momopoles and geodesics*, Commun. Math. Phys. **83** (1982), 579–602.

14. A. W. Knapp, *Introduction to representations in analytic cohomology*, in this volume, pp. 1–19.

15. C. R. LeBrun, *A finiteness theorem for quaternionic-Kähler manifolds with positive scalar curvature*, in this volume, pp. 89–101.

16. R. Penrose, *Twistor algebra*, Jour. Math. Phys. **8** (1967), 345–366.

17. _____, *Solutions of the zero rest-mass equations*, Jour. Math. Phys. **10** (1969), 38–39.

18. _____, *Non-linear gravitons and curved twistor theory*, Gen. Rel. Grav. **7** (1976), 31–52.

19. _____, *On the twistor description of massless fields*, in *Complex Manifold Techniques in Theoretical Physics*, Research Notes in Math., vol. 32, Pitman, 1979, pp. 55–91.

20. J. Radon, *Über die Bestimmung von Funktionen durch ihre Integralwerte längs gewisser Mannigfaltigkeiten*, Sächs. Akad. Wiss. Leipzig, Math.-Nat. Kl. **69** (1917), 262–277.

21. S. M. Salamon, *Quaternionic Kähler manifolds*, Invent. Math. **67** (1982), 143–171.

22. _____, *Topics in four-dimensional Riemannian geometry*, Geometry Seminar 'Luigi Bianchi', Lecture Notes in Math., vol. 1022, Springer, 1983, pp. 33–124.

23. W. Schmid, *Homogeneous complex manifolds and representations of semisimple Lie groups*, Ph.D. dissertation, University of California, Berkeley 1967, Representation Theory and Harmonic Analysis on Semisimple Lie Groups, Math. Surveys and Monographs, vol. 31, Amer. Math. Soc., 1989, pp. 223–286.

24. E. T. Whitakker, *On the general solution of Laplace's equation and the equation of wave motions, and on an undulatory explanation of gravity*, Monthly Notices Roy. Astron. Soc. **62** (1902), 617–620.

25. E. T. Whitakker and G. N. Watson, *A Course of Modern Analysis*, 3rd edition, Cambridge University Press, 1920.

26. H.-W. Wong, *Dolbeault cohomologies and Zuckerman modules*, in this volume, pp. 217–223.

DEPARTMENT OF PURE MATHEMATICS, UNIVERSITY OF ADELAIDE, SOUTH AUSTRALIA 5005

E-mail address: meastwoo@spam.maths.adelaide.edu.au

Contemporary Mathematics
Volume **154**, 1993

Strongly Harmonic Differential Forms on Elliptic Orbits

L. BARCHINI

ABSTRACT. This paper can be thought as an introduction based on examples to the works "Intertwining Operators in Dolbeault Cohomology Representations" [**BKZ**] and "Szegö Mapping, Harmonic Forms and Dolbeault Cohomology" [**B**]. We consider elliptic orbits of the form G/L with G semi-simple Lie Group and L the centralizer of a torus. We construct an explicit intertwining operator S, from derived functor modules, realized in the Langlands classification, into Dolbeault cohomology on G/L. This operator produces strongly harmonic forms. Since the Dolbeault cohomology representation was proved to be irreducible [**W**], we conclude that every K-finite class in cohomology admits a strongly harmonic representative which is in the image of the operator S. An understanding of the space of strongly harmonic forms is desirable in order to unitarize the cohomology.

§1. Dolbeault Cohomology

To describe our main results more precisely, we introduce some notation.

Let G be linear connected semi-simple with complexification $G^{\mathbb{C}}$, let K be maximal compact with Cartan involution θ, let T be a torus in K, and L the centralizer of T in G. We denote Lie algebras of Lie groups by $\mathfrak{g}_0$, $\mathfrak{k}_0$, $\mathfrak{l}_0$, etc. and their complexifications by, $\mathfrak{g}$, $\mathfrak{k}$, $\mathfrak{l}$, etc.

The quotient G/L has a number of invariant complex structures, and we pick one in the following way. Let $\mathfrak{q} = \mathfrak{l} \oplus \mathfrak{u}$ be a θ-stable parabolic subalgebra of $\mathfrak{g}$ containing $\mathfrak{l}$. If Q denotes the analytic subgroup of $G^{\mathbb{C}}$ with Lie algebra $\mathfrak{q}$, then G/L imbeds as an open subset of $G^{\mathbb{C}}/Q$ and inherits a complex structure in which $\mathfrak{u}$ is the antiholomorphic tangent space at the identity coset.

1991 *Mathematics Subject Classification*. Primary 22E45, 22E46.

Research supported in part by NSF Grant DMS 91 00383.

The final detailed version of this paper will appear in Journal of Functional Analysis [**B**].

A similar construction with $\mathfrak{q} \cap \mathfrak{k}$ makes $K/L \cap K$ into a compact complex submanifold of G/L. We set $s = \dim_{\mathbb{C}}(K/L \cap K) = \dim(\mathfrak{u} \cap \mathfrak{k})$.

Let ξ be a one dimensional representation of L, and let $\xi^{\sharp} = \xi \otimes \wedge^{\text{top}}\mathfrak{u}$. The complex line bundle $G \times_L \mathbb{C}_{\xi^{\sharp}} \to G/L$, with L acting on $\mathbb{C}$ via $\xi^{\sharp}$, canonically becomes a holomorphic line bunde [**TW**]. The bundle valued $(0, m)$ forms are the smooth sections of

$$(1.1) \qquad\qquad G \times_L \left(\mathbb{C} \otimes (\wedge^m \mathfrak{u})^* \right)$$

where $(\cdot)^*$ denotes dual. We write the space of sections as $C^{0,m}\left(G/L, \xi^{\sharp}\right)$ or

$$(1.2) \qquad \begin{aligned} C^{\infty}\left(G/L, \mathbb{C} \otimes (\wedge^m \mathfrak{u})^*\right) = \{f \, : \, & G \to \mathbb{C} \otimes (\wedge^m \mathfrak{u})^* \text{ is of class } C^{\infty} \\ & \mid f(x\ell) = \left(\xi^{\sharp}(\ell)^{-1} \otimes \text{Ad}(\ell)^{-1}\right) f(x)\}. \end{aligned}$$

Relative to the standard

$$(1.3) \qquad\qquad \bar{\partial} : \, C^{0,m}\left(G/L, \xi^{\sharp}\right) \longrightarrow C^{0,m+1}\left(G/L, \xi^{\sharp}\right)$$

the space of Dolbeault cohomology is

$$(1.4) \qquad\qquad H^{0,m}\left(G/L, \xi^{\sharp}\right) = \ker \bar{\partial}/\text{Image}\,\bar{\partial}.$$

There is a formal adjoint $\bar{\partial}^*$ to $\bar{\partial}$ (in (1.3)), denoted by [**BKZ**],

$$\bar{\partial}^* : \, C^{0,m+1}\left(G/L, \xi^{\sharp}\right) \longrightarrow C^{0,m}\left(G/L, \xi^{\sharp}\right).$$

We say that a bundle valued differential form ω is <u>strongly harmonic</u> if it satisfies $\bar{\partial}\omega = 0$ and $\bar{\partial}^*\omega = 0$.

§2. Positive Root System

We write the Cartan decomposition of $\mathfrak{g}_0$ relative to θ as $\mathfrak{g}_0 = \mathfrak{k}_0 \oplus \mathfrak{p}_0$. Extend $\mathfrak{t}_0$ to a maximal abelian subspace $\mathfrak{b}_0$ of $\mathfrak{k}_0$, and let $B = \exp \mathfrak{b}_0$. The centralizer $\mathfrak{h}_0$ of $\mathfrak{b}_0$ in $\mathfrak{g}_0$ is of the form $\mathfrak{h}_o = \mathfrak{b}_0 \oplus \mathfrak{a}'_0$ with $\mathfrak{a}'_0 \subset \mathfrak{p}_0$ and it is a maximally compact Cartan subalgebra of $\mathfrak{g}_0$. Let $\Delta = \Delta(\mathfrak{g}, \mathfrak{h})$ be the roots of $\mathfrak{g}$ with respect to $\mathfrak{h}$.

We choose Δ^+ a positive system for Δ such that $\Delta^+(\mathfrak{g}, \mathfrak{h}) = \Delta^+(\mathfrak{l}, \mathfrak{h}) \cup \Delta(\mathfrak{u}, \mathfrak{h})$ where $\Delta^+(\mathfrak{l}, \mathfrak{h})$ is given along with a compatible sequence $\alpha_1, \cdots, \alpha_{\ell}$. See [**BKZ**] p. 310. In particular $\mathfrak{a}''_0 = \sum \mathbb{R}(E_{\alpha_j} + E_{-\alpha_j})$ has $\mathfrak{a}_0 = \mathfrak{a}''_0 + \mathfrak{a}'_0$ maximal abelian in $\mathfrak{l} \cap \mathfrak{p}_0$.

Let

$$\mathfrak{b}''_0 = \sum i\mathbb{R}\, H_{\alpha_j} \subset \mathfrak{b}_0$$

$$\mathfrak{b}^-_0 = \text{orthocomplement of } \mathfrak{b}''_0 \text{ in } \mathfrak{b}_0.$$

We form a new Cartan subalgebra $\mathfrak{h}' = \mathfrak{b}^- \oplus \mathfrak{a}'' \oplus \mathfrak{a}'$ and define a positive root system $\Delta^+(\mathfrak{g}, \mathfrak{h}')$ in the following way. We list $H_{\alpha_1} \cdots H_{\alpha_{\ell}}$ as an ordered orthogonal basis in $i\mathfrak{b}''_0$ and extend it to an orthogonal basis of $i\mathfrak{b}''_0 \oplus \mathfrak{a}'_0$ by adjoining elements at the end. We use this basis in lexicographic fashion to determine positivity for members of $\Delta(\mathfrak{g}, \mathfrak{h}')$ that do not vanish identically on $\mathfrak{a}'' \oplus \mathfrak{a}'$. The positivity of the roots supported on $\mathfrak{b}^-$ is determined by $\Delta^+(\mathfrak{g}, \mathfrak{h})$.

Let Σ_L and Σ_G be the sets of restricted roots of L and G, and let Σ_L^+ and Σ_G^+ be the subsets of positive elements. Define ρ_L and ρ_G to be half sums of members of Σ_L^+ and Σ_G^+ with multiplicities counted. In [**BKZ**] it is shown that $\Delta^+(\mathfrak{l}, \mathfrak{h})$ and compatible $\alpha_1, \cdots, \alpha_\ell$ exist for which ρ_L is Σ_G^+-dominant.

§3. A Generalization of Borel-Weil-Bott Theorem

For the one dimensional representation ξ of L in §1, let $\lambda + \nu$ be the unique weight relative to $\mathfrak{h}$. Here λ is the part on $\mathfrak{b}$, and ν is the part on $\mathfrak{a}'$. The representation $\xi^\sharp = \xi \otimes \wedge^{\text{top}}\mathfrak{u}$ of L has weight $(\lambda + 2\delta(\mathfrak{u})) + \nu$.

We impose a dominance condition of ξ,

$$(3.1) \qquad \langle \lambda, \alpha \rangle \geq 0 \quad \text{for all} \quad \alpha \in \Delta(\mathfrak{u}).$$

The analytic construction of Dolbeault Cohomology has an algebraic analog based on a construction using derived functors. These are the Zuckerman modules $A_{\mathfrak{q}}(\lambda)$. Under the assumption (3.1) it was proven that both constructions coincide.

More precisely

THEOREM 3.1. *With the notation just introduced and under assumption* (3.1)

The Harish-Chandra Module of $H^s(G/L, \xi^\sharp)$ is $A_{\mathfrak{q}}(\lambda)$.

When G is a compact group, Theorem 3.1 is the Borel-Weil-Bott Theorem. If G is non-compact but L is compact, the theorem was proved by W. Schmid [**S**]. The general version can be found in [**W**].

Our main objective is to build an explicit intertwining operator between these modules. We observe that $A_{\mathfrak{q}}(\lambda)$ is an irreducible admissible $(\mathfrak{g}, K)$-module whose parameters in the Langlands classification are known. Thus, we can "identify" it with the Langlands quotient of an appropriate Principal Series. In fact, let A be the analytic subgroup of L with Lie algebra $\mathfrak{a}_0$ and let $P = MAN$ be a corresponding cuspidal parabolic subgroup of G with N chosen suitably. The Principal Series under consideration is $\text{ind}_{MAN}^G(\sigma \otimes \rho_L \otimes 1)$, where the discrete series σ of M is constructed by the Vogan algorithm from the minimal K-type of $A_{\mathfrak{q}}(\lambda)$.

Under the above identification, the operator $\mathcal{S}$ we are looking for is so that
$$(3.2)$$
$$\mathcal{S} : \text{ind}_{MAN}^G (\sigma \otimes \rho_L \otimes 1) \longrightarrow \ker \bar{\partial} \hookrightarrow \text{ind}_L^G (C \otimes (\wedge^s(\mathfrak{u})^*) \longrightarrow H^s(G/L, \xi^\sharp)$$

We observe that Theorem 3.1 guarantees the irreducibility of $H^s(G/L, \xi^\sharp)$. Thus, to show that the last map in (3.2), when restricted to K-finite vectors, is onto it is enough to prove that the image of $\mathcal{S}$ when viewed in cohomology is not zero. Moreover, our construction of $\mathcal{S}$ produces strongly harmonic forms. We have

$$\operatorname{ind}_{MAN}^{G}\left(\sigma \otimes \rho_{L} \otimes 1\right) \xrightarrow{\ \mathcal{S}\ } \ker \bar{\partial} \cap \ker \bar{\partial}^{*} \hookrightarrow \operatorname{ind}_{L}^{G}\left(\mathbb{C} \otimes (\wedge^{s}\mathfrak{u})^{*}\right) \longrightarrow H^{s}(G/L, \xi^{\sharp}).$$

As a consequence all non-trivial K-finite classes in cohomology admit a strongly harmonic representative which is in the image of the operator $\mathcal{S}$. It is of interest to study the space of strongly harmonic forms in order to describe the unitary structure of $H^{s}(G/L, \xi^{\sharp})$ from an analytic standpoint. See, for example, [**RSW**].

Our construction of $\mathcal{S}$ uses the theory of standard intertwining operators.

§4. Basic Intertwining Operators and Langlands Quotients

We revise some very well-known results on intertwining operators. We start by studying one example.

Let $G = SL(2, R)$. The Principal Series $P^{+,z}$, of $SL(2, R)$, is a representation in $L^{2}(\mathbb{R}, (1 + x^{2})^{\operatorname{Re} z} dx)$ with $z \in \mathbb{C}$. The representation $P^{+,z}$ is given by

$$\left[P^{+,z}\begin{pmatrix} a & b \\ c & d \end{pmatrix} f\right](x) = |-bx + d|^{-1-z} f\left(\frac{a\,x - c}{-bx + d}\right).$$

Some information about irreducibility and unitarity of $P^{+,z}$ is summarized in the following diagram.

(4.1)

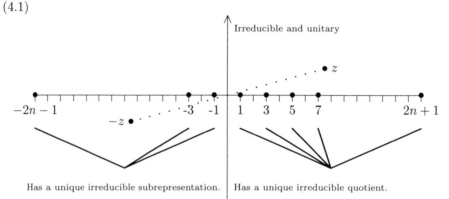

$P^{+,i\nu}$, $\nu \in \mathbb{R}$ is irreducible and unitary.

$P^{+,\pm(2n-1)}$ is reducible.

$P^{+,z}$ $z \neq \pm(2n-1)$ is an irreducible representation.

An operator

(4.2) $$A(+, z): \ P^{+,z} \longrightarrow P^{+,-z}$$

that satisfies

(4.3) $A(+, z) \ P^{+,z} \ = \ P^{+,-z} \ A(+, -z)$

is called an intertwining operator. A first example of such an operator is given by

$A(+, -(2n - 1)): \ P^{+,-(2n-1)} \longrightarrow P^{+,2n-1}$, where

(4.4) $[A(+, -(2n - 1))f](x) \ = \ \left[\left(\dfrac{d}{dx} \right)^{2n-1} f \right](x).$

If, Rez > 0, the Kunze-Stein Intertwining operator is given by an integral formula, i.e.

$$A(+, z): \ P^{+,z} \longrightarrow P^{+,-z}$$

(4.5) $[A(+, z) \, f] \, (x) \ = \ \displaystyle\int_{-\infty}^{\infty} \frac{f(x - y)}{|y|^{1-z}} \, dy.$

The operator in (4.5) can be meromorphically continued in z and identity (4.3) holds as an equality of meromorphic functions. Of course, for the values of z for which $P^{+,z}$ is irreducible, $A(+, z)$ provides an explicit isomorphism between $P^{+,z}$ and $P^{+,-z}$.

It is not difficult to show that in the points of reducibility of the Principal Series $A(+, 2n - 1)$ is so that Image of $A(+, 2n - 1)$ is infinitesimal equivalent to the unique irreducible quotient of $P^{+,2n-1}$ $(n \geq 1)$.

Using group notation

$$P^{+,z} \equiv \left\{ f: G \longrightarrow \mathbb{C} \ \middle| \ f \left(x \begin{pmatrix} a & b \\ 0 & a^{-1} \end{pmatrix} \right) = e^{-(1+z) \log |a|} f(x) \right\}$$

and the group G acts by left translation, i.e.,

(4.6) $\left[P^{+,z}(g)f \right] (x) \ = \ f(g^{-1}x).$

One can write down a corresponding expression for $A(+, z)$ in this picture, namely

(4.7) $[A(+, z)f] \, (x) \ = \ \displaystyle\int_{\overline{N}} f \left(x \begin{pmatrix} 0 & 1 \\ -1 & 0 \end{pmatrix} \bar{n} \right) d\bar{n}$

[Recall in $SL(2, R)$, $\overline{N} = \{ \begin{pmatrix} 1 & 0 \\ x & 1 \end{pmatrix} \}$]. Formula (4.7) can be thought as a composition of two operators $f \longrightarrow \int_{\overline{N}} f(x \, \bar{n}) \, d\bar{n}$ and right translation by $\begin{pmatrix} 0 & 1 \\ -1 & 0 \end{pmatrix}$

$$\mathrm{ind}_{MAN}^{G} \left(1 \otimes e^{z \log |a|} \otimes 1 \right) \longrightarrow \mathrm{ind}_{M\overline{AN}}^{G} \left(1 \otimes e^{z \log |a|} \otimes 1 \right)$$

$$\longrightarrow \mathrm{ind}_{MAN}^{G} \left(1 \otimes e^{-z \log |a|} \otimes 1 \right).$$

This expression suggests generalization and is the starting point for a systematic study of intertwining operators. [See [K1, ch. VII]].

If G is a connected semi-simple Lie group, then let $P = MAN$ be a parabolic subgroup of G. We set $\overline{P} = MA\overline{N}$ $(\overline{N} = \theta N)$. We consider the Principal Series $\text{ind}_{MAN}^G(\sigma \otimes e^\nu \otimes 1)$, where σ is a tempered unitary representation of M and $\nu \in \mathfrak{a}^*$ is such that $\text{Re}\nu$ is in an open positive Weyl chamber. The intertwining operator

$$A(\overline{P}, P, \sigma, \nu): \ \text{ind}_{MAN}^G(\sigma \otimes e^\nu \otimes 1) \longrightarrow \text{ind}_{MA\overline{N}}^G(\sigma \otimes e^\nu \otimes 1)$$

is formally given by an integral formula, i.e.

$$(4.8) \qquad\qquad \left(A(\overline{P}, P, \sigma, \nu) f\right)(x) = \int_{\overline{N}} f(x\bar{n})\, d\bar{n}.$$

The relation between intertwining operators and Langlands quotients observed in the example of $SL(2, R)$ also generalizes.

LEMMA 4.1.   *Under the assumptions already introduced, the Principal Series $\text{ind}_{MAN}^G(\sigma \otimes e^\nu \otimes 1)$ has a unique irreducible quotient $J(P, \sigma, \nu)$, and $J(P, \sigma, \nu)$ is infinitesimaly equivalent to the image of the intertwining operator $A(\overline{P}, P, \sigma, \nu)$. [Equivalently, the image of $A(\overline{P}, P, \sigma, \nu)$ is infinitesimaly equivalent to the unique irreducible subrepresentation of $\text{ind}_{MA\overline{N}}^G(\sigma \otimes e^\nu \otimes 1)$].*

Proof: See, for example, [**K1**, Theorem 7.24].

Fix a parabolic subgroup $P = MAN$ of G and form Σ_G the set of roots of $(\mathfrak{g}, \mathfrak{a})$. The choice of a positive system Σ_G^+ in Σ_G determines N. If we change the positive system, we obtain a new parabolic subgroup $P' = MAN'$. We can write down a formal expression for an intertwining operator $A(P', P, \sigma, \nu)$ with

$$(4.9) \qquad A(P', P, \sigma, \nu): \ \text{ind}_{MAN}^G(\sigma \otimes e^\nu \otimes 1) \longrightarrow \text{ind}_{MAN'}^G(\sigma \otimes e^\nu \otimes 1)$$

namely

$$(4.10) \qquad\qquad A(P', P, \sigma, \nu) f(x) = \int_{\overline{N} \cap N'} f(x\bar{n})\, d\bar{n}$$

where $\overline{N} = \theta N$. We can identify

$$\overline{N} \cap N' \equiv \frac{P'}{P' \cap \overline{P}}$$

and choose an appropriate P'-invariant measure on $P'/P' \cap P$ so that

$$(4.11) \qquad\qquad A(P', P, \sigma, \nu) f = \int_{P'/P' \cap P} f(xp')\, dp'.$$

Formula (4.11) suggests a philosophy that will allow us to treat more general situations. This is what Tony Knapp in [**K2**] calls the

Heuristic Principle.

Let H_1 and H_2 be two closed subgroups of G. Assume $H_1 \cap H_2 \neq \phi$ and consider $\text{ind}_{H_1}^G(\pi_1)$ and $\text{ind}_{H_2}^G(\pi_2)$. Here π_1 and π_2 may act on different Hilbert spaces. A natural intertwining operator $\mathcal{S}$ from $\mathrm{ind}_{H_1}^{G}(\pi_1)$ to $\mathrm{ind}_{H_2}^{G}(\pi_2)$ is given by some interpretation of

$$\mathcal{S}f(x) = \int_{H_2/H_1 \cap H_2} \pi_2(h)\,[T\,(f(xh)]\,dh$$

where

$$T:\ V^{\pi_1} \longrightarrow V^{\pi_2}\quad \text{is a }\ H_1 \cap H_2 - \text{map.}$$

This formula, in the case of parabolic induction, together with an analytic continuation, accounts for the standard Kunze-Stein operator.

§5. Construction of the Operator $\mathcal{S}$ for some Examples

Let $G = SO(4,1)$, so that $\mathfrak{g}$ consists of 5-by-5 complex matrices that are skew in the first four rows and columns, are symmetric in the last row and column, and are 0 on the diagonal. Thus,

(5.1)
$$\mathfrak{g}_0 = \left\{ \begin{pmatrix} \cdot & \cdot & | & \cdot & | & \cdot \\ \cdot & \cdot & | & \cdot & | & \cdot \\ \cdot & \cdot & | & \cdot & | & \cdot \\ \cdot & \cdot & | & \cdot & | & \cdot \end{pmatrix} \right\}$$

admits a Cartan decomposition $\mathfrak{g}_0 = \mathfrak{k}_0 \oplus \mathfrak{p}_0$. In (5.1) $\mathfrak{k}$ corresponds to the upper 4-by-4 rectangle and $\mathfrak{p}_0$ to the last row and column.

We choose $\mathfrak{a}_0$, the maximal abelian subspace of $\mathfrak{p}_o$, as indicated in the picture (5.1), i.e. $\mathfrak{a}_0 = R(E_{45} + E_{54})$. As usual E_{ij} denotes the matrix that is 1 in the $(i,j)^{\text{th}}$ entry and 0 elsewhere. If α denotes the unique positive restricted root, then $\rho_G = \frac{3}{2}\alpha$.

We take $\mathfrak{t}_0 \subset \mathfrak{k}_0$ to be the torus given by $\mathfrak{t}_0 = R\,(E_{12} - E_{21})$. (This corresponds, in the picture (5.1), to the 2-by-2 upper left square). $\mathfrak{l}_0$, the centralizer of $\mathfrak{t}_0$ in $\mathfrak{g}_0$ is $\mathfrak{l}_0 = \mathfrak{t}_0 \oplus SO(2,1)$. (i.e., the upper 2-by-2 left square plus the lower right 3-by-3 square). Here $\rho_L = \alpha/2$.

The corresponding root diagram is

$$\underset{e_1 - e_2}{\overset{\circ}{}} \!\!=\!\!=\!\!=\!\! \underset{\underset{\Delta(\ell)}{e_2}}{\overset{\bullet}{}}$$

We consider the quotient G/L with the complex structure given by the θ-stable parabolic subalgebra of $\mathfrak{g}$, $\mathfrak{q} = \mathfrak{l} \oplus \mathfrak{u}$, where $\Delta(\mathfrak{u}) = \{e_1 - e_2,\, e_1 + e_2,\, e_1\}$. Here, $s = \dim(\mathfrak{u} \cap \mathfrak{k}) = 2$.

Take a one-dimensional representation ξ of L with weight λ, acting on $\mathbb{C}_\lambda$. Denote $C_\lambda \otimes \wedge^{top}\mathfrak{u}$ by $C_\lambda^\sharp$. The interesting cohomology occurs in degree 2. We write the space of $(0,2)$-bundle valued forms variously as

$$C^{0,2}\left(G/L, C_\lambda^\sharp\right) = \left\{ f : G \longrightarrow C_\lambda^\sharp \otimes (\wedge^2\mathfrak{u})^* \,\middle|\, \begin{array}{l} \text{transformation law under } L \\ \text{on the right holds} \end{array} \right\}$$

or

$$\operatorname{ind}_L^G \left(C_\lambda^\sharp \otimes (\wedge^2\mathfrak{u})^* \right).$$

We observe that $C_\lambda^\sharp \otimes (\wedge^2\mathfrak{u})^*$ is the space of an irreducible three dimensional representation of L.

For expository purposes and in order to simplify the example, we assume that λ has $\lambda + 2\delta(\mathfrak{u} \cap \mathfrak{p}) = 0$ (observe that this parameter does not satisfy (3.1)). With this choice of λ, Vogan's construction [V] leads to the trivial representation of M. In other words, we need to build our intertwining operator $\mathcal{S}$ so that

(5.2) $\mathcal{S} : \operatorname{ind}_{MAN}^G (1 \otimes e^{\rho_L} \otimes 1) \longrightarrow \operatorname{ind}_L^G \left(C_\lambda^\sharp \otimes (\wedge^2\mathfrak{u})^* \right).$

We make this construction in three steps.

(i) First we observe that functions f in $\operatorname{ind}_{MAN}^G(1 \otimes e^{\rho_L} \otimes 1)$ transform according to the law

$$f(x\,m\,a\,n) = e^{-(\frac{1}{2}\alpha + \frac{3}{2}\alpha) \log a} f(x) \quad \text{for} \quad m\,a\,n \in MAN.$$

Here the term $\frac{3}{2}\alpha = \rho_G$ in the exponent is required for normalized induction. The same functions transform under $m\,a\,n \in L \cap MAN$ according to the law

$$f(x\,m\,a\,n) = e^{-(\frac{3}{2}\alpha + \frac{1}{2}\alpha) \log a} f(x) \quad \text{for} \quad m\,a\,n \in L \cap MAN.$$

Thus, we have a natural inclusion

$$i : \operatorname{ind}_{MAN}^G (1 \otimes e^{\rho_L} \otimes 1) \longrightarrow \operatorname{ind}_{L \cap MAN}^G (1 \otimes e^{\rho_G} \otimes 1).$$

(ii) By Lemma 4.1, we know that $\operatorname{ind}_{L \cap MA\overline{N}}^L(1 \otimes e^{\rho_G} \otimes 1)$ has a unique irreducible subrepresentation. In fact, such representation can be identified with the L-action on $C_\lambda^\sharp \otimes (\wedge^2\mathfrak{u})^*$. Again, applying Lemma 4.1, we have that the Kunze-Stein intertwining operator

$$A(\overline{P}, P, 1, \rho_G) : \operatorname{ind}_{L \cap MAN}^L (1 \otimes e^{\rho_G} \otimes 1) \longrightarrow \operatorname{ind}_{L \cap MA\overline{N}}^L (1 \otimes e^{\rho_G} \otimes 1)$$

has its image "isomorphic" to the irreducible subrepresentation of $\operatorname{ind}_{L \cap MA\overline{N}}^L (1 \otimes e^{\rho_G} \otimes 1)$, i.e. $C_\lambda^\sharp \otimes (\wedge^2\mathfrak{u})^*$.

(iii) Using the double induction formula we obtain an operator

$$\operatorname{ind} A(\overline{P}, P, 1, \rho_G) : \operatorname{ind}_L^G \left(\operatorname{ind}_{L \cap MAN}^L (1 \otimes e^{\rho_G} \otimes 1) \right) \longrightarrow \operatorname{ind}_L^G \left(C_\lambda^\sharp \otimes (\wedge^2\mathfrak{u})^* \right)$$

where

$$\operatorname{ind} A(\overline{P}, P, 1, \rho_G) f(x) = \int_{L \cap \overline{N}} f(x\bar{n})\, d\bar{n}.$$

The operator S we have been looking for is now just the restriction to $\operatorname{ind}_{MAN}^{G}(1 \otimes e^{\rho_L} \otimes 1)$ of $\operatorname{ind} A(\overline{P}, P, 1, \rho_G)$.

Using explicit formulas for $\bar\partial$ and $\bar\partial^*$ in terms of root vectors [**GS**], we calculate and check that Image $S \subset \ker \bar\partial \cap \ker \bar\partial^*$.

The second example will be to consider the case when M is compact (enough to assure real rank $L=$ real rank G and σ an irreducible representation of M). The construction of S in this case is very close to the one in the case treated before. I briefly indicate it here. For details see [**BKZ**].

Let μ be an irreducible representation of K with highest weight $\lambda + 2\delta(\mathfrak{u} \cap \mathfrak{p})$, acting in a space V^μ and having ϕ for a nonzero highest weight vector.

PROPOSITION 5.1. *The cyclic span of ϕ in V^μ under M is irreducible under M. Namely,*

(a) *ϕ is a highest weight vector under $\mathfrak{m}$, with highest weight $\lambda + 2\delta(\mathfrak{u} \cap \mathfrak{p})|_{\mathfrak{b}^-}$.*
(b) *$\mathbb{C}\phi$ is a one-dimensional subspace stable under the group $L \cap K$.*

Proof: Sèe [**BKZ**, Prop. 3.1].

We let σ be the irreducible representation of M acting in the M-cyclic span V^σ of ϕ, and we let τ be the one-dimensional representation of $L \cap K$ acting on $\mathbb{C}\phi$.

We construct the operator S in the following way.

(i)

$$i:\ \operatorname{ind}_{MAN}^{G}(\sigma \otimes e^{\rho_L} \otimes 1) \longrightarrow \operatorname{ind}_{L \cap MAN}^{G}(\tau|_{L \cap M} \otimes e^{\rho_G} \otimes 1)$$
$$f \longmapsto \tilde{f}(x) = \langle f(x), \phi \rangle\, \phi.$$

(ii) By Lemma 4.1, we know that $\operatorname{ind}_{L \cap MA\overline{N}}^{L}(\tau|_{L \cap M} \otimes e^{\rho_G} \otimes 1)$ has a unique irreducible subrepresentation, V^π. By Lepowsky-Wallach [**LW**], the representation (V^π, π) has highest weight $\rho_G - \rho_L$ and the highest weight space is acted upon by $L \cap M$ according to $\tau|_{L \cap M}$. Moreover, by Lemma 4.1 the standard Kunze-Stein operator for L

$$A(\overline{P}, P, 1, \rho_G):\ \operatorname{ind}_{L \cap MAN}^{L}(\tau|_{L \cap M} \otimes e^{\rho_G} \otimes 1) \to \operatorname{ind}_{L \cap MA\overline{N}}^{L}(\tau|_{L \cap M} \otimes e^{\rho_G} \otimes 1)$$

has its image "isomorphic" to (V^π, π).

One important point in [**BKZ**] is the construction of an L-irreducible subrepresentation $(\pi', V^{\pi'})$ of $C_\lambda^\sharp \otimes (\wedge^s \mathfrak{u})^*$, such that it is equivalent to (V^π, π).

(iii) Use double induction to obtain

$$\operatorname{ind} A(\overline{P}, P, 1, \rho_G) : \operatorname{ind}_L^G \left(\operatorname{ind}_{L\cap MA}^L (\tau|_{L\cap M} \otimes e^{\rho_G} \otimes 1) \right) \to$$

$$\operatorname{ind}_L^G (V^\pi) \to \operatorname{ind}_L^G (V^{\pi'}) \hookrightarrow \operatorname{ind}_L^G \left(C_\lambda^\sharp \otimes (\wedge^s \mathfrak{u})^* \right).$$

Under identifications the operator can be written as

$$(5.3) \qquad \mathcal{S}f(x) = \int_{L\cap K} \langle f(xk), \phi \rangle \, \pi'(k) \, (1 \otimes \omega_S) \, dk$$

for $f \in \operatorname{ind}_{MAN}^G (\sigma \otimes e^{\rho_G} \otimes 1)$. Here $1 \otimes \omega_S$ is a particular vector in $V^{\pi'} \subset C_\lambda^\sharp \otimes (\wedge^s \mathfrak{u})^*$.

In [**BKZ**] the invariance properties of the operator $\mathcal{S}$ are studied and a distribution argument shows that Image $\mathcal{S} \subset \ker \bar{\partial} \cap \ker \bar{\partial}^*$.

We make some observations about the vector $1 \otimes \omega_S$ (see [**BKZ**]).

(a) $1 \otimes \omega_S$ is a highest weight vector for $(\pi', V^{\pi'})$.
(b) The $L\cap M$-cyclic span of $1 \otimes \omega_S$ is a one-dimensional $L\cap M$ representation. Moreover,

$$(\langle L \cap M \ 1 \otimes \omega_S \rangle, L \cap M) \cong \tau|_{L\cap M}.$$

(c)

$$(5.4) \qquad (\langle L \cap M \ 1 \otimes \omega_S \rangle, \ L \cap M) \cong \tilde{C}_{\lambda + 2\delta(\mathfrak{u}\cap\mathfrak{p})} \otimes \wedge^{top}(\mathfrak{u} \cap \mathfrak{m} \cap \mathfrak{k})^*$$

where $\tilde{C}_{\lambda + 2\delta(\mathfrak{u}\cap\mathfrak{p})}$ denotes the space of realization of the one-dimensional representation of $L \cap M$ with weight $\lambda + 2\delta(\mathfrak{u}\cap\mathfrak{p}) + 2\delta(\mathfrak{u}\cap\mathfrak{m}\cap\mathfrak{k})$. Via the isomorphism (5.4) $1 \otimes \omega_S$ maps to a vector $1 \otimes \omega_1$ with $1 \in \tilde{C}_{\lambda + 2\delta(\mathfrak{u}\cap\mathfrak{p})}$ and ω_1 a non-trivial element of $\wedge^{top}(\mathfrak{u} \cap \mathfrak{m} \cap \mathfrak{k})^*$.

§6. The Operator $\mathcal{S}$ and the Heuristic Principle

According to the Heuristic Principle in §4., the operator $\mathcal{S}$, with

$$\mathcal{S} : \operatorname{ind}_{MAN}^G (\sigma \otimes e^{\rho_L} \otimes 1) \longrightarrow \operatorname{ind}_L^G (V^{\pi'}) \hookrightarrow \operatorname{ind}_L^G \left(C_\lambda^\sharp \otimes (\wedge^s \mathfrak{u})^* \right)$$

should be given by an integral formula of the form

$$(6.1) \qquad \mathcal{S}f(x) = \int_{L/L\cap MAN} \pi'(\ell) \, [T(f(x\ell)] \, d\ell = \int_{L\cap K} \pi'(\ell) \, [T\,(f(x\ell)] \, d\ell$$

with

$$T : V^\sigma \longrightarrow V^{\pi'} \subset C_\lambda^\sharp \otimes (\wedge^s \mathfrak{u})^* \quad \text{an} \quad L \cap M \quad \text{map}.$$

In the case when M is compact, we compare (5.3) and (6.1). Thus, we can take T to be given by

$$T: V^\sigma \longrightarrow V^{\pi'}$$
$$v \longmapsto \langle v, \phi \rangle \cdot 1 \otimes \omega_S$$
$$f(x) \longmapsto \langle f(x), \phi \rangle \cdot 1 \otimes \omega_S.$$

T is an $L \cap M$-map. If we use the identification mentioned in (5.4), we may say that $T(f(x)) = \langle f(x), \phi \rangle \cdot 1 \otimes \omega_1$, where $1 \otimes \omega_1 \in \tilde{C}_{\lambda + 2\delta(\mathfrak{u} \cap \mathfrak{p})} \otimes \wedge^{top}(\mathfrak{u} \cap \mathfrak{m} \cap \mathfrak{k})^*$. But the last formula reminds us of the isomorphism that implements the Borel-Weil-Bott Theorem in the direction

$$V^\sigma \longrightarrow H^{s_1}\left(M/L \cap M, \ \tilde{C}_{\lambda + 2\delta(\mathfrak{u} \cap \mathfrak{p})}\right).$$

In fact

$$V^\sigma \longrightarrow C^\infty\left(M/L \cap M, \ \tilde{C}_{\lambda + 2\delta(\mathfrak{u} \cap \mathfrak{p})} \otimes \wedge^{top}(\mathfrak{u} \cap \mathfrak{m} \cap \mathfrak{k})^*\right)$$
$$v \longmapsto \langle \sigma(m)\, v, \ \phi \rangle \quad 1 \otimes \omega_1$$

In other words, looking from the point of view of the Heuristic Principle, in the case M compact we obtain a method of lifting strongly harmonic forms for $M/L \cap M$ to strongly harmonic forms for G/L. With this same philosophy we can treat the general case (real rank $L \neq$ real rank G). In this case (σ, M) is a Discrete Series. A lifting from strongly harmonic forms for $M/L \cap M$ to strongly harmonic forms for G/L can be also obtained via the Heuristic Principle. There are some technical difficulties, due for example, to the fact that now M is not connected and of course V^σ is now a Hilbert space not finite dimensional. The details of the construction can be found in [**B**]. The philosophy is that an understanding of the theory for G/L lies in the understanding of the discrete series of $M/L \cap M$.

§7. Non-vanishing of the Image of $\mathcal{S}$ in Cohomology

In [**BKZ**] a Penrose transform $\mathcal{P}$ mapping $C^{0,s}(G/L, C_{\lambda^\sharp})$ into $C^\infty(G/K, V^\mu)$ was defined. Moreover, $\mathcal{P}$ annihilates co-boundaries. Thus, $\mathcal{P}$ is well defined on $H^s(G/L, C_{\lambda^\sharp})$. An explicit construction of the operator $\mathcal{P}$ can be found in [**BKZ**]. In [**BKZ**] we show commutativity of the following diagram.

$$\mathrm{ind}_{MAN}^G (\sigma \otimes e^{\rho_L} \otimes 1) \longrightarrow H^s(G/L, C_{\lambda^\sharp})$$

$$\mathcal{P} \circ \mathcal{S} \searrow \qquad \qquad \downarrow \mathcal{P}$$

$$C^\infty(G/K, V^\mu)$$

where $\mathcal{P} \circ \mathcal{S}$ is an Eisenstein integral operator that can be easily proved not to be zero.

An argument on the same lines works in the general case [**B**].

REFERENCES

[B]   L. Barchini, *Szegö Mappings, Harmonic Forms and Dolbeault Cohomology*, J. Func. Anal., to appear..

[BKZ]   L. Barchini, A. W. Knapp, and R. Zierau, *Intertwining operators into Dolbeault cohomology representations*, J. Func. Anal. **107** (1992), 302–341.

[GS]   P. Griffiths and W. Schmid, *Locally homogeneous complex manifolds*, Acta Math. **123** (1969), 253–302.

[K1]   A. W. Knapp, *Lie Groups, Lie Algebras, and Cohomology*, Princeton University Press, Princeton, 1988.

[K2]   _____, *Imbedding discrete series in $L^2(G/H)$*, Harmonic Analysis on Lie Groups (Sandbjerg Estate, August 26-30, 1991), Report Series, No. 3, Copenhagen University Mathematics Institute, 1991, pp. 27–29.

[LW]   J. Leposky and N. R. Wallach, *Finite and infinite dimensional representations of linear semisimple groups*, Trans. Amer. Math. Soc. **184**, 223–246.

[RSW]   J. W. Rawnsley, W. Schmid, and J. A. Wolf, *Singular unitary representations and indefinite harmonic theory*, J. Func. Anal. **51** (1983), 1–114.

[S]   W. Schmid, *Homogeneous complex manifolds and representations of semisimple Lie groups*, Ph.D. thesis, University of California at Berkeley, 1967, Representation Theory and Harmonic Analysis on Semisimple Lie Groups, Mathematical Surveys and Monographs **31** (eds. P. Sally and D. Vogan), 223–286, Amer. Math. Soc., 1988.

[TW]   J. A. Tirao and J. A. Wolf, *Homogeneous holomorphic vector bundles*, Indiana Univ. Math. Jour. **20** (1970), 15–31.

[V]   D. A. Vogan, *The algebraic structure of the representations of semisimple Lie groups*, I, Ann. of Math. **109** (1979), 1–60.

[W]   H.-W. Wong, *Dolbeault cohomologies and Zuckerman modules associated with finite rank representations*, Ph.D. thesis, Harvard University, 1991.

DEPARTMENT OF MATHEMATICS, TEMPLE UNIVERSITY, PHILADELPHIA, PENNSYLVANIA 19122, USA

E-mail address: barchini@euclid.math.temple.edu

Contemporary Mathematics
Volume **154**, 1993

A Finiteness Theorem for Quaternionic-Kähler Manifolds with Positive Scalar Curvature

CLAUDE LEBRUN

ABSTRACT. We study the topology and geometry of those compact Riemannian manifolds (M^{4n}, g), $n \geq 2$, with positive scalar curvature and holonomy in $Sp(n) \cdot Sp(1)$. Up to homothety, we show that there are only finitely many such manifolds of any dimension $4n$.

> ... che tu mi sie di tuoi prieghi cortese,
> in Fano, sì che ben per me s'adori
> pur ch'i' possa purgar le gravi offese.
>
> Dante Alighieri
> *Purgatorio V, 70-72*

Let (M^ℓ, g) be a connected Riemannian manifold. If $x \in M$ is an arbitrary base-point, one defines the *holonomy group* $\mathcal{H}(M, g, x) \subset End(T_x M)$ of (M, g, x) to be the set of linear maps $T_x M \to T_x M$ obtained by Riemannian parallel transport of tangent vectors around piece-wise smooth loops based at x; the *reduced holonomy group* $\mathcal{H}_0(M, g, x)$ is similarly defined, but now using only null-homotopic loops. The latter is automatically a connected Lie subgroup of the orthogonal transformations of the tangent space $T_x M$, and so may be identified with a Lie subgroup of $SO(\ell)$ by choosing an orthonormal basis for $T_x M$; and the conjugacy class of this subgroup is uneffected by a change of basis or base-point x. As was first pointed out by Berger, relatively few groups can actually arise in this way. If we exclude the so-called *locally reducible* manifolds, meaning those which are locally Riemannian Cartesian products of lower-dimensional manifolds, and the *locally symmetric* manifolds, meaning those for which the curvature tensor

---

1991 *Mathematics Subject Classification*. Primary 53C25; Secondary 14J45, 32J27.

Supported in part by NSF Grant DMS 92 04093.

A revised version of this paper constitutes part of [**16**], which has been submitted for publication elsewhere.

is covariantly constant, the only possibilities [6], [21] for the reduced holonomy of M^ℓ are those which appear on the following list:

ℓ	$\mathcal{H}_0$	geometry
ℓ	$SO(\ell)$	generic
$2m \geq 4$	$U(m)$	Kähler
$2m \geq 4$	$SU(m)$	Ricci-flat Kähler
$4n \geq 8$	$Sp(n)$	hyper-Kähler
$4n \geq 8$	$Sp(n) \times Sp(1)/\mathbb{Z}_2$	quaternionic-Kähler
7	G_2	imaginary Cayley
8	$Spin(7)$	octonionic

In particular, the universal cover of any complete Riemannian manifold is the product of globally symmetric spaces and manifolds whose holonomy appears on this list. Each of these possible holonomies may therefore be thought of as representing a "fundamental geometry," each of which has its own peculiar flavor.

Of these fundamental geometries, the *quaternionic-Kähler* or $Sp(n) \cdot Sp(1) := Sp(n) \times Sp(1)/\mathbb{Z}_2$ possibility in in some ways the most enigmatic. It is the only holonomy geometry which forces the metric to be Einstein but not Ricci-flat. In particular, the scalar curvature of a quaternionic-Kähler manifold is constant, and the sign of this constant turns out to have a decisive influence on the nature of manifold in question. Thus, while complete, non-compact, non-symmetric quaternionic-Kähler manifolds of negative scalar curvature exist in profusion [2], [7], [14], a complete quaternionic-Kähler manifold of positive scalar curvature must be compact, and the only known examples of such manifolds are symmetric. Indeed, Poon and Salamon [19], generalizing earlier work of Hitchin [10], have proved that there are no others in dimension 8, and it is therefore tempting to conjecture that this situation persists in higher dimensions. This article will indicate some new evidence supporting such a conjecture. In particular, a complete proof the following result is given:

THEOREM A (FINITENESS THEOREM). *For any n, there are, modulo isometries and rescalings, only finitely many compact quaternionic-Kähler 4n-manifolds of positive scalar curvature.*

Of course, this doesn't predict that every such manifold is symmetric. But the other main result which will be described in detail herein does allow one to draw such a conclusion for certain topological types:

THEOREM B (STRONG RIGIDITY). *Suppose (M, g) is a compact quaternionic-Kähler manifold of positive scalar curvature. Then $\pi_1(M) = 0$ and*

$$\pi_2(M) = \begin{cases} 0 & (M, g) = \mathbb{HP}_n \\ \mathbb{Z} & (M, g) = Gr_2(\mathbb{C}^{n+2}) \\ finite \supset \mathbb{Z}_2 & otherwise. \end{cases}$$

It might be particularly emphasized that these result rely very heavily on the global aspects of the problem. In particular, if one chooses to consider compact *orbifolds* rather than manifolds, both fail [8], even though the more general metrics involved are still smooth everywhere—at least when viewed from the skewed perspective of certain multi-valued coordinate charts.

Many of the results described herein were obtained as part of a joint research project with Simon Salamon, and further details will appear in our joint paper [16].

1. Preliminaries

DEFINITION 1. *Let (M, g) be a connected Riemannian $4n$-manifold, $n \geq 2$. We will say that (M, g) is a quaternionic-Kähler manifold iff the holonomy group $\mathcal{H}(M, g)$ is conjugate to $H \cdot Sp(1)$ for some Lie subgroup $H \subset Sp(n) \subset SO(4n)$.*

Example. The quaternionic projective spaces

$$\mathbb{HP}_n = Sp(n+1)/(Sp(n) \times Sp(1))$$

are quaternionic-Kähler manifolds. So are the complex Grassmannians

$$Gr_2(\mathbb{C}^{n+2}) = SU(n+2)/S(U(n) \times U(2))$$

and the oriented real Grassmannians

$$\tilde{Gr}_4(\mathbb{R}^{n+4}) = SO(n+4)/(SO(n) \times SO(4)).$$

In fact, these examples very nearly exhaust the compact homogeneous examples of quaternionic-Kähler manifolds. Indeed [1], every such homogeneous space is a symmetric space, and [23] there is exactly one such symmetric space for each compact simple Lie algebra. They can be constructed as follows: let G be a compact simple centerless group, and let $Sp(1)$ be mapped to G so that its root vector is mapped to a root of highest weight. If H is the centralizer of this $Sp(1)$, then the symmetric space $M = G/(H \cdot Sp(1))$ is quaternionic-Kähler, and every compact homogeneous quaternionic-Kähler manifold arises this way.

How typical are these symmetric examples? One geometric feature of any irreducible symmetric space is that it must be Einstein, with non-zero scalar curvature. This, it turns out, also happens for quaternionic-Kähler manifolds:

PROPOSITION 1 (BERGER). *Every quaternionic-Kähler manifold is Einstein, with non-zero scalar curvature.*

For details, see [6]. In particular, a complete quaternionic-Kähler manifold has constant scalar curvature.

DEFINITION 2. *We will say that a quaternionic-Kähler manifold is* positive *if it is complete and has positive scalar curvature.*

It is now an an immediate consequence of Myers' theorem that a positive quaternionic-Kähler manifold is compact and has finite fundamental group. Unfortunately, however, the only known positive quaternionic-Kähler manifolds are
the previously mentioned symmetric spaces! The main objective of the present
article will be to explain why this situation is hardly surprising. The main tool
in our investigation will be the next result:

THEOREM 1 (SALAMON [20]; BÉRARD-BERGERY [5]). *Suppose* (M^{4n}, g) *is a
quaternionic-Kähler manifold. Then there is a complex manifold* (Z, J) *of complex dimension* $2n + 1$, *called the* twistor space *of* (M, g), *such that*

- *there is a smooth fibration* $\wp : Z \to M$ *with fiber* S^2;
- *each fiber of* $\wp$ *is a complex curve in* (Z, J) *with normal bundle holomorphically isomorphic to* $[\mathcal{O}(1)]^{\oplus 2n}$, *where* $\mathcal{O}(1)$ *is the point-divisor
  line bundle on* $\mathbb{CP}_1$; *and*
- *there is a complex-codimension 1 holomorphic sub-bundle* $D \subset TZ$ *which
  is maximally non-integrable and transverse to the fibers of* $\wp$.

Moreover, if (M, g) *is* positive, *then* Z *carries a Kähler-Einstein metric of positive scalar curvature such that*

- $\wp$ *is a Riemannian submersion;*
- D *is the orthogonal complement of the vertical tangent bundle of* $\wp$; *and*
- *the induced metric on each fiber of* $\wp$ *has constant curvature.*

If (M, g) *is instead* negative, *there is an* indefinite *Kähler-Einstein pseudo-metric
on* Z *with all these properties.*

In particular, the twistor space Z of a positive quaternionic-Kähler manifold
is *Fano*:

DEFINITION 3. *A Fano manifold is a compact complex manifold* Z *such that
$c_1(Z)$ can be represented by a positive* $(1, 1)$-*form.*

That is, a Fano manifold is a compact complex manifold which admits Kähler
metrics of positive Ricci curvature. Every Fano manifold is simply connected,
since $c_1 > 0 \Rightarrow \chi(\mathcal{O}) = h^0(\mathcal{O}) = 1$ by the Kodaira vanishing theorem, thus
forbidding the possibility that the manifold might have a finite cover. Applying
the exact homotopy sequence of $Z \to M$, we now conclude the following:

PROPOSITION 2. *Any positive quaternionic-Kähler manifold is compact and
simply connected.*

A completely different and extremely important feature of our twistor spaces
is the holomorphic hyperplane distribution D, which gives a so-called *complex
contact structure* to Z. Such structures will be discussed systematically in §2.

Our definition of quaternionic has carefully avoided the case of $n = 1$; after
all, $Sp(1) \cdot Sp(1)$ is all of $SO(4)$, so such a holonomy restriction says nothing
at all. Instead, we choose the our definition in order to insure that Theorem 1
remains valid:

DEFINITION 4. *A Riemannian 4-manifold is called quaternionic-Kähler if it is Einstein, with non-zero scalar curvature, and half-conformally flat.*

2. Complex Contact Manifolds

DEFINITION 5. *A complex contact manifold is a pair (X, D), where X is a complex manifold and $D \subset TX = T^{1,0}X$ is a codimension-one holomorphic sub-bundle which is maximally non-integrable in the sense that the O'Neill tensor*

$$D \times D \quad \to \quad TX/D$$
$$(v, w) \quad \mapsto \quad [v, w] \bmod D$$

is everywhere non-degenerate.

Example. Let Y_{n+1} be any complex manifold, and let $X_{2n+1} = \mathbb{P}(T^*Y)$ be its projectived holomorphic cotangent bundle; dually stated, X is the Grassmann bundle of complex n-planes in TY. Let $\pi : X \to Y$ be the canonical projection, and let $D \subset TX$ be the sub-bundle defined by $D|_P := \pi_*^{-1}(P)$ for all complex n-planes $P \subset TY$. Then D is a complex contact structure on X.

The condition of non-integrability has a very useful reformulation, which we shall now describe. Given a codimension-one holomorphic sub-bundle $D \subset TX$, let $L := TX/D$ denote the quotient line bundle. Letting $\theta : TX \to L$ be the tautological projection, we may think of θ as a line-bundle-valued 1-form

$$\theta \in \Gamma(X, \Omega^1(L)) \, ,$$

and so attempt to form its exterior derivative $d\theta$. Unfortunately, this ostensibly depends on a choice of local trivialization; for if ϑ is any 1-form, $d(f\vartheta) = f d\vartheta + df \wedge \vartheta$. However, it is now clear that $d\theta|_D$ *is* well defined as a section of $L \otimes \wedge^2 D^*$, and an elementary computation, which we leave to the reader, shows that $d\theta|_D$, thought of in this way, is exactly the O'Neill tensor mentioned above. Now if the skew form $d\theta|_D$ is to be non-degenerate, D must have positive even rank $2n$, so that X must have odd complex dimension $2n + 1 \geq 3$. Moreover, the non-degeneracy exactly requires that

$$\theta \wedge (d\theta)^{\wedge n} \in \Gamma(X, \Omega^{2n+1}(L^{n+1}))$$

is nowhere zero. But this provides a bundle isomorphism between $L^{\otimes(n+1)}$ and the anti-canonical line bundle $\kappa^{-1} = \wedge^{2n+1} T^{1,0}X$.

Conversely, let X be a simply-connected compact complex $(2n + 1)$-manifold, and suppose that $c_1(X)$ is divisible by $n+1$. Then there is a unique holomorphic line bundle $L := \kappa^{-1/(n+1)}$ such that $L^{\otimes(n+1)} \cong \kappa^{-1}$. If we are then given a twisted holomorphic 1-form

$$\theta \in \Gamma(X, \Omega^1(\kappa^{-1/(n+1)}))$$

we may then construct

$$\theta \wedge (d\theta)^{\wedge n} \in \Gamma(X, \Omega^{2n+1}(\kappa^{-1})) = \Gamma(X, \mathcal{O}) = \mathbb{C} \ .$$

If this constant is non-zero, $D = \ker \theta$ is then a complex contact structure.

This simple observation has powerful consequences:

PROPOSITION 3. *Let X_{2n+1} be a simply connected compact complex manifold, and let $\mathcal{G}$ denote the identity component of the group of biholomorphisms $X \to X$. Then $\mathcal{G}$ acts transitively on the set of complex contact structures on X.*

Proof. We may assume that there is at least one complex contact structure on X, since otherwise there is nothing to prove. In this case, the canonical line bundle κ has a root $\kappa^{1/(n+1)}$, and there is only one such root because $H^1(X, \mathbb{Z}_{n+1}) = 0$. Thus any complex contact structure is determined by a class $[\theta] \in \mathbb{P}\Gamma(X, \Omega^1 \otimes (\kappa^{-1/(n+1)}))$ satisfying $\theta \wedge (d\theta)^n \neq 0$. The group $\mathcal{G}$ acts on this projective space $\cong \mathbb{P}_m$ in a manner preserving the hypersurface S defined by $\theta \wedge (d\theta)^n = 0$, and so partitions $\mathbb{P}_m - S$ into orbits; since $\mathbb{P}_m - S$ is connected, it therefore suffices to prove that each orbit is open, and for this it would be enough to prove that the holomorphic vector fields generating the action of the the Lie algebra of $\mathcal{G}$ on $\mathbb{P}_m - S$ span the tangent space at each point.

To prove the last statement, let $\theta \in \Gamma(X, \Omega^1 \otimes (\kappa^{-1/(n+1)}))$ be any contact form, and let $\phi \in \Gamma(X, \Omega^1 \otimes (\kappa^{-1/(n+1)}))$ be any other section. If D denotes the kernel of θ, $d\theta|_D : D \to D \otimes \kappa^{-1/(n+1)}$ is an isomorphism of holomorphic vector bundles, so we can define a holomorphic vector field $v \in \Gamma(X, \mathcal{O}(D))$ by $v = (d\theta|_D)^{-1}(\phi)$. We then have $\pounds_\xi \theta \equiv \xi \lrcorner d\theta \equiv \phi \mod \theta$, so that action of the Lie algebra of $\mathcal{G}$ spans the tangent space of $\mathbb{P}\Gamma(X, \Omega^1 \otimes (\kappa^{-1/(n+1)}))$ at $[\theta]$, thus proving the proposition. ∎

COROLLARY 1. *Two simply-connected compact complex manifolds are complex contact isomorphic iff the underlying complex manifolds are biholomorphically equivalent.*

This will now yield a result which is crucial for our purposes.

DEFINITION 6. *We will say that two Riemannian manifolds (M_1, g_1) and (M_2, g_2) are homothetic if there exists a diffeomorphism $\Phi : M_1 \to M_2$ such that $\Phi^* g_2 = c g_1$ for some constant $c > 0$. Such a map Φ will be called a homothety.*

PROPOSITION 4. *Two positive quaternionic-Kähler manifolds are homothetic iff their twistor spaces are biholomorphic.*

Proof. Let (M, g) and $(\tilde{M}, \tilde{g})$ be two given quaternionic-Kähler manifolds, $\wp : Z \to M$ and $\tilde{\wp} : \tilde{Z} \to \tilde{M}$ their twistor spaces, h and $\tilde{h}$ the Kähler-Einstein metrics of Z and $\tilde{Z}$. We also suppose that a biholomorphism $\Phi : Z \to \tilde{Z}$ is given to us. For some positive constant $c > 0$, h and $c\tilde{h}$ have the same scalar curvature; and notice that replacing $\tilde{h}$ with $c\tilde{h}$ just corresponds to replacing $\tilde{g}$ with $c\tilde{g}$. Now $\Phi^*c\tilde{h}$ is a Kähler-Einstein metric on Z with the same scalar curvature as h, and the Bando-Mabuchi theorem [**4**] on the uniqueness of Kähler-Einstein metrics now asserts that there exists a biholomorphism $\Psi : Z \to Z$ such that $\Psi^*(\Phi^*c\tilde{h}) = h$.

Let $N \subset \Gamma(Z, \Omega^1(\kappa^{-1/(n+1)}))$ be defined by $\phi \wedge (d\phi)^{\wedge n} = 1$. Proposition 3 implies that a finite connected cover $\mathcal{G}$ of the connected component of the automorphism group of (Z, J) acts transitively on N, since, in the notation of the proof of that proposition, $N \to \mathbb{P}_m - S$ is a finite covering. Because h is Kähler-Einstein, with positive scalar curvature, the Killing fields are a real form of the algebra of holomorphic vector fields, and a finite cover G of the connected component of the isometry group of (Z, h) is therefore a compact real form of $\mathcal{G}$. Morse theory now predicts that one orbit of the action of G on N is precisely the set of critical points of the G-invariant strictly plurisubharmonic function $f : N \to \mathbb{R}$ given by $\phi \mapsto \|\phi\|^2_{L^2, h}$. On the other hand, the derivative of f at a contact form ϕ in the direction of a real-holomorphic vector field ξ on Z is given by

$$df(\xi)|_\phi = \tfrac{1}{(2n)!} \int_X d(\xi \lrcorner \, \omega) \wedge |\phi|^2_h \left(\beta_\phi - \tfrac{n+2}{n+1} \omega^{2n} \right),$$

where ω is the Kähler form of (Z, J, h) and β_ϕ is the $(2n, 2n)$ form obtained by orthogonally extending the restriction of ω^{2n} from $D = \ker \phi$ to TZ. For the canonical contact form θ associated with the quaternionic-Kähler metric g by the twistor construction, $|\theta|_h$ is constant and $\beta_\theta = \wp^*(2n)! dvol_g$ is closed, so that θ is a critical point of f; but the same argument applies equally to the contact from of $\tilde{Z}$, and hence to the pull-back of this contact form via the holomorphic isometry $\Phi\Psi$. Hence there is a holomorphic isometry $\Xi : Z \to Z$ sending the first of these contact structures to the second, and $\Phi\Psi\Xi : Z \to \tilde{Z}$ is a then biholomorphism which sends h to $c\tilde{h}$ and D to $\tilde{D}$. Since the vertical tangent spaces of $\wp$ and $\tilde{\wp}$ are the orthogonal complements of D and $\tilde{D}$ with respect to h and $\tilde{h}$, respectively, it follows that $\Phi\Psi\Xi$ sends fibers of $\wp$ to fibers of $\tilde{\wp}$, and so covers a diffeomorphism $F : M \to \tilde{M}$. Moreover, since the $\wp$ and $\tilde{\wp}$ are Riemannian submersions, one has $F^*c\tilde{g} = g$, and F is thus a homothety between (M, g) and $(\tilde{M}, \tilde{g})$. ∎

For less precise but more broadly applicable theorems on the invertibility of the twistor construction, cf. [**13**], [**3**].

DEFINITION 7. *Let (X_{2n+1}, D) be a complex contact manifold. An n-dimensional submanifold $\Sigma_n \subset X_{2n+1}$ is called* Legendrian *if $T\Sigma \subset D$.*

LEMMA 1. *Let (X_{2n+1}, D) be a complex contact manifold, and let $\pi : X \to Y_{n+1}$ be a proper holomorphic submersion with Legendrian fibers. Then $X \cong \mathbb{P}(T^*Y)$ as complex contact manifolds.*

Proof. Define $\Psi : X \to Gr_n(TY)$ by $x \to \pi_*(D_x)$. This map preserves the contact structure; and since the pull-back of the contact form of $Gr_n(TY)$ via Ψ is the contact form of X, Ψ^* induces an isomorphism between forms of top degree. In other words, Ψ is a submersion onto its image, and, in particular, induces a submersion from each fiber of X onto its image in the fiber of $Gr_n(TY) = \mathbb{P}(T^*Y)$. By the properness assumption, Ψ is fiber-wise therefore a covering map. But the fibers of $\mathbb{P}(T^*Y)$ are projective spaces, and so simply connected. Hence Ψ is an injective holomorphic submersion, and so biholomorphic. ∎

DEFINITION 8. *If (X, D) is a complex contact manifold such that X is Fano, we will say that (X, D) is a Fano contact manifold.*

LEMMA 2. *Let $\varpi : \mathcal{X} \to \mathcal{B}$ be a holomorphic family of Fano contact manifolds with smooth connected parameter space—that is, let $\mathcal{B}$ be a connected complex manifold, ϖ a proper holomorphic submersion with Fano fibers, and assume that $\mathcal{X}$ is equipped with a maximally non-integrable, complex codimension 1 subbundle $D \subset \ker \varpi_*$ of the vertical tangent bundle. Then any two fibers $(X_0, D|_{X_0})$ and $(X_t, D|_{X_t})$ are isomorphic as complex contact manifolds.*

Proof. Since any two points in $\mathcal{B}$ can be joined by a finite chain of holomorphic images of the unit disk $\Delta \subset \mathbb{C}$, it suffices to prove the lemma when the base $\mathcal{B}$ is a disk Δ.

We now proceed as in [12]. By Darboux's theorem, any complex contact structure in dimension $2n + 1$ is locally complex-contact isomorphic to the one on $\mathbb{C}^{2n+1}$ determined by the 1-form

$$\vartheta = dz^{2n+1} + \sum_{j=1}^{n} z^j dz^{n+j} \ ,$$

so we may cover our family $\varpi : \mathcal{X} \to \Delta$ by Stein sets U_j on which we have holomorphic charts $\Phi_j : U_j \hookrightarrow \mathbb{C}^{2n+1} \times \Delta$ such that the last coordinate is given by ϖ and the fiber-wise contact structure on $\mathcal{X}$ agrees with that induced by $\Phi_j^* \vartheta$. Letting t denote the standard complex coordinate on Δ, we lift d/dt to each U_j as the vector field $v_j (\Phi_j^{-1})_* d/dt$, and observe that the t-dependent vertical vector field $w_{jk} := v_j - v_k$ satisfies $\pounds_{w_{jk}} \theta \propto \theta$.

Let $f_{jk} := \theta(w_{jk}) \in \Gamma(U_j \cap U_k, \mathcal{O}(L))$, and notice that the collection $\{f_{jk}\}$ is a Čech cocycle representing an element of $H^1(\mathcal{X}, \mathcal{O}(L))$. On the other hand, since L is a fiber-wise $(n + 1)^{st}$-root of the vertical anti-canonical bundle κ^{-1}, and since each fiber X_t of ϖ is assumed to be a Fano manifold, the bundle $\kappa^{-1} \otimes L$

is fiber-wise positive, and $H^1(X_t, \mathcal{O}(L)) = 0 \; \forall t \in \Delta$ by the Kodaira vanishing theorem. Thus the first direct image sheaf $\varpi_*^1 \mathcal{O}(L)$ is zero. Since Δ is Stein, the Leray spectral sequence now yields $H^1(\mathcal{X}, \mathcal{O}(L)) = 0$. Hence there exist sections $h_j \in \Gamma(U_j \mathcal{O}(L))$ such that $f_{jk} = h_j - h_k$ on $U_j \cap U_k$.

On U_j there is now a unique vertical holomorphic vector field u_j such that $\theta(u_j) = h_j$ and $\pounds_{u_j} \theta \propto \theta$. Indeed, taking a local trivialization of L so as to locally represent θ by a holomorphic 1-form ϑ, a vector field u satisfies $\pounds_u \theta \propto \theta$ iff

$$u \lrcorner \, d\vartheta \equiv -d(u \lrcorner \vartheta) \bmod \vartheta \; ,$$

so that such a field is uniquely determined by an arbitrary local function $f = \vartheta(u) = u \lrcorner \vartheta$. We therefore conclude that $v_j - v_k = w_{jk} = u_j - u_k$ on $U_j \cap U_k$, and the vector field $v = v_j - u_j$ is therefore globally defined. Since $\pounds_{v_j} \vartheta = 0$ and $\pounds_{u_j} \vartheta \equiv 0 \pmod{\vartheta, dt}$, the flow of $v = v_j - u_j$ preseves the fiberwise contact structure on $\mathcal{X}$. And since ϖ is a proper map, we can now integrate the flow of our lift v of d/dt to produce a fiber-wise contact biholomorphism between $\mathcal{X}$ and $X_0 \times \Delta$. In particular, any fiber X_t is complex-contact equivalent to the central fiber X_0. ∎

3. Mori Theory

Mori's theory of extremal rays [17] has led to a startling series of advances in the classification of complex algebraic varieties, especially in the Fano case which interests us. One beautiful consequence of this is the so-called *contraction theorem*: if X is a Fano manifold, there is always a map $\Upsilon : X \to Y$ to some other variety Y which decreases the second Betti number b_2 by one, and where the kernel of $\Upsilon_* : H_2(X, \mathbb{R}) \to H_2(Y, \mathbb{R})$ is generated by the class of a rational holomorphic curve $\mathbb{CP}_1 \subset X$. (The positive half of such a one-dimensional subspace $\ker \Upsilon_* \subset H_2(X, \mathbb{R})$ is called an "extremal ray".) If $b_2(X) = 1$, this tells us next to nothing, because we can take Y to be a point; but for $b_2(X) \geq 2$, it is quite a powerful tool. In particular, it gives rise to the following very useful result of Wiśniewski [22]:

THEOREM 2 (WIŚNIEWSKI). *Let X be a Fano manifold of dimension $2r - 1$ for which $r | c_1$. Then $b_2(X) = 1$ unless X is one of the following: (i) $\mathbb{CP}_{r-1} \times Q_r$; (ii) $\mathbb{P}(T^* \mathbb{CP}_r)$; or (iii) $\mathbb{CP}_{2r-1}$ blown up along $\mathbb{CP}_{r-2}$.*

Here $Q_r \subset \mathbb{CP}_{r+1}$ denotes the r-quadric, while the projectivization of a bundle $E \to Y$ is defined by $\mathbb{P}(E) := (E - 0_Y)/(\mathbb{C} - 0)$. The essence of the proof is that, since the rational curves collapsed by the Mori contraction have, in these circumstances, normal bundles of rather large index, they are so mobile that they sweep out projective spaces of comparatively large dimension, and these must therefore be the fibers of the contraction map.

The following is now an easy consequence:

COROLLARY 2. *Let (X_{2n+1}, D) be a Fano contact manifold. If $b_2(X) > 1$, then $X = \mathbb{P}(T^*\mathbb{CP}_{n+1})$.*

Proof. Setting $r = n + 1$, we notice that the existence of a contact structure implies that $(n+1)|c_1$. We may therefore invoke Theorem 2. On the other hand, spaces (i) and (iii) aren't complex contact manifolds, since $\Gamma(\mathbb{CP}_{r-1}, \Omega^1(1)) = 0$ and therefore the obvious foliations by $\mathbb{CP}_{r-1}$'s would necessarily have Legendrian leaves, implying (by Lemma 1) that these spaces would then have to be of the form $\mathbb{P}(T^*Y)$, where Y is the leaf space Q_r or $\mathbb{CP}_r$—a contradiction. So the only candidate left is (ii), and this *is* in fact a contact manifold. ∎

THEOREM 3 ([**15**]). *Let (M, g) be a compact quaternionic-Kähler $4n$-manifold with $s > 0$. Then either*

    (1) $b_2(M) = 0$; *or else*

    (2) $M = Gr_2(\mathbb{C}^{n+2})$ *with its symmetric-space metric.*

Proof. By the Leray-Hirsch theorem on sphere bundles, the second Betti numbers of M^{4n} and its twistor space Z_{2n+1} are related by $b_2(Z) = b_2(M) + 1$. Since Z is a Fano contact manifold, $b_2(M) > 0 \Rightarrow Z = \mathbb{P}(T^*\mathbb{CP}_{n+1})$ by Corollary 2. But this is the twistor space of $Gr_2(\mathbb{C}^{n+2})$. The result therefore follows by Proposition 4. ∎

THEOREM B (STRONG RIGIDITY). *Let M be a compact quaternionic-Kähler manifold of positive scalar curvature. Then $\pi_1(M) = 0$ and*

$$H_2(M, \mathbb{Z}) = \begin{cases} 0 & M = \mathbb{HP}_n \\ \mathbb{Z} & M = Gr_2(\mathbb{C}^{n+2}) \\ \text{finite} \supset \mathbb{Z}_2 & \text{otherwise.} \end{cases}$$

Proof. If (M, g) is not homothetic to the symmetric space $Gr_2(\mathbb{C}^{n+2}) = SU(n+2)/S(U(n) \times U(2))$, $b_2(M) = 0$ by Theorem 3, so that $H^2(M, \mathbb{Z}) = 0$ and $H_2(M, \mathbb{Z})$ is finite. Since we also know that $H_1(M, \mathbb{Z}) = 0$, $H^2(M, \mathbb{Z}_2)$ is exactly the the 2-torsion of $H_2(M, \mathbb{Z})$ by the universal coefficients theorem. If, on the other hand, (M, g) is not homothetic to the symmetric space $\mathbb{HP}_n$, the class $\varepsilon \in H^2(M, \mathbb{Z}_2)$ must be non-zero [**20**], and the finite group $\pi_2(M) = H_2(M, \mathbb{Z})$ must therefore contain an element of order 2. ∎

4. The Finiteness Theorem

THEOREM 4. *Up to biholomorphism, there are only finitely many Fano contact manifolds of any given dimension $2n + 1$.*

Proof. By Wisniewski's theorem, we may restrict our attention to Fano manifolds with $b_2 = 1$. A theorem of Nadel [**18**] then asserts that there are only a finite number of deformation types of any fixed dimension.[1]

For any fixed deformation type, we may embed each Fano manifold in a fixed projective space $\mathbb{CP}_N$ in such a manner that the restriction of the generator $\alpha \in H^2(\mathbb{CP}_N, \mathbb{Z})$ is a fixed multiple $\ell c_1(Z)/q$ of the anti-canonical class, and we may freely choose the positive integers ℓ and q as long as $q|c_1$ and ℓ is sufficiently large. Thus, let $\mathcal{F}$ denote the set of all complex submanifolds $Z \subset \mathbb{CP}_N$ of some fixed dimension m and degree d, and with the additional property that, for fixed integers ℓ, q, the restriction of the hyperplane class is $\ell c_1(Z)/q$. Thus $\mathcal{F}$ is a Zariski-open subset in a component of the Chow variety, and so, in particular, is quasi-projective. There is now a tautological family

$$
\begin{array}{ccc}
\mathcal{Z} & \hookrightarrow & \mathcal{F} \times \mathbb{CP}_N \\
\varpi \downarrow & & \downarrow \\
\mathcal{F} & = & \mathcal{F}
\end{array}
$$

such that the fiber of $Z \in \mathcal{F}$ is the submanifold $Z \subset \mathbb{CP}_N$.

We now assume moreover that the dimension m is an odd number $2n + 1$, take $q = n + 1$, and choose $\ell \gg 0$ such that $\gcd(n + 1, \ell) = 1$. Letting $V \to \mathcal{Z}$ denote the vertical tangent bundle $\ker \varpi_*$, the vertical anti-canonical line bundle $\kappa^{-1} := \wedge^{2n+1}V$ has a consistent fiber-wise $(n+1)^{st}$-root $L \to \mathcal{Z}$; indeed, using the Euclidean algorithm to write $1 = a(n+1)+b\ell$, we may define L by $L = \kappa^{-a} \otimes \mathcal{H}^b$, where $\mathcal{H}$ is the pull-back of $\mathcal{O}(1)$ from $\mathbb{CP}_N$ to $\mathcal{Z}$.

Let $\mathcal{F}_j \subset \mathcal{F}$ denote the locus

$$
\mathcal{F}_j := \left\{ Z \in \mathcal{F} \mid h^0(Z, \Omega_Z^1 \otimes \kappa^{-1/(n+1)}) \geq j \right\}
$$

where the space of candidate contact forms has dimension at least j. Since $\Omega_Z^1 \otimes \kappa^{-1/(n+1)}$ is just the restriction of $V^* \otimes L$ to the appropriate fiber of ϖ, and since ϖ is a flat morphism, it follows from the semi-continuity theorem [**9**] that each $\mathcal{F}_j$ is a Zariski-closed subset of the quasi-projective variety $\mathcal{F}$, and so, in particular, has only finitely many components.

Let $\tilde{\mathcal{F}}_j := \mathcal{F}_j - \mathcal{F}_{j-1}, j \geq 1$. Since $\tilde{\mathcal{F}}_j$ is a quasi-projective variety, it is a finite union of irreducible strata $\mathcal{F}_{jk}$, each of which is a connected complex manifold. On each stratum $\mathcal{F}_{jk}$, define a vector bundle E_{jk} as the zero-th direct image $\mathcal{O}(E_{jk}) = \varpi_*^0 \mathcal{O}(V^* \otimes L)$ of the fiber-wise 1-forms with values in L. Let $\mathcal{L}$ denote the line bundle $\varpi_*^0 \mathcal{O}(L^{n+1} \otimes \wedge^{2n+1}V)$ on $\mathcal{F}$. Then $\theta \mapsto \theta \wedge (d\theta)^n$ is defines a canonical holomorphic section of the symmetric-product bundle $\mathcal{L} \otimes \bigodot^{n+1} E_{jk}^*$;

---

[1]It is now known [**11**] that this is true even *without* the restriction $b_2 = 1$.

let $\mathcal{E}_{jk}$ denote the open subset in the total space of $E_{jk} \to \mathcal{F}_{jk}$ where this homogeneous function is non-zero. Thus each $\mathcal{E}_{jk}$ is either a connected complex manifold or is empty. Each $\mathcal{E}_{jk}$ may now be viewed as the smooth parameter space of a connected family of Fano contact manifolds by taking the fiber over $\theta \in \Gamma(Z, \Omega^1(L))$ to be the pair $(Z, \ker \theta)$. On the other hand, every Fano contact manifold (Z, D), where Z is of the fixed deformation type, appears in one of these families—albeit many times. Applying Lemma 2, each of these families is of constant contact type. Since we must construct $\mathcal{F}$ only for a finite number of degrees in order to account for all Fano deformation types of the given dimension, and since, for each $\mathcal{F}$ we only have a finite number of contact families $\mathcal{E}_{jk}$, the result now follows. ∎

THEOREM A (FINITENESS THEOREM). *Up to homothety, there are only finitely many compact quaternionic-Kähler manifolds of positive scalar curvature in any given dimension $4n$.*

Proof. By Proposition 4, two positive quaternionic-Kähler manifolds are homothetic iff their twistor spaces are biholomorphic. Since the twistor space of any such manifold is a Fano contact manifold, the result now follows immediately from Theorem 4. ∎

5. Other Results

We have seen in §3 that the second homology of a positive quaternionic-Kähler manifold is far from arbitrary, and may by itself contain enough information to determine the metric up to isometry. Recent calculations of Salamon show that the higher homology groups are similarly constrained, in the following remarkable manner:

THEOREM 5 (SALAMON). *Let (M^{4n}, g) be a compact quaternionic-Kähler manifold with positive scalar curvature. Then the "odd" Betti numbers b_{2k+1} of M vanish, and the "even" Betti numbers $b_{2k} = b_{2(2n-k)}$ are subject to the linear constraint*

$$\sum_{k=0}^{n} a_k b_{2k} = 0 \; ,$$

where $a_k = \begin{cases} 1 + 2k + 2k^2 - 4n/3 - 2kn + n^2/3 & k < n \\ (n^2 - n)/6 & k = n. \end{cases}$

The proof of this result involves an intricate interplay between the Kodaira vanishing theorem and the Penrose transform. Details will appear elsewhere [16].

Acknowledgements. The author would like to thank Shigeru Mukai, Janoš Kollár, and Alan Nadel for their helpful explanations of Fano theory, and Simon Salamon for many, many helpful conversations.

REFERENCES

1. D. V. Alekseevskii, *Compact quaternion spaces*, Funct. Anal. Appl. **2** (1968), 106–114.
2. _____, *Classification of quaternionic spaces with transitive solvable groups of motions*, Math. USSR—Izv. **9** (1975), 297–339.
3. T. N. Baily and M. G. Eastwood, *Complex paraconformal manifolds—their differential geometry and twistor theory*, Forum Math. **3** (1991), 61–103.
4. S. Bando and T. Mabuchi, *Uniqueness of Kähler-Einstein metrics modulo connected group actions*, Algebraic Geometry, Sendai, 1985, (T. Oda, ed.), Adv. Stud. Math., vol. 10, North-Holland, 1987, pp. 11–40.
5. L. Bérard-Bergery, *Variétés Quaternioniennes*, unpublished lecture notes, Espalion, 1979.
6. A. Besse, *Einstein Manifolds*, Springer, 1987.
7. K. Galicki, *Generalization of the momentum mapping construction for quaternionic Kähler manifolds*, Comm. Math. Phys. **108** (1987), 117–138.
8. K. Galicki and H. B. Lawson, *Quaternionic reduction and quaternionic orbifolds*, Math. Ann. **282** (1988), 1–21.
9. R. Hartshorne, *Algebraic Geometry*, Springer, 1977.
10. N. J. Hitchin, *Kählerian Twistor Spaces*, Proc. Lond. Math. Soc. **43** (1981), 133–150.
11. J. Kollár, Y. Miyaoka and S. Mori, *Rational connectedness and boundedness of Fano manifolds*, J. Diff. Geom. **36** (1992), 765–779.
12. C. R. LeBrun, *A rigidity theorem for quaternionic-Kähler manifolds*, Proc. Am. Math. Soc. **103** (1988), 1205–1208.
13. _____, *Quaternionic-Kähler manifolds and conformal geometry*, Math. Ann. **284** (1989), 353–376.
14. _____, *On complete quaternionic-Kähler manifolds*, Duke Math. J. **63** (1991), 723–743.
15. _____, *On the topology of quaternionic manifolds*, Twistor Newsletter **32** (1991), 6–7.
16. C. R. LeBrun and S. Salamon, *Strong Rigidity of Positive Quaternion-Kähler Manifolds*, preprint.
17. S. Mori, *Hartshorne conjecture and extremal ray*, Sugaku Expositions **0** (1988), 15–37.
18. A. Nadel, *The boundedness of degree of Fano varieties with Picard number one*, J. Am. Math. Soc. **4** (1991), 681–692.
19. Y.-S. Poon and S. M. Salamon, *Eight-dimensional quaternionic-Kähler manifolds with positive scalar curvature*, J. Diff. Geom. (to appear).
20. S. M. Salamon, *Quaternionic-Kähler manifolds*, Inv. Math. **67** (1982), 143–171.
21. _____, *Riemannian Geometry and Holonomy Groups*, Pitman Research Notes in Mathematics vol. 201, Longman, 1989.
22. J. A. Wiśniewski, *On Fano manifolds of large index*, Manu. Math. **70** (1991), 145–152.
23. J. A. Wolf, *Complex homogeneous contact manifolds and quaternionic symmetric spaces*, J. Math. Mech. **14** (1965), 1033–1047.

DEPARTMENT OF MATHEMATICS, STATE UNIVERSITY OF NEW YORK, STONY BROOK, NEW YORK 11794-3651, USA
E-mail address: claude@math.sunysb.edu

Contemporary Mathematics
Volume **154**, 1993

Holomorphic Language for $\overline{\partial}$-cohomology and Representations of Real Semisimple Lie Groups

SIMON GINDIKIN

ABSTRACT. We announce a purely holomorphic language for $\overline{\partial}$-cohomology and show how it applies in some interesting cases, specifically the cases of hyperfunctions, of nonholomorphic discrete series for $SU(2,1)$, and of Speh representations of $SL(2m; \mathbb{R})$.

1. Choice of the cohomological language I—continuous Čech cohomology

There are two canonical cohomological languages for Cauchy-Riemann cohomology: Čech and Dolbeault cohomology. The standard Čech language is not quite natural for the case of coverings that depend continuously on parameters. There we use a continuous version of Čech cohomology.

Let M be a complex manifold. Let $\{M_\xi \mid \xi \in \Xi\}$ be a covering of M by Stein manifolds M_ξ, parameterized by a a real manifold Ξ. Let

$$X = \{(z,\xi) \mid \xi \in \Xi \text{ and } z \in M_\xi\}.$$

Then we have the double (C^∞) fibering

$$X$$

(1)
$$\pi \swarrow \quad \searrow \rho$$

$$M \qquad \Xi$$

with the natural projections π, ρ. The covering sets M_ξ correspond to the fibers $\{(z,\xi) \mid z \in M_\xi\}$, ξ fixed, of ρ. The $\Xi_z = \{(z,\xi) \mid z \in M_\xi\}$, z fixed, correspond to the fibers of π.

1991 *Mathematics Subject Classification.* Primary 32F10, 32M15, 22E46.

Research partially supported by NSF Grant DMS 92 02049.

This paper is in final form and no version of it will be submitted for publication elsewhere.

Let $\Phi^r = \Phi^r(M, \Xi)$ be the space of differential r-forms $\varphi(z|\xi, d\xi)$ on Ξ that are holomorphic in $z \in M_\xi$. In other words, Φ^r is the pullback under $X \hookrightarrow M \times \Xi$ of the tensor product of the space of holomorphic functions on M with the space of smooth r-forms on Ξ. Let

$$\cdots \to \Phi^{r-1} \to \Phi^r \to \Phi^{r+1} \to \cdots$$

be the complex for the differential d_ξ and

$$(2) \qquad H_I^{(r)}(M) = ker(\Phi^r \to \Phi^{r+1})/im(\Phi^{r-1} \to \Phi^r).$$

Then we can prove

PROPOSITION 1. *Let the double fibration* (1) *satisfy the condition*

(C)   *the fibers* $\Xi_z, z \in M$, *of* π *are contractible.*

Then $H_I^{(r)}(M) = H^{(r)}(M, \mathcal{O})$.

Here one can replace the coefficient sheaf $\mathcal{O}$ by the sheaf of germs of holomorphic sections of a holomorphic vector bundle.

If we consider $H^{(r)}(M, \mathcal{O})$ in its Dolbeault realization then we can obtain the isomorphism in the same way as the isomorphism between Dolbeault and sheaf cohomology. Let $r = 1$ and ω be a (0,1)-form on M with $\bar{\partial}\omega = 0$. Then $\omega = \bar{\partial}_z f(z|\xi), z \in M_\xi$ and $\varphi = d_\xi f \in \Phi^1$ because $\bar{\partial}_z \varphi = d_\xi \omega = 0$.

It is remarkable that there is another operator from H_I onto Dolbeault cohomology which does not use the solution of $\bar{\partial}$-equations.

PROPOSITION 2. *Under condition* (C) *let* γ *be a* (*smooth*) *section of the fibering* $\pi : X \to M$. *Then the operator on* Φ^r

$$(3) \qquad \varphi \mapsto \kappa\varphi = (-1)^r (\varphi|_{\xi=\gamma(z)})^{(0,r)}$$

induces an isomorphism between $H_I^{(r)}(M)$ *and* $H^{(0,r)}$.

So we consider the result of the restriction of φ to the section γ as an r-form on M and take the $(0,r)$-component of this form. Let us remark that for $r = 1$ if $\varphi \in \Phi^1, \varphi = d_\xi f(z|\xi)$, then $\kappa\varphi$ will differ from $\bar{\partial}_z f \in Z^{(0,1)}(M)$ by the $\bar{\partial}$-exact form $\bar{\partial} f(z|\gamma(z))$.

2. Example 1 (hyperfunctions on $\mathbb{R}^n$)

By definition, hyperfunctions on $\mathbb{R}^n$ are elements of $H^{(n-1)}(\mathbb{C}^n \backslash \mathbb{R}^n, \mathcal{O})$. So $M = \mathbb{C}^n \backslash \mathbb{R}^n$. Let $\Xi = S^{n-1}$ be the unit sphere in $\mathbb{R}^n$

$$(4) \qquad M_\xi = \{z \in \mathbb{C}^n | \text{Im}\langle \xi, z \rangle > 0, \xi \in S^{n-1}\}.$$

These are Stein manifolds, products of the halfplane and $\mathbb{C}^{n-1}$. We consider the space of $(n-1)$-forms $\varphi(z|\xi, d\xi)$, holomorphic in $z \in M_\xi$ (all such forms are closed) and divide by the subspace of d_ξ-exact forms. If $\varphi \in \Phi^{n-1}$ has boundary values in a sense for $z = x \in \mathbb{R}^n$ then the hyperfunction φ defines a function (still called φ) by the formula

$$(5) \qquad \varphi(x) = \int_{S^{n-1}} \varphi(x|\xi, d\xi), \qquad x \in \mathbb{R}^n.$$

Let $T = \mathbb{R}^n + iV$ be a tube domain with a convex sharp (does not contain straight lines, so the dual cone is empty) cone V as a base. It is known that any holomorphic function f in T has well defined hyperfunction boundary values. To find the boundary values of f as a hyperfunction in this language, we need to represent f as an average of holomorphic functions $f(z|\xi)$ in M_ξ for $M_\xi \supset T$.

There is an interesting generalization of this construction. Let $T = \mathbb{R}^n + iV$ be a tube domain where the cone V is not necessarily convex. Let $V_\xi, \xi \in \Xi$, be the family of maximal convex subcones in V. We suppose that $V = \bigcup V_\xi$ and all V_ξ are a product of $\mathbb{R}^q$ and a sharp cone in $\mathbb{R}^{n-q}$ and in Ξ there are q-dimensional cycles γ such that $V = \bigcup_{\xi \in \gamma} V_\xi$. We will call q the index of convexity. Let

$$T_\xi = \mathbb{R}^n + iV_\xi$$

and we have the Stein covering $T = \bigcup T_\xi$. Let us consider $H^{(q)}(T, \mathcal{O})$ using this covering. It turns out that:

There is a canonical operator (of boundary values)

$$H^{(q)}(T, \mathcal{O}) \to H^{(n-1)}(\mathbb{C}^{n-1} \backslash \mathbb{R}^{n-1}, \mathcal{O}).$$

So not only holomorphic functions in convex tube domains have hyperfunction boundary values on $\mathbb{R}^n$ but also $\bar{\partial}$-cohomology of a appropriate dimension q in nonconvex tube domains (q has the geometrical interpretation) have hyperfunction boundary values.

In particular if $\varphi(z|\xi, d\xi) \in \Phi^q(T)$ has regular boundary values for $z \in \mathbb{R}^n$ then the corresponding hyperfunction boundary values are realized by the regular function

$$\varphi(x) = \int_\gamma \varphi(x|\xi, d\xi), \qquad x \in \mathbb{R}^n.$$

Correspondingly, if $\mathbb{R}^n$ is represented as the closure of a union of cones V_j with different indexes q_j, then we can represent hyperfunctions on $\mathbb{R}^n$ as a sum of boundary values of $\bar{\partial}$-cohomology (different dimensions) in $T_j = \mathbb{R}^n + iV_j$.

The language of H_I-cohomology is convenient for definitions of different functional spaces of $\bar{\partial}$-cohomology. Hardy spaces of $\bar{\partial}$-cohomology in nonconvex tube domains were defined in such a way in [1].

The simplest cone for which these constructions make sense is the cone V in $\mathbb{R}^3$

$$y_1^2 - y_2^2 - y_3^2 < 0.$$

The cones V_ξ will be wedge-products of $\mathbb{R}^1$ and two-dimensional sharp angles. We investigate $H^{(1)}(\mathbb{R}^3 + iV, \mathcal{O})$. If the cones $V_\pm$ are defined by the conditions

$$y_1^2 - y_2^2 - y_3^2 > 0, \quad y_1 \gtrless 0$$

then any hyperfunction on $\mathbb{R}^3$ can be represented as the sum of boundary values of holomorphic functions on $\mathbb{R}^3 + iV_\pm$ and of cohomology from $H^{(1)}(\mathbb{R}^3 + iV, \mathcal{O})$.

3. Choice of cohomological language II—Stein coverings

Let Ξ in (1) be a complex manifold, let π and ρ be holomorphic, and suppose

(C′)   the $\Xi_z, z \in M$, are contractible Stein manifolds.

Then X is a complex manifold and one can prove

THEOREM 3. *Under condition* (C′) *the inclusion* $\{\Phi_h^r\} \hookrightarrow \{\Phi^r\}$, *of the subcomplex consisting of forms whose coefficients are holomorphic on X, induces isomorphisms in cohomology.*

So we have the Stein manifold X with holomorphic double fibering

$$X$$
$$\pi \swarrow \quad \searrow \rho$$
$$M \qquad \qquad \Xi$$

and we consider holomorphic r-forms $\varphi(z|\xi, d\xi)$ on X with differentials only on the fibers. M. Eastwood remarked that for the theorem it is important only that we have a Stein manifold X which is fibered over M. At this point it is not essential that we use the construction of X through a covering $\{M_\xi\}$, (i.e. $X \subset M \times \Xi$).

So we have a purely holomorphic definition of $\bar\partial$-cohomology. Proposition 2 shows that cohomology exists on the Stein covering $X \to M$ as holomorphic forms and only after restriction to a (nonholomorphic) section γ of $X \to M$ do we obtain nonholomorphic objects—the Dolbeault realization of $\bar\partial$-cohomology. In other words, the natural place for ("the life of") $\bar\partial$-cohomology is the Stein covering X ("skies") rather than the manifold M ("earth").

Maybe it is a little bit surprising that the existence of Stein coverings M is a fairly generic situation. Let us begin with compact manifolds.

For $M = \mathbb{C}P^1$ we can take as X the manifold of pairs (x, ξ) of different points of $\mathbb{C}P^1$. It will be a Stein manifold, $\Xi \cong M$, and M_ξ, Ξ_z are affine lines.

For $M = \mathbb{C}P^n$ let Ξ be the dual space $\mathbb{C}P^n$ of hyperplanes ξ, and take $X = \{(z, \xi) : z \notin \xi\}$. Then M_ξ, Ξ_z are isomorphic to $\mathbb{C}^n$.

We can observe in this example a general phenomenon: we take as Ξ the set of Stein submanifolds M_ξ such that $M = cl(M_\xi)$ (closure of M_ξ). We will see that such construction can be generalized to arbitrary flag manifolds.

4. Example 2 (complement of the ball in $\mathbb{C}P^2$)

Let M be the domain in $\mathbb{C}P^2$ (with homogeneous coordinates (z_0, z_1, z_2)):

$$(6) \qquad H(z, z) = |z_0|^2 - |z_1|^2 - |z_2|^2 < 0.$$

We will construct a Stein covering of M with $\Xi \cong M$. Namely, for $\xi \in M$ let M_ξ be the union of all (complex) lines in M going through ξ with removed ξ. It is clear that M_ξ will be the product of a disk and $\mathbb{C}^1$ so that it is a Stein manifold. We have

$$X = \{(z, \xi) | z \in M, \xi \in M, z \neq \xi, \text{line}(z, \xi) \subset M\}.$$

The manifold X is not homogeneous.

We will investigate $H^{(1)}(M, \mathcal{O}(-k))$ for $k \geq 2$ through the consideration of the holomorphic forms

$$\varphi(z|\xi, d\xi) \in \Phi_h^{(1)}, \quad z \in M_\xi; \quad \xi \in \Xi = M;$$

$$\varphi(\lambda z | \mu \xi, d(\mu \xi)) = \lambda^{-k} \varphi(z|\xi, d\xi).$$

PROPOSITION 4. *For $k \geq 2$ in any cohomology class $H_I^{(1)}(M, \mathcal{O}(-k))$ there is one and only one 1-form $\varphi(z|\xi, d\xi) \in \Phi^{(1)}(M, \mathcal{O}(-k))$ which is constant on z along lines of the fibering M_ξ (going through ξ).*

This proposition is only the reformulation of results from [2], [3] about the explicit reconstruction of $\bar{\partial}$-cohomology in M through its Penrose transform. In a sense it is an analog of the Hodge theorem for this homomorphic cohomological language.

We have the natural action of $SU(1,2)$ in $H^{(1)}(M, \mathcal{O}(-k))$. These representations will be equivalent to the representations of holomorphic discrete series. Let us give the construction of the invariant Hermitian form.

(i) Our first step is to choose a submanifold $Y \subset X$ transversal to the lines (z, ξ). Proposition 4 show that it is natural, in a sense, since cohomology is trivial along of such lines.

Let $L_\xi \subset M_\xi$ be defined by the condition: $z \in M_\xi$, $H(z, \xi) = 0$, so that

$$(7) \qquad L_\xi = \{z : H(z, \xi) = 0, H(z, z) < 0\}.$$

Each line (z, ξ) in M_ξ has a unique representative on L_ξ. Let

$$Y = \{(z, \xi), \xi \in \Xi, z \in L_\xi\}.$$

It is a 6-dimensional (over $\mathbb{R}$) homogeneous (noncomplex) submanifold (fibering over Ξ with disks as fibers).

(ii) For $\varphi \in \Phi^1$ we consider the $(2,2)$-form on Y

$$(8) \qquad \Lambda(\varphi) = \varphi \wedge \overline{\varphi} \wedge \Theta,$$

where

(9) $$\Theta = \frac{H(z,z))^{k-2}}{H(\xi,\xi)} det(z,\xi,dz) \wedge det(\overline{z},\overline{\xi},d\overline{z}).$$

The form Θ is independent of ξ on Y. It will be closed (φ is closed in ξ and Θ has maximal degree in z).

If $\gamma \subset \Xi$ let

$$Y(\gamma) = \{(z,\xi), \xi \in \gamma, z \in L_\xi\}.$$

Then we can prove the following statement in a number of cases, and we conjecture that it is true much more generally:

PROPOSITION 5. *If $\gamma \subset \Xi$ is a 2-dimensional cycle (over $\mathbb{R}$) and $k > 2$*

(10) $$\int\limits_{Y(\gamma)} \Lambda(\varphi) = c(\gamma)N(\varphi)$$

where $c(\gamma)$ depends only on γ and $N(\varphi)$ depends only on ϕ, with $\infty \geq N(\varphi) > 0$ if the cohomology class $\{\varphi\} \neq 0$; c and N are defined up to a scalar factor.

So we can use $N(\varphi)$ as $\|\varphi\|^2$ and consider the Hilbert space $H_2^{(1)}(M, \mathcal{O}(-k))$ (the invariance of this norm is evident); $H_2^{(1)}(M, \mathcal{O}(-k))$ consists of those φ such that $N(\varphi) < \infty$. It can be shown that $H_2^{(1)}(M, \mathcal{O}(-k))$ is nonzero. The proof will not be included here.

(iii) It is convenient to have some standard cycles with $c(\gamma) \neq 0$. We can take as γ a line $l \subset \Xi$. This construction admits the following interpretation. The covering Ξ of M is not minimal. We have the minimal covering if we take M_ξ for $\xi \in l$ (line $\subset \Xi$) and we can construct cohomology using this minimal covering $l \subset \Xi$. The condition of the closure of $\varphi \in \Phi^{(1)}$ on l is trivial.

So we can realize our representation in 1-forms $\varphi(z|\xi, d\xi)$, $\xi \in l \subset \Xi$, $z \in M_\xi$ (or $z \in L_\xi$). But l is not invariant under $SU(1;2)$ and we need to recompute cohomology for the covering $g(l), g \in SU(1;2)$ on the covering l. The operators of this representation will be integral operators of Bessel kind.

5. Example 3 (ladder representations of SU(p,q))

Let us briefly describe a multidimensional generalization of this construction. Let M be the domain in $\mathbb{C}P^n$

(11) $$H(z,z) = |z_0|^2 + \cdots + |z_q|^2 - |z_{q+1}|^2 - \cdots |z_n|^2 > 0.$$

We will investigate $H^{(q)}(M, \mathcal{O}(-k))$.

(i) Let Ξ be the domain in the Grassmanian $Gr(q, n+1)$ of $(q-1)$-planes which are contained in M. We will describe $\xi \in \Xi$ by a q-frame $(\xi^{(1)}, \cdots, \xi^{(q)})$.

Let X be the manifold of pairs $(z, \xi), z \in M, \xi \in \Xi$, such that $z \in \xi$, and the q-plane $\pi(z, \xi)$ is contained within M. So

(12)
$$M_\xi = \bigcup_{M \supset \pi \supset \xi} \pi \backslash \xi,$$

where we take the union of all q-planes π, with $M \supset \pi \supset \xi$.

We have
$$M_\xi \cong D_{n-q} \times \mathbb{C}^q,$$

where D_{n-q} is an $(n-q)$-dimensional ball.

So elements of $\Phi^k(M, \Xi)$ are holomorphic q-forms $\varphi(z|\xi, d\xi), \xi \in \Xi, \xi \in M_\xi$.

(ii) *Holomorphic Hodge forms.* We say the form $\varphi(z|\xi, d\xi)$ satisfies the condition (H) if

(1) φ as a function in z is constant on q-planes $\pi \supset \xi$,

(2) φ is a sum of differential monomials $d\xi_{i_1}^{(1)} \wedge d\xi_{i_2}^{(2)} \wedge \cdots \wedge d\xi_{i_q}^{(q)}$.

We could not yet see the condition (2) in example 2 (it was trivial for $q = 1$). It has a simple geometrical sense: φ must be dual to the Euler class (equal zero on other Schubert cells) as well as on each set of $(q-1)$-planes, which contain a general line.

It follows from [**2**], [**3**] that there is a unique form satisfying the condition (H) in each cohomology class.

(iii) Let us construct $Y \subset X$ which is "transversal" to $\xi \in \Xi$. Let $L_\xi, \xi \in \Xi$, be defined by the conditions

(13)
$$H(z, z) > 0, H(z, \xi^{(j)}) = 0, 1 \le j \le q.$$

Each q-plane in M_ξ has a unique representative in L_ξ and $L \cong D_{n-q}$. Denote

(14)
$$Y = \{(z, \xi), \xi \in \Xi, z \in L_\xi\}.$$

For $\gamma \subset \Xi$ let $Y(\gamma)$ denote $\{(z, \xi), \xi \in \gamma, z \in L_\xi\}$.

(iv) Let for $\varphi \in \Phi^q, d_\xi \varphi = 0, \Lambda(\varphi) = \varphi \wedge \overline{\varphi} \wedge \Theta$ where

$$\Theta = \frac{(H(z, z))^{k-n+q-1}}{\underset{j}{\Pi} H(\xi^{(j)}, \xi^{(j)})} det(z, \xi^{(1)}, \cdots, \xi^{(q)}, dz, \cdots, dz)$$
$$\wedge det(\overline{z}, \overline{\xi}^{(1)}, \cdots, \overline{\xi}^{(q)}, d\overline{z}, \cdots, d\overline{z})$$

is the $(n-q, n-q)$-form on z(of maximal degree) which (on Y) is independent of ξ. We use the exterior product of the dz_j in the calculation of the determinants.

Correspondingly,

(15)
$$\| \varphi \|^2 = \frac{1}{c(\gamma)} \int_{Y(\gamma)} \Lambda(\varphi), \qquad k > n - q + 1,$$

gives the invariant norm and we can take as $(2q)$-cycle $\gamma \subset \Xi$ the manifold $\gamma(\pi)$ of all $(q-1)$-planes ξ in a q-plane $\pi \subset M$. Then $\{M_\xi\}, \xi \in \gamma(\pi)$, is a (minimal) covering of M. So we obtain a Hilbert space $H_2^{(q)}(M, \mathcal{O}(-k))$ consisting of those

φ such that $\| \varphi \| < \infty$. The proof that $H_2^{(q)}(M, \mathcal{O}(-k))$ is nonzero will not be given here.

6. Structurization of the construction

Let us try to separate a structure in this construction. We worked in the examples with very special Stein coverings $\{M_\xi\}$ of a complex manifold M. They have strong connections with maximal compact submanifolds in M. Namely, we represent these submanifolds π as the closures $cl(\pi_s)$ of Stein submanifolds $\pi_s \subset \pi$ and M_ξ are disjoint unions of π_s ($cl(M_\xi)$ are unions of π). Roughly speaking, we take the unions of compact submanifolds with common "infinity". Let us consider some examples.

Flag manifolds. Let $M = F = G_u/U = G_\mathbb{C}/P$ be a flag manifold, G_u be a compact Lie group, U be the centralizer of a torus, $G_\mathbb{C}$ be the complexification of G_u and P be the corresponding parabolic subgroup. The nilpotent radical N of an opposite parabolic subgroup has only one open orbit $M_N \subset M$, which is a Stein manifold. Let Ξ be the manifold of all such N (it will be the dual flag manifold and it is very simple to describe it in root language), $M_\xi = M_N$ and

$$(16) \qquad\qquad X = \{(z, \xi); \xi \in \Xi, z \in M_\xi\}.$$

In this case M_ξ coincides with the Stein part of the compact submanifold and we do not need to take their unions.

Flag domains. Let G be a real form of $G_\mathbb{C}$ and M be an open orbit in the flag manifold $F = G_\mathbb{C}/P$ from the previous example, $M = G/H$.

If M is a Hermitian symmetric manifold (H is compact) then M is Stein. In other cases there are compact submanifolds $\pi \subset M$ which are isomorphic to $K/K \cap H = K_\mathbb{C}/P_K$ where K is the maximal compact subgroup in G. $K_\mathbb{C}$ is the complexification of K, P_K is a parabolic subgroup in $K_\mathbb{C}$. Let N be the unipotent radical in P_K.

Let us take as Ξ the manifold of all subgroups N_ξ in $G_\mathbb{C}$ which are conjugate to N and have orbits $\sigma(N_\xi)$ such that $(cl(\sigma(N_\xi)) \subset M$. We will see that it is convenient sometimes to consider a weaker version of this condition. Then Ξ will be a domain in $G_\mathbb{C}/\mathrm{Norm}(N)$. We have the natural action of G on Ξ. But, in general, Ξ will not be homogeneous for this action and this is the cause of many troubles.

Then M_ξ is the union of such orbits of N_ξ which are contained in M together with their closures.

It is clear that in Example 3 we considered a particular case of this construction. Usually it is not difficult to construct a section $Y \subset X$ transversal to cycles.

The Hodge condition must be the combination of two conditions. Firstly, we consider the forms $\varphi(z|\xi, d\xi)$ which are constant in z along cycles—constituents of M_ξ. Secondly, we consider cycles $\gamma \subset \Xi$ corresponding to minimal coverings and φ must be the "dual" to such cycles.

REMARK. We have connected our Stein coverings with compact submanifolds $\pi \subset M$. So it is natural to compare our representatives of cohomology $\varphi(z|\xi, d\xi)$ with the Penrose transform—the integration on π. It turns out that φ can be obtained from the Penrose transform by differential operators with constant coefficients. It is remarkable that differential equations describing the image of Penrose transform (generalized massless equations) have been transformed in such a way into the universal condition of the closure of the forms φ. In a sense, the holomorphic representation of $\bar{\partial}$-cohomology combines the direct and inverse Penrose transforms (cf. [2], [3]).

7. Example 4 (nonholomorphic discrete series for SU(2,1))

We will illustrate this structure on another example–the flag domain

$$M = SU(2;1)/S(U(1) \times U(1) \times U(1)).$$

It corresponds to the simplest nonholomorphic discrete series of $SU(2;1)$.

For the realization of M let us consider the domain D from example 2:

$$(6) \quad H(z,z) = |z_0|^2 - |z_1|^2 - |z_2|^2 < 0.$$

Then M consists of the flags (z, l), $z \in l$, $z \in D$, the line $l \not\subset cl(D)$. It will be a domain on the flag manifold $SU(3)/S(U(1) \times U(1) \times U(1) = SL(3;\mathbb{C})/B$, where B is a Borel subgroup.

We have dim $M = 3$ and there are 1-dimensional complex compact submanifolds in M (rational curves). They are parametrized by pairs (m, w), where the line $m \subset D, w \in \mathbb{C}P^2 \backslash cl(D)$ $(H(w,w) > 0)$. The curve $C(m, w)$ is

$$(17) \qquad C(m, w) = \{(z, l) \in M; z \in m, l \ni w\}.$$

It is evident that $C(m, w) \cong \mathbb{C}P^1$.

Now we need to build the Stein manifolds M_ξ covering M from these cycles. In our construction Ξ will coincide with M. Let $(\xi, \alpha) \in \Xi \cong M, \xi \in D$, line $\alpha \not\subset cl(\Xi), \xi \in \alpha$ and

$$(18) \qquad M_{(\xi,\alpha)} = \{(z, l) \in M; \text{line}(z, \xi) \subset D, l \cap \alpha \in \mathbb{C}^2 \backslash cl(D)\}$$

So $cl(M_{(\xi,\alpha)})$ is the union of $C(m, w) \ni (\xi, \alpha)$. Clearly, $M_{(\xi,\alpha)} \cong \mathbb{C}^1 \times D^1 \times D^1$ (D^1 is the disk) is a Stein manifold and $\{M_{(\xi,\alpha)}, (\xi, \alpha) \in \Xi\}$ is a (nonminimal) covering. Let

$$X = \{[(z, l), (\xi, \alpha)], (z, l) \in M_{(\xi,\alpha)}\}.$$

and

$$L_{(\xi,\alpha)} = \{(z, l) \in M_{(\xi,\alpha)}, H(z, \xi) = 0\},$$
$$Y = \{[(z, l), (\xi, \alpha)] \in X, H(z, \xi) = 0\};$$

$Y \subset X$ will be transversal to cycles $C(m, \omega)$.

We will change notation. The flag (z, l) we will be characterized by z and another point $w \in l$ (notation: $f(z|w)$).

If $\{(z,l),(\xi,\alpha)\} \in Y$ we will suppose $H(\xi,\eta) = 0, \eta \in \alpha, w \in l, w \neq z, w \in l$. So we can describe Y by the system

$$\{f(z|w), f(\xi|\eta)\}$$

(19)
$$H(z,z) < 0, H(z,\xi) = 0, H(\xi,\xi) < 0,$$
$$H(\xi,\eta) = 0, H(\eta,\eta) > 0, det(\xi,\eta,w) = 0.$$

The last condition means that $w \in \text{line}(\xi,\eta)$.

For the description of $H^{(1)}(M, \mathcal{O}(-r) \times \mathcal{O}(-s))$ we consider holomorphic 1-forms

$$\varphi(z, \omega | \xi, \eta, d\xi, d\eta)$$

of bihomogeneity $(-r, -s)$ on $(z, w), d_{(\xi,\eta)}\varphi = 0$ and

$$\Lambda(\varphi) = \varphi \wedge \overline{\varphi} \wedge \Theta$$

where Θ is (2,2)-form with respect to (z, w) on Y which is constant on (ξ, η):

$$\Theta = \frac{H(z,z)^{r-2} H(w,w)^{s-2} H(\eta,\eta)}{|det(\xi,\eta,z)|^2} \times$$
$$det(z, \xi, dz) \wedge det(\overline{z}, \overline{\xi}, d\overline{z}) \wedge det(w, z, dw) \wedge det(\overline{w}, \overline{z}, d\overline{w}).$$

Then

$$\int_{Y(\gamma)} \Lambda(\varphi) = c(\gamma) \| \varphi \|^2, \qquad r > 2, s > 2$$

where $\gamma \subset \Xi$ is a (real) 2-dimensional cycle in Ξ, and $Y(\gamma)$ is its preimage in Y. We can take the standard curves (17) as γ. They correspond to the minimal covering of M by $M_{(\xi,l)}$. In such a way we can define the invariant Hermitian form on the subspace $H_2^{(1)}(M, \mathcal{O}(-r,) \times \mathcal{O}(-s)), r > 2, s > 2$ of $H^{(1)}(M, \mathcal{O}(-r,) \times \mathcal{O}(-s))$ on which the the integral converges.

8. Example 5 (Speh representations for SL(2m;$\mathbb{R}$))

In our last example we will show that sometimes it is useful for the construction of the Stein manifolds M_ξ to use not only compact submanifolds $\pi \subset M$, but also π such that $\pi \not\subset M, \pi \subset cl(M)$ but the Stein part $\pi_S \subset M(cl(\pi_S) = \pi)$.

Let $M = SL(2m; \mathbb{R})/SL(m; \mathbb{C}) \times T$. This pseudo Hermitian symmetric manifold can be realized as the open orbit of $SL(2m; \mathbb{R})$ on the complex Grassmanian $Gr_\mathbb{C}(2m; m) = SU(2m)/S(U(m) \times U(m)) = SL(2m; \mathbb{C})/B$, where elements of B are block-triangular matrixes $g = \begin{pmatrix} \alpha & \beta \\ 0 & \gamma \end{pmatrix}$, $g \in SL(2m; \mathbb{C})$, $\alpha, \rho, \gamma \in M_\mathbb{C}(m; m)$ (the $m \times m$-matrices).

On $Gr_\mathbb{C}(2m; m)$ we can define the Stiefel coordinates $Z \in M_\mathbb{C}(2m, m)$, with $\text{rank} Z = m$:

$$Z \sim uZ, \qquad u \in Gl(m; \mathbb{C}).$$

The group $SL(2m; \mathbb{C})$ will act on equivalency classes by

$$Z \mapsto Zg, \qquad g \in SL(2m; \mathbb{C}).$$

The domain $M \subset Gr_\mathbb{C}(2m; m)$ given by

(20)
$$\frac{1}{i^m} det \begin{pmatrix} \overline{Z} \\ Z \end{pmatrix} > 0, \qquad Z \in M_\mathbb{C}(2m; m)$$

(here $\begin{pmatrix} \overline{Z} \\ Z \end{pmatrix} \in M_\mathbb{C}(2m, 2m)$) will be an $SL(2m; \mathbb{R})$-orbit which is isomorphic to $SL(2m; \mathbb{R})/SL(m; \mathbb{C}) \times T$. The Shilov boundary of M is isomorphic to the real Grassmannian $Gr_\mathbb{R}(2m; m)$. In the coordinate chart

(21)
$$Z = (I, z), z \in M_\mathbb{C}(m, m),$$

where I is unit matrix (it is a dense chart in Gr) the condition (21) becomes

(22)
$$det(Im\, z) > 0.$$

So we have in this chart the tube domain M^0 (Siegel domain of the 1st kind) with nonconvex cone V of $m \times m$ matrixes with the positive determinant (cf. [4] and §2). M^0 will be the Zariski open part of M.

There are compact complex submanifolds isomorphic to $SO(2m)/SU(m) = SO(2m; \mathbb{C})/\tilde{B}$ in M and we will construct corresponding Stein manifolds M_ξ for $\xi \in \Xi$. But in this case the complex manifold Ξ will not be homogeneous and we prefer to work with a (real) homogeneous manifold Ξ on the boundary of Ξ.

Correspondingly, we will consider the compact submanifolds $\pi \subset M$ for which only the Stein parts $\pi_S \subset M$.

Let us consider the orbits on Gr of the unipotent subgroup $N \subset \tilde{B}$ of matrices

(23)
$$\lambda = \begin{pmatrix} I & u \\ 0 & I \end{pmatrix}, \qquad u \in M_\mathbb{C}(m, m), u = -u^\top.$$

Some of its orbits are contained in M and moreover in M^0, namely the orbits (cf. (21)):

(24)
$$Im(z + z^\top) = const \gg 0, \qquad z \in M_\mathbb{C}(m, m),$$

(positively definite). The closures of (24) in $Gr(2m; n)$ will be isomorphic to $SO(2n)/SU(n)$ but they will all contain the boundary point $Z = (0, I)$ of M.

We take their union M_0 as one element of the covering. So

$$M_0 = \{z \in M_\mathbb{C}(m, m); Im(z + z^\top) \gg 0\}.$$

It is a convex tube domain and is, of course, a Stein manifold.

Let us consider the family of $M_{\tilde{\xi}} = M_0 \cdot g, g \in SL(2m; \mathbb{R}), \tilde{\xi} \in \tilde{\Xi} = SL(2m; \mathbb{R})/$ Norm$(N) : g \in$ Norm$(N) :$

$$g = \begin{pmatrix} a & u \\ 0 & a \end{pmatrix}, \qquad aa^\top = I, a, u \in M_\mathbb{C}(m, m).$$

For the technical reasons we consider the subgroup $N_1 \subset N$ of elements (23) with $u = u^\top$ and $\Xi = SL(2m; \mathbb{R})/N_1$. We have the canonical projection $\Xi \mapsto \tilde{\Xi}$ and will identify $M_\xi, \xi \in \Xi$, with $M_{\tilde{\xi}}$ for $\tilde{\xi} = pr\,\xi$ (so M_ξ will be the same for all ξ with equal $pr\,\xi$). We have $\{M_\xi, \xi \in \Xi\}$ is the covering of M. Let

(25)
$$
\begin{aligned}
&X = \{(Z, \xi), \xi \in \Xi, Z \in M_\xi\}, \\
&L_0 = (I, iI) \cdot N_1 = \{z \in M^0, z = z^\top, Im\,z \gg 0\}. \\
&L_\xi = L_0 \cdot g, \text{if}\, \xi = \xi_0 \cdot g, g \in SL(2n, \mathbb{R}) \\
&Y = \{(z, \xi), \xi \in L_\xi, \xi \in \Xi\}.
\end{aligned}
$$

Then

(26)
$$
\begin{aligned}
&L_\xi = \{ Z J_g Z^\top = 0, \frac{1}{i}(Z J_g \overline{Z}^\top) \gg 0\}, \text{where}\, \xi = \xi_o g, \\
&J_g = g J g^\top, g \in SL(m; \mathbb{R}), J = \begin{pmatrix} 0 & I \\ -I & 0 \end{pmatrix}
\end{aligned}
$$

and indeed L_ξ depends only on the image of g in $\Pi = SL(2m; \mathbb{R})/Sp(2m; \mathbb{R})$. So there is the natural map of Y on the fibering $\tilde{Y}$ with the symmetric space Π as the base and the Hermitian symmetric manifold $Sp(2m; \mathbb{R})/SU(m)$ (Siegel half-plane) as fiber. The homogeneous manifold $\tilde{Y}$ is $SL(2m; \mathbb{R})/SU(m)$.

REMARK. We may have $\pi(\xi') = \pi(\xi'')$ and so $M_{\xi'} = M_{\xi''}$ but $L_{\xi'} \neq L_{\xi''}$. This was the reason why we change from $\tilde{\Xi}$ to Ξ. Now L_ξ is homogeneous relative to the isotropy subgroup $G_\xi = N_1$ (moreover relative $Sp(m; \mathbb{R})$), and Y will be homogeneous relative to $G = SL(2m; \mathbb{R})$.

We will consider the cohomology $H^{(q)}(M, \mathcal{O}(-r)), q = \frac{m(m-1)}{2}$, where $\mathcal{O}(-r)$ is the line bundle on $Gr_\mathbb{C}(2m; m)$ whose sections will satisfy the condition

(27) $\qquad\qquad f(uZ) = (det\,u)^{-r} f(Z), \qquad u \in Gl(m; G)$

Correspondingly, we will consider holomorphic q-forms $\varphi(Z|\xi, d\xi)$ on Z for $\xi \in \Xi$, $Z \in M_\xi$, satisfying on Z the condition (27).

We have the natural action of the group $SL(2m; \mathbb{R})$ on such forms, which induces the representation of $SL(2m; \mathbb{R})$ in $H^{(q)}(M, \mathcal{O}(-r))$. This representation is unitarizable (on the subspace) if r is sufficiently large. The structure of the norm is similar to the norms from examples 2–4 but more complicated and we will not give the formula here. The corresponding representation will be equivalent to the Speh representation of $SL(2m; \mathbb{R})$.

REMARK. As we saw the manifold M contains the Zariski open part M^0 (22) which is biholomorphicaly equivalent to the nonconvex tube domain (22). Our covering $\{M_\xi\}$ contains the subcovering $\{M_\xi, \xi \in \Xi^0\}$ of M^0 by the convex tube domains. They correspond to $\xi = \xi^0 \cdot P$ where P is the parabolic subgroup in $SL(2m; \mathbb{R})$—the isotropy subgroup of the boundary point $(0, I)$ of M. We obtain in such a way the description of the cohomology in tube domain M^0 as in §2. It is important that the cohomology can be extended on M and we realize the Speh representation in the cohomology M^0. This realization has a strong connection with the Sahi-Stein realization of the Speh representation [5].

I thank Michael Eastwood for many useful discussions and Andrew Leahy for help in the preparation of this manuscript.

REFERENCES

1. S. Gindikin, *Hardy spaces and Fourier transform of $\bar{\partial}$-cohomology in tube domains*, C. R. Acad. Sci. Paris **315** (1992), 1139–1143.
2. S. Gindikin and G. Henkin, *Transformation de Radon pour la d''-cohomologie des domaines q-linearement concaves*, C.R. Acad. Sci. Paris, **4** (1978), A209–A212.
3. S. Gindikin and G. Henkin, *Integral geometry for $\bar{\partial}$-cohomology in q-linearly concave domains in $\mathbb{C}P^n$*, Functional Anal. Appl. (Russian) **12** (1978), 6–23.
4. J. E. D'Atri and S. Gindikin, *Siegel domain realization of pseudo-Hermitian symmetric manifolds*, Geometrica Dedicata **46** (1993), 91–125..
5. S. Sahi and E. Stein, *Analysis in matrix space and Speh's representation*, Invent. Math. **101** (1990), 379–393.

DEPARTMENT OF MATHEMATICS, RUTGERS UNIVERSITY, NEW BRUNSWICK, NEW JERSEY 08903, USA

E-mail address: gindikin@math.rutgers.edu

Contemporary Mathematics
Volume **154**, 1993

Twistor Theory for Indefinite Kähler Symmetric Spaces

EDWARD G. DUNNE AND ROGER ZIERAU

1. Introduction
2. Standard twistor theory
3. Maximal compact subvarieties: the three cases
4. Generalized conformal structures
5. Examples

1. Introduction

Let $D = G/H$ be an open orbit of the real semisimple Lie group G in a generalized flag variety, $X = G^{\mathbf{C}}/P$, where $G^{\mathbf{C}}$ is the complexification of G. For instance, X could be $\mathbf{P}^n$, $G = U(n,1)$ and D could be the set of positive lines in $\mathbf{P}^n$. Then, $V_0 = K.eH \cong K^{\mathbf{C}}.eH \cong K/(K \cap H)$ is a maximal compact subvariety in D, for a suitable choice of a maximal compact subgroup K in G. Wolf [**30**] has shown that the parameter space M_D of all $G^{\mathbf{C}}$-translates of V_0 in D is a Stein manifold.

The cohomology $H^s(D, \mathcal{O}(\mathcal{V}))$ is a natural place to look for certain representations of the group G. (See, also, the talks of Knapp [**15**] and Wong [**31**].) When $G = U$ is a compact group, the Borel-Weil-Bott theorem tells us that all irreducible, finite-dimensional unitary representations of U arise on such cohomology spaces. If G is non-compact and has a compact Cartan subgroup, $H = T$, then Schmid's thesis [**23**] and his proof of the Langlands' conjecture [**24**], [**25**] show how the cohomology $H^s(G/T, \mathcal{O}(\mathcal{L}))$ gives rise to the discrete series for G, where, now, $\mathcal{L}$ is an appropriate line bundle. When $D = G/H$ is an indefinite Kähler symmetric space satisfying a 'holomorphicity' condition (the natural maps $G/H \xleftarrow{\tau} G/(H \cap K) \xrightarrow{\pi} G/K$ should form a G-invariant holomorphic double fibration), then there are techniques for realizing the unitary

---

1991 *Mathematics Subject Classification.* Primary 22E46, 53C10; Secondary 53A30, 53C55.
This paper is in final form and no version of it will be submitted for publication elsewhere.

structure on the cohomology of certain vector bundles on D. See, for example [20], [21], or [18].

Penrose and MacCallum [20] wrote down the inner product for the case in standard twistor theory, $H^1(\mathbf{P}^+, \mathcal{O}(-n-2))$. (See §2 for notation.) This method uses what became known as the *twistor transform*. The twistor transform was then generalized to construct the inner product on the ladder representations of $SU(p,q)$, which arise via line bundles on the open $SU(p,q)$-orbit in $\mathbf{CP}^{p+q-1}$ consisting of positive lines (cf. [5]). This method relies in an essential way on establishing an isomorphism between cohomology on the space D and the solutions of certain differential equations on G/K, which, in this case, is our space M_D. (Cf. §3.) Moreover, the differential equations that arise are invariants of the conformal structure.

One wishes to generalize the above situation. If D does not satisfy the holomorphicity condition, a Penrose transform is defined, but is more difficult to study in detail. The twistor transform has not been very much explored. Moreover, one is also interested in the more general case where D need not be a symmetric space, but only an open G-orbit in a generalized flag variety. In such cases, M_D can serve as a replacement for the space G/K in part of the picture. Indeed, in the holomorphic case, the parameter space *is* G/K. In the non-holomorphic case, M_D is not a homogenous space for any subgroup of $G^{\mathbf{C}}$. It does, though, carry a generalized conformal structure, in the sense of Baston [2] or Gindikin [10] and Goncharov [12]. An interesting open question is: What role does the generalized conformal structure play in the integral geometry of the double fibration

$$
(1) \qquad \begin{array}{ccc} & Y_D & \\ {\scriptstyle\tau}\swarrow & & \searrow{\scriptstyle\pi} \\ D & & M_D \end{array}
$$

where Y_D is the correspondence space $Y_D = \{(z, V_x) \in D \times M_D \mid z \in V_x\}$? The conjectural answer is that there is an analogue of the Penrose transform that relates the cohomology $H^s(D, \mathcal{O}(\mathcal{V}))$ to the kernel of a differential operator $D : \mathcal{V}' \to \mathcal{V}''$ that is an invariant of the generalized conformal structure. Here, $\mathcal{V}'$ and $\mathcal{V}''$ are holomorphic vector bundles on M_D, depending on $\mathcal{V}$.

2. Standard twistor theory

In this section we give a brief sketch of some of the ideas of twistor theory for compact, complex Minkowski space. (Fuller descriptions are given in [3] and [5].) This provides an example of what we would like to be true in a more general setting, as explained in the introduction and in §4. We set the following notation:

Let $U(2,2) = \{g \in GL(4, \mathbf{C}) \mid g$ preserves the Hermitian form $\Phi\}$ where Φ has signature $(+ + --)$. We fix a maximal compact subgroup, $K \subset U(2,2)$,

$K \cong U(2) \times U(2)$. Furthermore, let

$$
\begin{aligned}
\mathbf{P} &= \mathbf{P}^3 = GL(4, \mathbf{C})/P_1 \\
\mathbf{P}^+ &= \{z \in \mathbf{P} \mid \Phi|_z \gg 0\} \\
&= U(2,2)/(U(1) \times U(1,2)) \\
\mathbf{M} &= G_2(\mathbf{C}^4) = \text{Grassmannian of 2-planes in } \mathbf{C}^4 \\
\mathbf{M}^+ &= \{w \in \mathbf{M} \mid \Phi|_w \gg 0\} \\
\mathbf{F} &= \{(z,w) \mid z \in \mathbf{P}, w \in \mathbf{M}, \text{ and } z \subset w\} \\
\mathbf{F}^+ &= \{(z,w) \mid z \in \mathbf{P}^+, w \in \mathbf{M}^+, \text{ and } z \subset w\}
\end{aligned}
$$

The subgroup P_1 is a parabolic subgroup in $GL(4, \mathbf{C})$ that is the isotropy group of a basepoint in $\mathbf{P}$. The space $\mathbf{F}$ is the correspondence space that sits between $\mathbf{P}$ and $\mathbf{M}$ in the *Penrose correspondence*, described below. That is to say, there is a holomorphic double fibration:

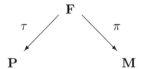

Since $\mathbf{P}^+$ is an open $U(2,2)$-orbit in $\mathbf{P}$, it carries a natural $U(2,2)$-invariant complex structure. We pick Φ to be the diagonal matrix with $(1,1,-1,-1)$ for diagonal, and choose K to be $\begin{pmatrix} U(2) & 0 \\ 0 & U(2) \end{pmatrix}$. We use $x_0 = [1,0,0,0]$ as the basepoint of $\mathbf{P}^+ \subset \mathbf{P}$, where the square brackets indicate homogeneous coordinates. Then, $K.x_0$ is a maximal compact subvariety and a simple computation shows that, as homogeneous space, it is just $(U(2) \times U(2))/(U(1) \times U(1) \times U(2)) \cong \mathbf{P}^1$. We are interested in the translates of $K.x_0$ in $\mathbf{P}^+$ by elements of $GL(4, \mathbf{C})$. We use L to denote a general maximal compact subvariety in $\mathbf{P}$ of this type, (i.e., an embedded $\mathbf{P}^1$).

The geometry of the lines in $\mathbf{P}$ is closely related to the geometry of $\mathbf{M}$. For instance, by considering $\mathbf{M}$ as a smooth quadric in $\mathbf{P}^5$ (via the Plücker embedding) and $\mathbf{P} = \mathbf{P}^3$ as the parameter space for one of the two rulings of $\mathbf{M}$ by 2-planes, we can deduce that every point, $w \in \mathbf{M}$ corresponds to a unique line, $L_w \cong \mathbf{P}^1$, in $\mathbf{P}$. Moreover, every line in $\mathbf{P}$ is of this form. Thus, $\mathbf{M}$ parametrizes the lines in $\mathbf{P}$. We can also see that the lines in $\mathbf{P}^+$ are parametrized by $\mathbf{M}^+$. This correspondence encodes the conformal geometry of $\mathbf{M}$. Specifically, $\mathbf{M}$ carries an invariant complex metric such that two points are null-separated if and only if the corresponding lines in $\mathbf{P}$ intersect [19]. These facts constitute what is known as the *Penrose correspondence*, as depicted in Figure 1. The picture is meant to illustrate the fact that w_2 lies on the light cone of w_1 if and only if the lines L_{w_1} and L_{w_2} intersect non-trivially.

One may show that if $\mathcal{L}$ is a sufficiently negative holomorphic line bundle on $\mathbf{P}^+$ then $H^q(\mathbf{P}^+, \mathcal{L})$ is zero if $q > 1$ and is zero if $q = 0$. When $q = 1$,

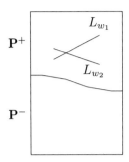
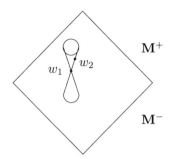

FIGURE 1. The Penrose correspondence.

by choosing an appropriate line bundle, one gets an infinite-dimensional space which carries a natural action of $U(2,2)$. These provide realizations of singular representations of $U(2,2)$ which are unitarizable (that is, representations which do not occur in the decomposition of $L^2(U(2,2))$). In particular, this is true if we choose $\mathcal{L} = \mathcal{O}(-n-2)$ where $n \geq 0$.

Under the same 'sufficiently negative' assumption, there is a general procedure for moving the data of the cohomology on D to data on $\mathbf{M}^+$, which goes as follows (see [**3**], [**5**], [**6**], or [**8**]): By Theorem 1 in §3, $H^1(\mathbf{P}^+, \mathcal{O}(-n-2))$ is isomorphic to $H^1(\mathbf{F}^+, \tau^{-1}\mathcal{O}(-n-2))$, where $\tau^{-1}\mathcal{O}(-n-2))$ is the topological inverse image sheaf (not to be confused with the pull-back sheaf, which is this tensored over $\tau^{-1}\mathcal{O}_{\mathbf{P}^+}$ with $\mathcal{O}_{\mathbf{F}^+}$). The topological inverse image sheaf is *not* an $\mathcal{O}_{\mathbf{F}^+}$-module, hence not coherent. One needs to resolve it by coherent $\mathcal{O}_{\mathbf{F}^+}$-modules, in order to compute its cohomology. One particularly good resolution is the (geometric) relative Bernstein-Gel'fand-Gel'fand resolution of $\tau^{-1}\mathcal{O}(-n-2)$. This is a good resolution because it is shorter than the relative de Rham resolution and the terms are geometric versions of the terms in the Bernstein-Gel'fand-Gel'fand resolution for Verma modules. However, this is not necessarily an acyclic resolution. So one needs to use the hypercohomology spectral sequence to compute $H^s(\mathbf{F}^+, \tau^{-1}\mathcal{O}(-n-2))$ from this resolution. The typical term in this spectral sequence is the cohomology group of a line bundle. This cohomology may be computed by pushing down the data to $\mathbf{M}^+$ using the Leray spectral sequence for the direct images of the line bundles. The Borel-Weil-Bott theorem allows us to identify the direct images, many of which are zero. Because $\mathbf{M}^+$ is Stein, the spectral sequence collapses, leaving us with an injection:

$$H^1(\mathbf{P}^+, \mathcal{O}(-n-2)) \hookrightarrow H^0(\mathbf{M}^+, \mathcal{V}_n)$$

where $\mathcal{V}_n$ is a homogeneous vector bundle on $\mathbf{M}^+$. It is the nth symmetric power of the primed spin bundle. The image of $H^1(\mathbf{P}^+, \mathcal{O}(-n-2))$ is characterized as the kernel of a $U(2,2)$-invariant operator, which is comes from the d-operator in

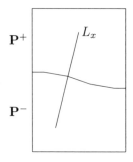
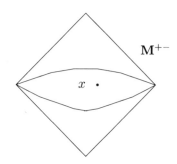

FIGURE 2. Correspondence for $\mathbf{M}^{+-}$.

the Leray spectral sequence. When $n \geq 1$ this operator is first order; it is second order for $n = 0$. These are the massless field operators, a sort of generalized Dirac operator, for the line bundles with $n \geq 1$, and it is the wave operator for $n = 0$. The final result, then, is an isomorphism from $H^1(\mathbf{P}^+, \mathcal{O}(-n-2))$ to the space of holomorphic solutions of the zero-rest-mass equations on $\mathbf{M}^+$.

2.1. A variation. Now let $D = \mathbf{M}^{+-}$ be the space of 2-planes x in $\mathbf{C}^4$ such that $\Phi|_x$ has signature $(1,1)$. Then, $\mathbf{M}^{+-}$ is an open $U(2,2)$-orbit in $\mathbf{M}$, which is isomorphic to $SU(2,2)/S(U(1,1) \times U(1,1))$. Again, we may interpret $\mathbf{M}^{+-}$ as a certain set of lines in $\mathbf{P}^3$. Namely, it is the collection of lines that meet both $\mathbf{P}^+$ and $\mathbf{P}^-$, as in Figure 2. Let x_0 be the 2-plane $\{(z_1, 0, z_2, 0)\}$. What is $K.x_0 = V_0$? It is the set of all planes spanned by λ_1 and λ_2 where λ_1 lies in $L_1 = \{(*, *, 0, 0)\}$ and λ_2 lies in $L_2 = \{(0, 0, *, *)\}$. Thus,

$$\begin{aligned} V_0 &= \{\text{lines in } \mathbf{P}^3 \text{ that meet both } L_1 \text{ and } L_2\} \\ &\cong L_1 \times L_2 \\ &\cong \mathbf{P}^1 \times \mathbf{P}^1. \end{aligned}$$

It is now fairly easy to see that $M_D = \{L_1 \subset \mathbf{P}^+\} \times \{L_2 \subset \mathbf{P}^-\}$. But the lines in $\mathbf{P}^+$ are just $\mathbf{M}^+$ and the lines in $\mathbf{P}^-$ are just $\mathbf{M}^-$. Therefore,

$$\begin{aligned} M_D &\cong \mathbf{M}^+ \times \mathbf{M}^- \\ &\cong G/K \times \overline{G/K}. \end{aligned}$$

The space M_D again carries a conformal structure, defined by the incidence relation. Note that (x_1, y_1) is null-connected to (x_2, y_2) in $\mathbf{M}^+ \times \mathbf{M}^-$ if and only if L_{x_1} meets L_{x_2} and L_{y_1} meets L_{y_2}. One could define a degenerate conformal structure by replacing the *and* by *or*.

3. Maximal compact subvarieties

Let $D = G/H$ be an open G-orbit in a generalized flag variety for $G^{\mathbf{C}}$. Let $V_0 = K.eH$ be our base maximal compact subvariety. We will call any $G^{\mathbf{C}}$-translate of V_0 a c.s.v.. We note that not every compact subvariety of maximal dimension must be a c.s.v.. For example, D could be the exterior of a ball in $\mathbf{P}^n \cong U(n,1)/U(1) \times U(n-1,1)$ with $V_0 \cong \mathbf{CP}^{n-1}$. Then D contains compact $(n-1)$-dimensional subvarieties that are not homologous to V_0, hence cannot be translates of V_0 by an element of $G^{\mathbf{C}}$.

The general procedure for constructing a Penrose transform for D and M_D is similar to that for $\mathbf{P}^3$ and Minkowski space outlined in §2. However, there are several points that need to be checked in general.

First, the isomorphism $H^s(D, \mathcal{O}(\mathcal{L})) \cong H^s(Y_D, \tau^{-1}\mathcal{O}(\mathcal{L}))$ depends on the topology of the fibers of τ. The 'Buchdahl conditions' for when one does have an isomorphism are encapsulated in the theorem:

THEOREM 1. (Buchdahl [4]) *Let* X, Y *be complex manifolds,* $f : Y \to X$ *be a surjective holomorphic mapping of maximal rank and* $\mathcal{V}$ *a holomorphic vector bundle on* X. *If, for some* $N \geq 0$, $H^p(f^{-1}(x), \mathbf{C}) = 0$ *for* $p = 0, 1, \ldots, N$ *and all* $x \in X$, *then the canonical homomorphism* $H^q(X, \mathcal{O}(\mathcal{V})) \to H^q(Y, f^{-1}\mathcal{O}(\mathcal{V}))$ *is an isomorphism for* $q = 0, 1, \ldots, N$ *and an injection for* $q = N + 1$. *(For* $N = 0$, *the condition is that the fibers must be connected.)*

Thus, one would like to have $H^p(\tau^{-1}(z), \mathbf{C}) = 0$ for $p = 0, 1, \ldots, s$. It is often the case that the fibers are contractible, as in the holomorphic case below.

Second, the inverse image sheaf $\tau^{-1}\mathcal{O}(\mathcal{V})$ is not coherent, so we need an effective resolution by $\mathcal{O}_Y$-modules. The relative de Rham resolution is always available, but, because we are expecting irreducible representations to occur, it is better if we can work with a geometric version of the relative Bernstein-Gel'fand-Gel'fand resolution, as in the standard case. This sequence is shorter and its terms are easier to relate to well-understood objects. (See [3].) If the fibers are not homogeneous, this is not defined, however.

When pushing-down the information on Y_D to M_D, using the Leray spectral sequence, we can still take advantage of the Borel-Weil-Bott theorem because the fibers of π are precisely the c.s.v.'s, i.e., copies of $K/(K \cap H)$, which are generalized flag varieties for $K^{\mathbf{C}}$. Moreover, Wolf's theorem [30] guarantees that the parameter space is a Stein manifold. Therefore, the Leray spectral sequence will collapse, as in the standard twistor case.

We distinguish three cases, as in [28] and [30].

Case 1: The holomorphic case. Suppose the natural maps

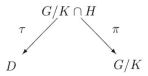

are holomorphic. Then G/K carries an invariant Hermitian structure. It turns out that G/K with this complex structure is M_D.

In this case, everything works as in standard twistor theory: The fibers of τ are contractible. They are just $H/(H \cap K)$. There is a relative Bernstein-Gel'fand-Gel'fand resolution that computes the cohomology of the inverse image sheaf. Then, proceed as in the standard twistor case. Once everything is sorted out, one is left with an isomorphism between $H^s(D, \mathcal{O}(\mathcal{V}))$ and the kernel of an invariant differential operator between homogeneous vector bundles on G/K. For the details, see [3] and [21], Corollary 5.32. (There are some difficult 'unitary' questions about the cohomology here, which we will not address.)

Case 2: The Hermitian case. Suppose that G/K is a Hermitian symmetric space, but that there is no G-invariant *holomorphic* double fibration

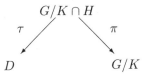

Then, M_D is an open subset of $G^{\mathbf{C}}/K^{\mathbf{C}}$ (see [28]). In particular, it cannot be G/K because the dimensions don't match. One example is $D = \mathbf{M}^{+-}$, as in §2.1. Interestingly, all the examples of this case, so far, have $M_D \cong G/K \times \overline{G/K}$.

Case 3: The non-Hermitian case. Here we assume G/K is not a Hermitian symmetric space. Hence, a G-invariant holomorphic double fibration is not possible. Again, we can view M_D as an open subset of $G^{\mathbf{C}}/K^{\mathbf{C}}$. Examples of such spaces are given in §5.

4. Generalized conformal structures

In his talk, Michael Eastwood [7] alluded to the *conformal structure* on $\mathbf{M}$, which is part of the correspondence between points of $\mathbf{M}$ and lines in $\mathbf{P}$. Generally, a conformal structure on a complex manifold, M, is an equivalence class of (holomorphic) Riemannian metrics, $[g]$, where two metrics g and $\tilde{g}$ are equivalent if $\tilde{g} = \lambda g$ for some non-vanishing holomorphic function, λ. In his nonlinear graviton paper [19], Penrose shows that such a structure is equivalent to a specification of quadratic cones in the tangent spaces. These cones become the null-cones of the conformal structure. Thus, to specify a conformal structure on a complex manifold, it is enough to specify at each point p, which points

q in a neighborhood of p are null-connected to p, i.e., are connected to p by null-geodesics. For $\mathbf{M}$, two points x_1, x_2 are null-connected if and only if the corresponding lines L_{x_1} and L_{x_2} in $\mathbf{P}$ intersect non-trivially.

One can make a similar definition for points in M_D: Two points $\xi_1, \xi_2 \in M_D$ are 'null-connected' if and only if the corresponding c.s.v.'s, V_{ξ_1}, V_{ξ_2}, intersect non-trivially. Now, however, one need not obtain a conformal structure on M_D because the corresponding cones defined in the tangent spaces need not be quadratic. This leads to the following two definitions (cf. [**12**]).

Let $X = G^{\mathbf{C}}/K^{\mathbf{C}}P^+$ be a compact Hermitian symmetric space. Let $P_x^{\mathbf{C}} = K_x^{\mathbf{C}}P_x^+$ be the stabilizer of $x \in X$. Let $x_0 = eK^{\mathbf{C}}P^+$ be a basepoint. Set $K(X)$ to be the cone of highest weight vectors in the $K^{\mathbf{C}}$-module $T_{x_0}X$.

DEFINITION 1. *Let $\mathcal{X}$ be a complex manifold. Suppose X is a compact Hermitian symmetric space of rank greater than 1. A generalized conformal structure of type X is given on $\mathcal{X}$ if there exist cones $\mathcal{K}_x \subset T_x\mathcal{X}$, depending holomorphically on x, such that each $\mathcal{K}_x$ is $\mathbf{C}$-linearly equivalent to $K(X)$.*

The case of a rank 1 Hermitian symmetric space corresponds to a *projective structure*. These are not defined by cones in the tangent spaces. Rather, one requires that there be an equivalence class of connections on the manifold, such that two connections are equivalent if they have the same geodesics as unparametrized curves.

As an example, let X be a non-degenerate (smooth) quadric in $\mathbf{P}^{n+1}$. $\mathcal{K}_x$ is the non-degenerate quadratic cone consisting of lines in $\mathbf{P}^{n+1}$ passing through x and lying completely in X. This is the 'standard' flat conformal structure. The quadric, X, may be viewed as the conformal compactification of $\mathbf{C}^n$.

A more general notion is that of a Frobenius structure:

DEFINITION 2. *Let $\mathcal{X}$ be an n-dimensional complex manifold. Suppose at every point $x \in \mathcal{X}$ there is given a family $\mathcal{F}(x)$ of k-dimensional subspaces in $T_x\mathcal{X}$ ($k \le n$) that depends holomorphically on x. Thus, $\mathcal{F}(x) \subset Gr_k(T_x\mathcal{X})$. Then $\mathcal{F}$ is a Frobenius structure on $\mathcal{X}$.*

An important example is where Z is a complex manifold, V_0 is a compact subvariety in Z and Γ is the space of all deformations of V_0 in Z. Let $Y = \{(z, V) \mid z \in V \subset Z, V \in \Gamma\}$. Then there is a holomorphic double fibration:

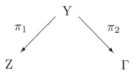

Define $Z_V = \pi_1\pi_2^{-1}(V)$ and $\Gamma_z = \pi_2\pi_1^{-1}(z)$. Then Z_V is (biholomorphic to) the subvariety V. Γ_z consists of all the subvarieties V containing z. Finally,

$$\{T_V(\Gamma_z) \mid z \in V\}$$

defines a Frobenius structure on Γ.

Example. Let

$$D = \frac{U(p,q)}{U(r) \times U(p-r,q)}$$

with $r < p$, be the space of positive r-planes in $\mathbf{C}^{p,q}$. It is an open $U(p,q)$-orbit in $\mathrm{Gr}_r(\mathbf{C}^{p+q})$. This is an example in the holomorphic case of §3. So, $M_D \cong U(p,q)/(U(p) \times U(q)) \subset \mathrm{Gr}_p(\mathbf{C}^{p+q})$, the space of all positive p-planes. Pick the base compact subvariety to be

$$
\begin{aligned}
V_0 \;&=\; \{r\text{-planes in } (z_1, z_2, \ldots, z_p, 0, 0, \ldots, 0)\} \\
&\cong\; \mathrm{Gr}_r(\mathbf{C}^p) \,.
\end{aligned}
$$

We think of an element $\xi \in M_D$ as an element of $\mathrm{Gr}_p(\mathbf{C}^{p+q})$. The corresponding subvariety is

$$
\begin{aligned}
V_\xi \;&=\; \{ \text{ all } r\text{-planes in the given positive } p\text{-plane } \xi\} \\
&=\; \mathrm{Gr}_r(\xi)
\end{aligned}
$$

The natural Frobenius structure says that ξ and ξ' are null-connected in M_D if V_ξ and $V_{\xi'}$ meet non-trivially. However, this condition misses some of the K-invariance inherent in the picture. Choose coordinates in $\mathrm{Gr}_p(\mathbf{C}^{p+q})$ where $\xi = \begin{bmatrix} I_p \\ Z \end{bmatrix}$ represents the p-plane spanned by the columns of the matrix, I_p is the $(p \times p)$ identity matrix and Z is any $(q \times p)$ matrix. In these coordinates, the base c.s.v. corresponds to V_{ξ_0} where $\xi_0 = \begin{bmatrix} I_p \\ 0_{q \times p} \end{bmatrix}$. For any other ξ, $\dim(V_{\xi_0} \cap V_\xi)$ is related to the rank of Z:

$$V_{\xi_0} \cap V_\xi \cong \mathrm{Gr}_r(\mathbf{C}^{p-k}) \,,$$

where $k = \mathrm{rk}(Z)$.

The $K^{\mathbf{C}}$-action on the parameter space is by

$$\begin{pmatrix} A & 0 \\ 0 & D \end{pmatrix} \cdot \begin{bmatrix} I_p \\ Z \end{bmatrix} = \begin{bmatrix} A \\ DZ \end{bmatrix} \equiv \begin{bmatrix} I_p \\ DZA^{-1} \end{bmatrix}$$

and $\mathrm{rk}(DZA^{-1}) = \mathrm{rk}(Z)$. Thus, we can partition the Frobenius structure at ξ_0 into $K^{\mathbf{C}}$-invariant strata according to $\dim(V_{\xi_0} \cap V_\xi)$. The same holds, of course, at any point in M_D. Because of this stratification, as Toby Bailey puts it, "some vectors are 'nuller' than others."

Using the coordinates above, we can identify the tangent space at ξ_0 with the space of $(q \times p)$ matrices. The action of $K^{\mathbf{C}}$ is again $Z \mapsto DZA^{-1}$. The highest weight vectors for the standard Borel of upper triangular matrices in $K^{\mathbf{C}}$ are

$$\begin{bmatrix} 0 & 0 & \cdots & 0 & z \\ 0 & 0 & \cdots & 0 & 0 \\ & & \cdots & & \\ 0 & 0 & \cdots & 0 & 0 \end{bmatrix}_{q \times p}$$

which has rank 1. Clearly any other rank 1 matrix is obtained under the action of $K^{\mathbf{C}}$. Thus, the points in M_D that are null-connected to ξ_0 in the generalized conformal structure are $\xi = \begin{bmatrix} I_p \\ Z \end{bmatrix}$ where Z has minimal rank. Hence, the corresponding subvarieties are those that intersect V_0 in the maximal possible dimension. We see, then, that the generalized conformal structure comes from the lowest-dimensional stratum in the Frobenius structure.

5. Examples

We have already pointed out that for the holomorphic case, one has $M_D \cong G/K$. Also, G/K is a noncompact Hermitian symmetric space, sitting inside its compact dual. Thus, the generalized conformal geometry for these examples is straightforward. We now give some examples of the nonholomorphic cases (cases 2 and 3 of §3).

Example 1. (Novak [17]) An interesting family of examples where G/K is Hermitian symmetric, but there is no holomorphic double fibration between D and G/K, is the case where $G = Sp(n, \mathbf{R})$ and H is one of $H_i = SU(i, n - i)$ for some $i = 1, 2, \ldots, n - 1$. We consider $\mathbf{C}^{2n}$ with a symplectic form ω and a Hermitian form $\langle \cdot, \cdot \rangle$ of signature (n, n). Then $D_i = G/H_i$ is the set of all isotropic (with respect to ω) n-planes of signature $(i, n - i)$ in $\mathbf{C}^{2n}$. Each D_i is an open G-orbit in the flag variety for $Sp(n, \mathbf{C})$ consisting of all isotropic n-planes in $\mathbf{C}^{2n}$.

Let M_{D_i} be the connected component of $K.x_i$ in the parameter space of all $G^{\mathbf{C}}$-translates of $K.x_i$ that lie in D_i, where x_i is a basepoint in D_i, fixed once and for all. Then the main result of [17] is:

THEOREM 2. (Novak) *The space* M_{D_i} *is isomorphic to* $G/K \times \overline{G/K}$, *where* $\overline{G/K}$ *denotes* G/K *with the opposite complex structure.*

Furthermore, Novak has shown that the fibers of the map $\tau : Y_{D_i} \to D_i$ from the correspondence space to D_i are contractible. Thus, one is able to apply Buchdahl's result. Moreover, she has obtained good descriptions of the fibers and the correspondence space, which should allow one to construct a useful resolution of the inverse-image sheaves on Y_{D_i}.

Example 2. Let $G = SO(n, 1)$ be the orthogonal group in $\mathbf{R}^{n+1}$ preserving a non-degenerate symmetric form $\epsilon(\cdot, \cdot)$ with signature $(n, 1)$. Assume $n \geq 4$. For

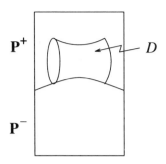

FIGURE 3. The quadric and D for $G = SO(n, 1)$.

this calculation we take:

$$\epsilon(z, w) = z_1 w_z + z_2 w_2 + \ldots + z_n w_n - z_{n+1} w_{n+1} .$$

We also consider the complexification of $\epsilon(\cdot, \cdot)$ on $\mathbf{C}^{n+1}$ and denote it by the same symbol. Then, $G^{\mathbf{C}} = SO(n+1, \mathbf{C})$, $K = SO(n) \times \mathbf{Z}_2$ and $K^{\mathbf{C}} = SO(n, \mathbf{C})$. We denote the associated Hermitian form on $\mathbf{C}^{n+1}$ by

$$\langle z, w \rangle = \epsilon(z, \overline{w}) .$$

We say that a point z is *null* if $\epsilon(z, z) = 0$. We say z is *isotropic* if $\langle z, z \rangle = 0$.

Let X be the smooth quadric in $\mathbf{P}^n$ defined by $\epsilon(z, z) = 0$. The manifold X is a generalized flag variety for $G^{\mathbf{C}}$. There is precisely one open G-orbit:

$$
\begin{aligned}
D &= \{ z \in X \mid \langle z, z \rangle > 0 \} \\
&= \{ z \in X \mid \langle z, z \rangle \neq 0 \}
\end{aligned}
$$

Also, notice that D lies in the closure of $\mathbf{P}^+$. A rough picture is given in Figure 3. Since G/K is not Hermitian symmetric, this is an example of the third case, as defined in §3.

Let $x_0 = [1, i, 0, 0, \ldots, 0, 0] \in X$ be the basepoint. Then our base compact subvariety V_0 is the K- (equivalently, the $K^{\mathbf{C}}$-) orbit of x_0. This is the $(n-2)$-dimensional quadric obtained by intersecting X with the hyperplane $\{z_{n+1} = 0\}$. Let $\xi_0 = [0, 0, \ldots, 0, 1]$ and let $Z^\perp$ denote the orthogonal complement of a subspace $Z \subset \mathbf{C}^{n+1}$ with respect to $\epsilon(\cdot, \cdot)$. Then the hyperplane defining V_0 is $\xi_0^\perp$. It is easy to see that any $G^{\mathbf{C}}$-translate of V_0 is of the form

$$V = \xi^\perp \cap X$$

for some $\xi \in \mathbf{P}^n$. Thus we may parametrize the c.s.v.'s by elements of $\xi \in \mathbf{P}^n$. (Actually, they are elements of the dual projective space, which we have identified with $\mathbf{P}^n$ via the non-degenerate form $\epsilon(\cdot, \cdot)$.)

Wolf's theorem implies that M_D is Stein. A concrete description of M_D is given as follows (with thanks to David Vogan):

First, note that it is sufficient to check that $\xi^\perp$ does not contain any points that are both null and isotropic. If $z \in \mathbf{P}^n$ is null and isotropic, then we may take it to be of the form $z = (v; 1)$ where $v \in \mathbf{C}^n$. It is a simple calculation to show that v must actually lie in $S^{n-1} \subset \mathbf{R}^n \subset \mathbf{C}^n$. It is also easy to see that any $\xi = [\xi_1, \xi_2, \ldots, \xi_n, \xi_{n+1}]$ in M_D must have $\xi_{n+1} \neq 0$. Since ξ is an element of projective space, we may work with the element $(\xi_1, \xi_2, \ldots, \xi_n, 1)$, which we will again call ξ. Write α and β for the real and imaginary parts of $(\xi_1, \xi_2, \ldots, \xi_n) \in \mathbf{C}^n$.

We now look for conditions on α and β that guarantee that no null, isotropic $(v; 1)$ lies in $\xi^\perp$. Let (a, b) denote the standard inner product on $\mathbf{C}^n$ (and on $\mathbf{R}^n$). If $(\alpha, \alpha) < 1$, then $(\alpha, v) < 1$ and $\epsilon(\xi, z) < 0$. Thus, every such ξ is in M_D.

If $(\alpha, \alpha) = 1$, the real part of $\epsilon(\xi, z)$ is zero only if α and v are co-linear. The imaginary part is zero only if $(\alpha, \beta) = 0$. Thus, the corresponding ξ is in M_D if and only if $(\alpha, \beta) = 0$.

If $(\alpha, \alpha) > 0$, then a little work shows that, for $\beta \neq 0$, the condition that $(\alpha + i\beta; 1)$ is in M_D is

$$|\alpha|^2 \left(1 - \frac{(\alpha, \beta)^2}{|\alpha|^2 |\beta|^2}\right) < 1.$$

For $\beta = 0$, the condition is $|\alpha|^2 < 1$.

We can combine the three cases and rewrite the final condition as

$$(2) \qquad\qquad\qquad |\alpha - P_\beta(\alpha)| < 1$$

where P_β is the orthogonal projection onto β and we take $P_0 \equiv 0$. As stated above, we know that this subset of $\mathbf{C}^n$ is Stein, though this is difficult to show directly.

We now want to consider the incidence relation for two c.s.v.'s, V_1 and V_2. Each is an $(n-2)$-dimensional smooth quadric in the smooth quadric X. Provided $n > 2$, these will *always* intersect non-trivially! Also, M_D inherits a generalized conformal structure because it sits naturally inside $\mathbf{P}^n$. But this is the 'degenerate' case of a projective structure where the cone in the tangent space consists of everything, except the origin.

One might ask for finer information about the intersection of two c.s.v.'s, as in §4. Now, the dimension of the intersection is invariant, but the intersection may be singular. We consider the base c.s.v., V_0 and any other c.s.v.V_ξ in M_D. The intersection is characterized by the system of equations:

$$(3) \qquad \begin{aligned} z_1^2 + z_2^2 + \ldots + z_n^2 - z_{n+1}^2 &= 0 \\ z_{n+1} &= 0 \\ \xi_1 z_1 + \xi_2 z_2 + \ldots + \xi_n z_n - \xi_{n+1} z_{n+1} &= 0 \end{aligned}$$

We are looking for places where the derivative of this map from $\mathbf{C}^{n+1}$ to $\mathbf{C}^3$ drops rank. This occurs precisely at

$$(4) \qquad\qquad (\xi_1, \xi_2, \ldots, \xi_n, \xi_{n+1}) \text{ where } \xi_1^2 + \xi_2^2 + \ldots + \xi_n^2 = 0.$$

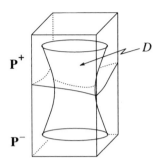

FIGURE 4. The quadric and D for $G = SO(p, q)$.

By coupling condition (2) with Eq. 4 we know that there is a complex hyper-surface worth of such points in M_D. Notice also that this singularity condition is invariant under the action of K (and of $K^{\mathbf{C}}$) on the parameter space.

Example 3. We now look at a variation on the previous example. Let $G = SO(p, q)$ with $p \geq q \geq 2$, but we exclude the case $p = q = 2$. Then, $\epsilon(z, w) = z_1 w_1 + \ldots + z_p w_p - z_{p+1} w_{p+1} - \ldots - z_{p+q} w_{p+q}$ and $\langle z, w \rangle = \epsilon(z, \overline{w})$. Let X be the smooth quadric in $\mathbf{P}^{p+q-1}$ defined by $\epsilon(z, z) = 0$. It is a generalized flag variety for $G^{\mathbf{C}} = SO(p + q, \mathbf{C})$. Define

$$D = \{[z] \in X \mid \langle z, z \rangle > 0\}.$$

Pick the basepoint $[z_0] = [1, i, 0, 0, \ldots, 0]$. Then $H \cong SO(2) \times SO(p-2, q)$. Let $K = SO(p) \times SO(q) \times \mathbf{Z}_2$ be the fixed maximal compact subgroup of G and $K^{\mathbf{C}} = SO(p, \mathbf{C}) \times SO(q, \mathbf{C})$.

When $q = 2$, this is an example of Case 2, the Hermitian case, since G/K is Hermitian symmetric, but there is no holomorphic double fibration between G/H and G/K. Otherwise, this falls into Case 3, since G/K is not Hermitian symmetric.

Our base compact subvariety is

$$\begin{aligned} V_0 &= K.[z_0] \\ &= [z_1, z_2, \ldots, z_p, 0, 0, \ldots, 0] \mid z_1^2 + z_2^2 + \ldots + z_p^2 = 0\} \\ &= \xi_0^{\perp} \cap X \end{aligned}$$

where ξ_0 is the q-plane $\{z_1 = 0, z_2 = 0, \ldots, z_p = 0\}$. Choose coordinates in $\mathrm{Gr}_q(\mathbf{C}^{p+q})$ so that $\xi = \begin{bmatrix} Z \\ I_q \end{bmatrix}$ represents the q-plane spanned by the columns of the matrix, I_q is the $(q \times q)$ identity matrix and Z is any $(p \times q)$ matrix. Then $\xi_0 = \begin{bmatrix} o \\ I_q \end{bmatrix}$. Any $G^{\mathbf{C}}$-translate of V_0 is also realizable as the intersection of X with a q-plane, ξ. Moreover, the q-plane must be non-degenerate with respect to $\epsilon(\cdot, \cdot)$. Thus, the parameter space of *all* $G^{\mathbf{C}}$-translates of V_0 in X is the space of all non-degenerate q-planes in $\mathbf{C}^{p+q}$, i.e., $M_X = G^{\mathbf{C}}/K^{\mathbf{C}}$. This is an open subset of the Grassmannian of all q-planes in $\mathbf{C}^{p+q}$. As mentioned in §3, M_D is an open subset in M_X, hence an open subset of the space of all non-degenerate q-planes.

Because we are assuming that $q \geq 2$, M_D inherits a Frobenius structure and a generalized conformal structure (which is not a projective structure) from the Grassmannian, $\mathrm{Gr}_q(\mathbf{C}^{p+q})$. These two structures are exactly as given in the example in §4.

There are two stratifications of the Frobenius structure. The first is by the rank of Z and is just as in the example of §4. The second is by the rank of Z and the rank of $Z^\mathsf{T} Z$. The latter condition measures the degeneracy of Z with respect to $\epsilon(\cdot, \cdot)$, which in turn measures the singularity of the intersection, $(V_0 \cap V_\xi)$. For instance, if $p \geq 2q$, then Z could span a totally null, q-dimensional subspace in $\xi_0^\perp$ along which (after projectivization) $V_0 \cap V_\xi$ would be singular. This happens precisely when all the columns of Z are null, i.e., when $\mathrm{rk}(Z^\mathsf{T} Z) = 0$.

For $k = \begin{pmatrix} A & 0 \\ 0 & D \end{pmatrix} \in SO(p) \times SO(q)$ and $\xi = \begin{bmatrix} Z \\ I_q \end{bmatrix}$, we have

$$k.\xi = \begin{bmatrix} AZD^{-1} \\ I_q \end{bmatrix} = \begin{bmatrix} AZD^\mathsf{T} \\ I_q \end{bmatrix}.$$

And,

$$
\begin{aligned}
\mathrm{rk}((AZD^\mathsf{T})(DZ^\mathsf{T} A^\mathsf{T})) &= \mathrm{rk}(AZZ^\mathsf{T} A^\mathsf{T}) \\
&= \mathrm{rk}(AZZ^\mathsf{T} A^{-1}) \\
&= \mathrm{rk}(ZZ^\mathsf{T})
\end{aligned}
$$

since $D^\mathsf{T} = D^{-1}$ and $A^\mathsf{T} = A^{-1}$. Therefore, this finer stratification is preserved by K and $K^\mathbf{C}$, but not by $GL(p, \mathbf{C}) \times GL(q, \mathbf{C})$, the group preserving the first stratification.

As before, the extreme stratum (in the first stratification) defined by $\mathrm{rk}(Z) = 1$ coincides with the generalized conformal structure on the Grassmannian. However, the second stratification does not have an extreme stratum corresponding to any generalized conformal structure.

The stratification of the Frobenius structures is reminiscent of the stratification of the cones that arise in the constructions of unitary highest weight representations in [14] for $SU(2, 2)$ and in [22] for the highest weight modules coming from a line bundle on G/K. When $X = G^\mathbf{C}/K^\mathbf{C} P^+$, Goncharov [12] attaches a representation of the complex Lie algebra $\mathbf{g}$ of minimal Kirillov dimension to the cone of highest weight vectors $K(X)$ in $T_{x_0} X$ (which is the minimal $K^\mathbf{C}$-orbit in $\mathbf{p}^+$) by mapping $\mathbf{g}$ into the algebra of regular differential operators on the singular variety $K(X)$ using a Fourier transform. Unlike [14] and [22], Goncharov does not address the issue of the unitarity of the representation.

These results make us hopeful of further connections between the geometry of the space M_D and representations of the group G. In particular, we would expect there to be a way of attaching representations to the other strata in the Frobenius structure.

With regard to the description of the parameter space, we can show for all open $U(p,q)$-orbits D in any generalized flag variety for $G^{\mathbf{C}} = GL(p+q, \mathbf{C})$ that M_D is $G/K \times \overline{G/K}$, except in the holomorphic cases, where $M_D \cong G/K$.

REFERENCES

1. D. N. Ahiezer and S. G. Gindikin, *On Stein extensions of real symmetric spaces*, Math. Ann. **286** (1990), 1–12.

2. R. J. Baston, *Almost Hermitian symmetric manifolds*, I and II, Duke Math. Jour. **63** (1991), 81–112 and 113–138.

3. R. J. Baston and M. G. Eastwood, *The Penrose Transform; its Interaction with Representation Theory*, Oxford University Press, 1989.

4. N. Buchdahl, *On the relative de Rham sequence*, Proc. A.M.S. **87** (1983), 363–366.

5. E. G. Dunne and M. G. Eastwood, *The twistor transform*, Twistors in Mathematics and Physics, Lond. Math. Soc. Lecture Note Series, vol. 156, Cambridge University Press, 1990, pp. 110–128.

6. M. G. Eastwood, The Penrose transform, Twistors in Mathematics and Physics, Lond. Math. Soc. Lecture Note Series, vol. 156, Cambridge University Press, 1990, pp. 87–103.

7. _____, *Introduction to the Penrose transform*, in this volume, pp. 71–75.

8. M. G. Eastwood, R. Penrose and R. O. Wells, Jr., *Cohomology and massless fields*, Commun. Math. Phys. **78** (1981), 305–351.

9. M. Flensted-Jensen, *Analysis on non-Riemannian Symmetric Spaces*, Reg. Conf. Ser. Math. **61**, Amer. Math. Soc., Providence, 1986.

10. S. G. Gindikin, *Generalized conformal structures*, Twistors in Mathematics and Physics, Lond. Math. Soc. Lecture Note Series, vol. 156, Cambridge University Press, 1990, pp. 36–52.

11. S. G. Gindikin, *Integral geometry and twistors*, Lecture Notes in Math., vol. 970, Springer, 1982, pp. 2–42.

12. A. B. Goncharov, *Generalized conformal structures on manifolds*, Selecta Mathematica Sovietica **6** (1987), 307–340.

13. S. Helgason, *Differential Geometry, Lie Groups and Symmetric Spaces*, Academic Press, 1978.

14. H. P. Jakobsen and M. Vergne, *Wave and Dirac operators and representations of the conformal group*, J. Funct. Anal. **24** (1977), 52–106.

15. A. W. Knapp, *Introduction to representations in analytic cohomology*, in this volume, pp. 1–19.

16. K. Kodaira, *On stability of compact submanifolds of complex manifolds*, Amer. J. Math. **85** (1963), 79–94.

17. J. Novak, *Parametrizing maximal compact subvarieties*, preprint, Oklahoma State University, 1992.

18. C. M. Patton and H. Rossi, *Unitary structures on cohomology*, Trans. A.M.S. **290** (1985), 235–258.

19. R. Penrose, *Nonlinear gravitons and curved twistor theory*, General Relativity and Gravitation **7** (1976), 31–52.

20. R. Penrose and M. A. H. MacCallum, *Twistor theory: an approach to the quantisation of fields and space-time*, Physics Reports (Section C of Physics Letters) **6** (1972), 241–316.

21. J. Rawnsley, W. Schmid and J. A. Wolf, *Singular unitary representations and indefinite harmonic theory*, J. Func. An. **51** (1983), 1–114.

22. H. Rossi and M. Vergne, *Analytic continuation of the holomorphic discrete series*, Acta Math. **136** (1976), 1–59.

23. W. Schmid, *Homogeneous complex manifolds and representations of semisimple Lie groups*, Ph.D. dissertation, University of California, Berkeley 1967, Representation Theory

and Harmonic Analysis on Semisimple Lie Groups, Math. Surveys and Monographs, vol. 31, Amer. Math. Soc., 1989, pp. 223–286.

24. _____, *On a conjecture of Langlands*, Ann. of Math. **93** (1971), 1–42.

25. _____, L^2-*cohomology and the discrete series*, Ann. of Math. **103** (1976), 375–394.

26. W. Schmid and J. A. Wolf, *A vanishing theorem for open orbits on complex flag manifolds*, Proc. A.M.S. **92** (1984), 461–464.

27. R. O. Wells, Jr., *Complex manifolds and mathematical physics*, Bull. A.M.S. **1** (1979), 296–336.

28. R. O. Wells, Jr. and J. A. Wolf, *Poincaré series and automorphic cohomology on flag domains*, Ann. of Math. **105** (1977), 397–448.

29. J. A. Wolf, *The action of a real semisimple group on a complex flag manifold I: orbit structure and holomorphic components*, Bull. A.M.S. **75** (1969), 1121–1237.

30. _____, *The Stein condition for cycle spaces of open orbits on complex flag manifolds*, Ann. of Math. **136** (1982), 541–555.

31. H. W. Wong, *Dolbeault cohomologies and Zuckerman modules*, in this volume, pp. 217–223.

DEPARTMENT OF MATHEMATICS, OKLAHOMA STATE UNIVERSITY, STILLWATER, OKLAHOMA 74078-0613, USA

E-mail address: dunne@math.okstate.edu

DEPARTMENT OF MATHEMATICS, OKLAHOMA STATE UNIVERSITY, STILLWATER, OKLAHOMA 74078-0613, USA

E-mail address: zierau@math.okstate.edu

Contemporary Mathematics
Volume **154**, 1993

Algebraic $\mathcal{D}$-modules and Representation Theory of Semisimple Lie Groups

DRAGAN MILIČIĆ

ABSTRACT. This expository paper represents an introduction to some aspects of the current research in representation theory of semisimple Lie groups. In particular, we discuss the theory of "localization" of modules over the enveloping algebra of a semisimple Lie algebra due to Alexander Beilinson and Joseph Bernstein [**1**], [**2**] and the work of Henryk Hecht, Wilfried Schmid, Joseph A. Wolf and the author on the localization of Harish-Chandra modules [**7**], [**8**], [**13**], [**17**], [**18**]. These results can be viewed as a vast generalization of the classical theorem of Armand Borel and André Weil on geometric realization of irreducible finite-dimensional representations of compact semisimple Lie groups [**3**].

1. Introduction

Let G_0 be a connected semisimple Lie group with finite center. Fix a maximal compact subgroup K_0 of G_0. Let $\mathfrak{g}$ be the complexified Lie algebra of G_0 and $\mathfrak{k}$ its subalgebra which is the complexified Lie algebra of K_0. Denote by σ the corresponding Cartan involution, i.e., σ is the involution of $\mathfrak{g}$ such that $\mathfrak{k}$ is the set of its fixed points. Let K be the complexification of K_0. The group K has a natural structure of a complex reductive algebraic group.

Let π be an admissible representation of G_0 of finite length. Then, the submodule V of all K_0-finite vectors in this representation is a finitely generated module over the enveloping algebra $\mathcal{U}(\mathfrak{g})$ of $\mathfrak{g}$, and also a direct sum of finite-dimensional irreducible representations of K_0. The representation of K_0 extends uniquely to a representation of the complexification K of K_0, and it is also a direct sum of finite-dimensional representations.

We say that a representation of a complex algebraic group K in a linear space V is *algebraic* if V is a union of finite-dimensional K-invariant subspaces V_i,

---

1991 *Mathematics Subject Classification*. Primary 22E46.

This paper is in final form and no version of it will be submitted for publication elsewhere.

$i \in I$, and for each $i \in I$ the action of K on V_i induces a morphism of algebraic groups $K \to \mathrm{GL}(V_i)$.

This leads us to the definition of a *Harish-Chandra module* V:

  (i) V is a finitely generated $\mathcal{U}(\mathfrak{g})$-module;
  (ii) V is an algebraic representation of K;
  (iii) the actions of $\mathfrak{g}$ and K are compatible, i.e.,
      (a) the action of $\mathfrak{k}$ as the subalgebra of $\mathfrak{g}$ agrees with the differential of the action of K;
      (b) the action map $\mathcal{U}(\mathfrak{g}) \otimes V \to V$ is K-equivariant (here K acts on $\mathcal{U}(\mathfrak{g})$ by the adjoint action).

A morphism of Harish-Chandra modules is a linear map which intertwines the $\mathcal{U}(\mathfrak{g})$- and K-actions. Harish-Chandra modules and their morphisms form an abelian category. We denote it by $\mathcal{M}(\mathfrak{g}, K)$.

Let $\mathcal{Z}(\mathfrak{g})$ be the center of the enveloping algebra of $\mathcal{U}(\mathfrak{g})$. If V is an irreducible Harish-Chandra module, the center $\mathcal{Z}(\mathfrak{g})$ acts on V by multiples of the identity operator, i.e., $\mathcal{Z}(\mathfrak{g}) \ni \xi \to \chi_V(\xi) 1_V$, where $\chi_V : \mathcal{Z}(\mathfrak{g}) \to \mathbb{C}$ is the *infinitesimal character* of V. In general, if a Harish-Chandra module V is annihilated by an ideal of finite codimension in $\mathcal{Z}(\mathfrak{g})$, it is of finite length.

Since the functor attaching to admissible representations of G_0 their Harish-Chandra modules maps irreducibles into irreducibles, the problem of classification of irreducible admissible representations is equivalent to the problem of classification of irreducible Harish-Chandra modules. This problem was solved in the work of R. Langlands [11], Harish-Chandra, A.W. Knapp and G. Zuckerman [10], and D. Vogan [19]. Their proofs were based on a blend of algebraic and analytic techniques and depended heavily on the work of Harish-Chandra.

In this paper we give an exposition of the classification using entirely the methods of algebraic geometry [8], [14]. In §2, we recall the Borel-Weil theorem. In §3, we introduce the localization functor of Beilinson and Bernstein, and sketch a proof of the equivalence of the category of $\mathcal{U}(\mathfrak{g})$-modules with an infinitesimal character with a category of $\mathcal{D}$-modules on the flag variety of $\mathfrak{g}$. This equivalence induces an equivalence of the category of Harish-Chandra modules with an infinitesimal character with a category of "Harish-Chandra sheaves" on the flag variety. In §4, we recall the basic notions and constructions of the algebraic theory of $\mathcal{D}$-modules. After discussing the structure of K-orbits in the flag variety of $\mathfrak{g}$ in §5, we classify all irreducible Harish-Chandra sheaves in §6. In §7, we describe a necessary and sufficient condition for vanishing of cohomology of irreducible Harish-Chandra sheaves and complete the geometric classification of irreducible Harish-Chandra modules. The final section 8, contains a discussion of the relationship of this classification with the Langlands classification, and a detailed discussion of the case of the group $\mathrm{SU}(2,1)$.

2. The Borel-Weil theorem

First we discuss the case of a connected compact semisimple Lie group. In this situation $G_0 = K_0$, and we denote by G the complexification of G_0. In this case, the irreducible Harish-Chandra modules are just irreducible finite-dimensional representations of G.

For simplicity, we assume that G_0 (and G) is simply connected. Denote by X the flag variety of $\mathfrak{g}$, i.e., the space of all Borel subalgebras of $\mathfrak{g}$. It has a natural structure of a smooth algebraic variety. Since all Borel subalgebras are mutually conjugate, the group G acts transitively on X. For any $x \in X$, the differential of the orbit map $g \longmapsto g \cdot x$ defines a projection of the Lie algebra $\mathfrak{g}$ onto the tangent space $T_x(X)$ of X at x. Therefore, we have a natural vector bundle morphism from the trivial bundle $X \times \mathfrak{g}$ over X into the tangent bundle $T(X)$ of X. If we consider the adjoint action of G on $\mathfrak{g}$, the trivial bundle $X \times \mathfrak{g}$ is G-homogeneous and the morphism $X \times \mathfrak{g} \to T(X)$ is G-equivariant. The kernel of this morphism is a G-homogeneous vector bundle $\mathcal{B}$ over X. The fiber of $\mathcal{B}$ over $x \in X$ is the Borel subalgebra $\mathfrak{b}_x$ which corresponds to the point x. Therefore, we can view $\mathcal{B}$ as the "tautological" vector bundle of Borel subalgebras over X. For any $x \in X$, denote by $\mathfrak{n}_x = [\mathfrak{b}_x, \mathfrak{b}_x]$ the nilpotent radical of $\mathfrak{b}_x$. Then $\mathcal{N} = \{(x, \xi) \mid \xi \in \mathfrak{n}_x\} \subset \mathcal{B}$ is a G-homogeneous vector subbundle of $\mathcal{B}$. We denote the quotient vector bundle $\mathcal{B}/\mathcal{N}$ by $\mathcal{H}$. If B_x is the stabilizer of x in G, it acts trivially on the fiber $\mathcal{H}_x$ of $\mathcal{H}$ at x. Therefore, $\mathcal{H}$ is a trivial vector bundle on X. Since X is a projective variety, the only global sections of $\mathcal{H}$ are constants. Let $\mathfrak{h}$ be the space of global sections of $\mathcal{H}$. We can view it as an abelian Lie algebra. The Lie algebra $\mathfrak{h}$ is called *the (abstract) Cartan algebra* of $\mathfrak{g}$. Let $\mathfrak{c}$ be any Cartan subalgebra of $\mathfrak{g}$, R the root system of the pair $(\mathfrak{g}, \mathfrak{c})$ in the dual space $\mathfrak{c}^*$ of $\mathfrak{c}$, and R^+ a set of positive roots in R. Then $\mathfrak{c}$ and the root subspaces of $\mathfrak{g}$ corresponding to the roots in R^+ span a Borel subalgebra $\mathfrak{b}_x$ for some point $x \in X$. We have the sequence $\mathfrak{c} \to \mathfrak{b}_x \to \mathfrak{b}_x/\mathfrak{n}_x = \mathcal{H}_x$ of linear maps, and their composition is an isomorphism. On the other hand, the evaluation map $\mathfrak{h} \to \mathcal{H}_x$ is also an isomorphism, and by composing the previous map with the inverse of the evaluation map, we get the canonical isomorphism $\mathfrak{c} \to \mathfrak{h}$. Its dual map is an isomorphism $\mathfrak{h}^* \to \mathfrak{c}^*$ which we call *a specialization* at x. It identifies an (abstract) root system Σ in $\mathfrak{h}^*$, and a set of positive roots Σ^+, with R and R^+. One can check that Σ and Σ^+ do not depend on the choice of $\mathfrak{c}$ and x. Therefore, we constructed *the (abstract) Cartan triple* $(\mathfrak{h}^*, \Sigma, \Sigma^+)$ of $\mathfrak{g}$. The dual root system in $\mathfrak{h}$ is denoted by $\Sigma^{\check{}}$.

Let $P(\Sigma)$ be the weight lattice in $\mathfrak{h}^*$. Then to each $\lambda \in P(\Sigma)$ we attach a G-homogeneous invertible $\mathcal{O}_X$-module $\mathcal{O}(\lambda)$ on X. We say that a weight λ is *antidominant* if $\alpha^{\check{}}(\lambda) \leq 0$ for any $\alpha \in \Sigma^+$. The following result is the celebrated Borel-Weil theorem. We include a proof inspired by the localization theory.

2.1. THEOREM (BOREL-WEIL). *Let λ be an antidominant weight. Then*

(i) $H^i(X, \mathcal{O}(\lambda))$ *vanish for $i > 0$.*

(ii) $\Gamma(X, \mathcal{O}(\lambda))$ *is the irreducible finite-dimensional G-module with lowest weight λ.*

PROOF. Denote by F_λ the irreducible finite-dimensional G-module with lowest weight λ. Let $\mathcal{F}_\lambda$ be the sheaf of local sections of the trivial vector bundle with fibre F_λ over X. Clearly we have

$$H^i(X, \mathcal{F}_\lambda) = H^i(X, \mathcal{O}_X) \otimes_{\mathbb{C}} F_\lambda \text{ for } i \in \mathbb{Z}_+.$$

Since X is a projective variety, the cohomology groups $H^i(X, \mathcal{O}_X)$ are finite dimensional.

Let Ω be the Casimir element in the center of the enveloping algebra $\mathcal{U}(\mathfrak{g})$ of $\mathfrak{g}$. Then for any local section s of $\mathcal{O}(\mu)$, Ωs is proportional to s. In fact, if we denote by $\langle \cdot, \cdot \rangle$ the natural bilinear form on $\mathfrak{h}^*$ induced by the Killing form of $\mathfrak{g}$, by a simple calculation using Harish-Chandra homomorphism we have $\Omega s = \langle \mu, \mu - 2\rho \rangle s$ for any section s of $\mathcal{O}(\mu)$. In particular, Ω annihilates $\mathcal{O}_X$, hence it also annihilates finite-dimensional $\mathfrak{g}$-modules $H^i(X, \mathcal{O}_X)$. Since finite-dimensional $\mathfrak{g}$-modules are semisimple, and Ω acts trivially only on the trivial irreducible $\mathfrak{g}$-module, we conclude that the action of $\mathfrak{g}$ on $H^i(X, \mathcal{O}_X)$ is trivial. Therefore, $\Omega - \langle \lambda, \lambda - 2\rho \rangle$ annihilates $H^i(X, \mathcal{F}_\lambda)$. On the other hand, the Jordan-Hölder filtration of F_λ, considered as a B-module, induces a filtration of $\mathcal{F}_\lambda$ by G-homogeneous locally free $\mathcal{O}_X$-modules such that $F_p \mathcal{F}_\lambda / F_{p-1} \mathcal{F}_\lambda$ is a G-homogeneous invertible $\mathcal{O}_X$-module $\mathcal{O}(\nu_p)$ for a weight ν_p of F_λ. This implies that $\prod_0^{\dim F_\lambda} (\Omega - \langle \nu_p, \nu_p - 2\rho \rangle)$ annihilates $\mathcal{F}_\lambda$.

Assume that $\langle \nu_p, \nu_p - 2\rho \rangle = \langle \lambda, \lambda - 2\rho \rangle$ for some weight ν_p. It leads to $\langle \nu_p - \rho, \nu_p - \rho \rangle = \langle \lambda - \rho, \lambda - \rho \rangle$, and, since λ is the lowest weight, we finally see that $\nu_p = \lambda$. Therefore, $\mathcal{F}_\lambda$ splits into the direct sum of the Ω-eigensheaf $\mathcal{O}(\lambda)$ for eigenvalue $\langle \lambda, \lambda - 2\rho \rangle$ and its Ω-invariant complement. Since cohomology commutes with direct sums, we conclude that

$$H^i(X, \mathcal{O}(\lambda)) = H^i(X, \mathcal{O}_X) \otimes_{\mathbb{C}} F_\lambda$$

for $i \in \mathbb{Z}_+$. Clearly, $\Gamma(X, \mathcal{O}_X) = \mathbb{C}$ and (ii) follows immediately. This implies that invertible $\mathcal{O}_X$-modules $\mathcal{O}(\lambda)$, for regular antidominant λ, are very ample. By a theorem of Serre, (i) follows for geometrically "very positive" λ (i.e., far from the walls in the negative chamber). Hence $H^i(X, \mathcal{O}_X) = 0$ for $i > 0$, which in turn implies (i) in general. $\square$

3. Beilinson-Bernstein equivalence of categories

Now we want to describe a generalization of the Borel-Weil theorem established by A. Beilinson and J. Bernstein.

First we have to construct a family of sheaves of algebras on the flag variety X. Let $\mathfrak{g}^\circ = \mathcal{O}_X \otimes_{\mathbb{C}} \mathfrak{g}$ be the sheaf of local sections of the trivial bundle $X \times \mathfrak{g}$. Denote by $\mathfrak{b}^\circ$ and $\mathfrak{n}^\circ$ the corresponding subsheaves of local sections of $\mathcal{B}$ and $\mathcal{N}$, respectively. The differential of the action of G on X defines a natural homomorphism τ of the Lie algebra $\mathfrak{g}$ into the Lie algebra of vector fields on X. We define a structure of a sheaf of complex Lie algebras on $\mathfrak{g}^\circ$ by putting

$$[f \otimes \xi, g \otimes \eta] = f\tau(\xi)g \otimes \eta - g\tau(\eta)f \otimes \xi + fg \otimes [\xi, \eta]$$

for $f, g \in \mathcal{O}_X$ and $\xi, \eta \in \mathfrak{g}$. If we extend τ to the natural homomorphism of $\mathfrak{g}^\circ$ into the sheaf of Lie algebras of local vector fields on X, $\ker \tau$ is exactly $\mathfrak{b}^\circ$. In addition, the sheaves $\mathfrak{b}^\circ$ and $\mathfrak{n}^\circ$ are sheaves of ideals in $\mathfrak{g}^\circ$.

Similarly, we define a multiplication in the sheaf $\mathcal{U}^\circ = \mathcal{O}_X \otimes_\mathbb{C} \mathcal{U}(\mathfrak{g})$ by

$$(f \otimes \xi)(g \otimes \eta) = f\tau(\xi)g \otimes \eta + fg \otimes \xi\eta$$

where $f, g \in \mathcal{O}_X$ and $\xi \in \mathfrak{g}$, $\eta \in \mathcal{U}(\mathfrak{g})$. In this way $\mathcal{U}^\circ$ becomes a sheaf of complex associative algebras on X. Evidently, $\mathfrak{g}^\circ$ is a subsheaf of $\mathcal{U}^\circ$, and the natural commutator in $\mathcal{U}^\circ$ induces the bracket operation on $\mathfrak{g}^\circ$. It follows that the sheaf of right ideals $\mathfrak{n}^\circ\mathcal{U}^\circ$ generated by $\mathfrak{n}^\circ$ in $\mathcal{U}^\circ$ is a sheaf of two-sided ideals in $\mathcal{U}^\circ$. Therefore, the quotient $\mathcal{D}_\mathfrak{h} = \mathcal{U}^\circ/\mathfrak{n}^\circ\mathcal{U}^\circ$ is a sheaf of complex associative algebras on X.

The natural morphism of $\mathfrak{g}^\circ$ into $\mathcal{D}_\mathfrak{h}$ induces a morphism of the sheaf of Lie subalgebras $\mathfrak{b}^\circ$ into $\mathcal{D}_\mathfrak{h}$ which vanishes on $\mathfrak{n}^\circ$. Hence there is a natural homomorphism ϕ of the enveloping algebra $\mathcal{U}(\mathfrak{h})$ of $\mathfrak{h}$ into the global sections $\Gamma(X, \mathcal{D}_\mathfrak{h})$ of $\mathcal{D}_\mathfrak{h}$. The action of the group G on the structure sheaf $\mathcal{O}_X$ and $\mathcal{U}(\mathfrak{g})$ induces a natural G-action on $\mathcal{U}^\circ$ and $\mathcal{D}_\mathfrak{h}$. On the other hand, triviality of $\mathcal{H}$ and constancy of its global sections imply that the induced G-action on $\mathfrak{h}$ is trivial. It follows that ϕ maps $\mathcal{U}(\mathfrak{h})$ into the G-invariants of $\Gamma(X, \mathcal{D}_\mathfrak{h})$. This implies that the image of ϕ is in the center of $\mathcal{D}_\mathfrak{h}(U)$ for any open set U in X. One can show that ϕ is actually an isomorphism of $\mathcal{U}(\mathfrak{h})$ onto the subalgebra of all G-invariants in $\Gamma(X, \mathcal{D}_\mathfrak{h})$. In addition, the natural homomorphism of $\mathcal{U}(\mathfrak{g})$ into $\Gamma(X, \mathcal{D}_\mathfrak{h})$ induces a homomorphism of the center $\mathcal{Z}(\mathfrak{g})$ of $\mathcal{U}(\mathfrak{g})$ into $\Gamma(X, \mathcal{D}_\mathfrak{h})$. Its image is also contained in the subalgebra of G-invariants of $\Gamma(X, \mathcal{D}_\mathfrak{h})$. Hence, it is in $\phi(\mathcal{U}(\mathfrak{h}))$. Finally, we have the canonical Harish-Chandra homomorphism $\gamma : \mathcal{Z}(\mathfrak{g}) \to \mathcal{U}(\mathfrak{h})$, defined in the following way. First, for any $x \in X$, the center $\mathcal{Z}(\mathfrak{g})$ is contained in the sum of the subalgebra $\mathcal{U}(\mathfrak{b}_x)$ and the right ideal $\mathfrak{n}_x\mathcal{U}(\mathfrak{g})$ of $\mathcal{U}(\mathfrak{g})$. Therefore, we have the natural projection of $\mathcal{Z}(\mathfrak{g})$ into

$$\mathcal{U}(\mathfrak{b}_x)/(\mathfrak{n}_x\mathcal{U}(\mathfrak{g}) \cap \mathcal{U}(\mathfrak{b}_x)) = \mathcal{U}(\mathfrak{b}_x)/\mathfrak{n}_x\mathcal{U}(\mathfrak{b}_x) = \mathcal{U}(\mathfrak{b}_x/\mathfrak{n}_x).$$

Its composition with the natural isomorphism of $\mathcal{U}(\mathfrak{b}_x/\mathfrak{n}_x)$ with $\mathcal{U}(\mathfrak{h})$ is independent of x and, by definition, equal to γ. The diagram

$$
\begin{array}{ccc}
\mathcal{Z}(\mathfrak{g}) & \xrightarrow{\ \gamma\ } & \mathcal{U}(\mathfrak{h}) \\
\Big\| & & \Big\downarrow{\phi} \\
\mathcal{Z}(\mathfrak{g}) & \longrightarrow & \Gamma(X, \mathcal{D}_\mathfrak{h})
\end{array}
$$

of natural algebra homomorphisms is commutative. We can form $\mathcal{U}(\mathfrak{g}) \otimes_{\mathcal{Z}(\mathfrak{g})} \mathcal{U}(\mathfrak{h})$, which has a natural structure of an associative algebra. There exists a natural algebra homomorphism

$$\Psi : \mathcal{U}(\mathfrak{g}) \otimes_{\mathcal{Z}(\mathfrak{g})} \mathcal{U}(\mathfrak{h}) \to \Gamma(X, \mathcal{D}_\mathfrak{h})$$

given by the tensor product of the natural homomorphism of $\mathcal{U}(\mathfrak{g})$ into $\Gamma(X,\mathcal{D}_{\mathfrak{h}})$ and ϕ. The next result describes the cohomology of the sheaf of algebras $\mathcal{D}_{\mathfrak{h}}$. Its proof is an unpublished argument due to Joseph Taylor and the author.

3.1. LEMMA.

(i) *The morphism*

$$\Psi : \mathcal{U}(\mathfrak{g}) \otimes_{\mathcal{Z}(\mathfrak{g})} \mathcal{U}(\mathfrak{h}) \to \Gamma(X,\mathcal{D}_{\mathfrak{h}})$$

   is an isomorphism of algebras.
(ii) $H^i(X,\mathcal{D}_{\mathfrak{h}}) = 0$ *for* $i > 0$.

SKETCH OF THE PROOF. First we construct a left resolution

$$\ldots \to \mathcal{U}^\circ \otimes_{\mathcal{O}_X} \textstyle\bigwedge^p \mathfrak{n}^\circ \to \ldots \to \mathcal{U}^\circ \otimes_{\mathcal{O}_X} \mathfrak{n}^\circ \to \mathcal{U}^\circ \to \mathcal{D}_{\mathfrak{h}} \to 0$$

of $\mathcal{D}_{\mathfrak{h}}$ (here $\bigwedge^p \mathfrak{n}^\circ$ is the pth exterior power of $\mathfrak{n}^\circ$). The cohomology of each component in this complex is given by

$$H^q(X,\mathcal{U}^\circ \otimes_{\mathcal{O}_X} \textstyle\bigwedge^p \mathfrak{n}^\circ) = H^q(X,\mathcal{U}(\mathfrak{g}) \otimes_{\mathbb{C}} \textstyle\bigwedge^p \mathfrak{n}^\circ) = \mathcal{U}(\mathfrak{g}) \otimes_{\mathbb{C}} H^q(X,\textstyle\bigwedge^p \mathfrak{n}^\circ).$$

Let $\ell : W \to \mathbb{Z}_+$ be the length function on the Weyl group W of Σ with respect to the set of reflections corresponding to simple roots Π in Σ^+. Let $W(p) = \{w \in W \mid \ell(w) = p\}$ and $n(p) = \operatorname{Card} W(p)$. By a lemma of Bott [5] (which follows easily from the Borel-Weil-Bott theorem),

$$H^q(X,\textstyle\bigwedge^p \mathfrak{n}^\circ) = 0 \text{ if } p \neq q;$$

and $H^p(X,\bigwedge^p \mathfrak{n}^\circ)$ is a linear space of dimension $n(p)$ with trivial action of G. Now, a standard spectral sequence argument implies that (ii) holds, and that $\Gamma(X,\mathcal{D}_{\mathfrak{h}})$ has a finite filtration such that the corresponding graded algebra is isomorphic to a direct sum of $\operatorname{Card} W$ copies of $\mathcal{U}(\mathfrak{g})$. Taking the G-invariants of this spectral sequence we see that the induced finite filtration of $\Gamma(X,\mathcal{D}_{\mathfrak{h}})^G = \mathcal{U}(\mathfrak{h})$ is such that the corresponding graded algebra is isomorphic to a direct sum of $\operatorname{Card} W$ copies of $\mathcal{Z}(\mathfrak{g})$. This implies (i). $\square$

Denote by ρ the half-sum of all positive roots in Σ. The enveloping algebra $\mathcal{U}(\mathfrak{h})$ of $\mathfrak{h}$ is naturally isomorphic to the algebra of polynomials on $\mathfrak{h}^*$, and therefore any $\lambda \in \mathfrak{h}^*$ determines a homomorphism of $\mathcal{U}(\mathfrak{h})$ into $\mathbb{C}$. Let I_λ be the kernel of the homomorphism $\varphi_\lambda : \mathcal{U}(\mathfrak{h}) \to \mathbb{C}$ determined by $\lambda + \rho$. Then $\gamma^{-1}(I_\lambda)$ is a maximal ideal in $\mathcal{Z}(\mathfrak{g})$, and, by a result of Harish-Chandra, for $\lambda, \mu \in \mathfrak{h}^*$,

$$\gamma^{-1}(I_\lambda) = \gamma^{-1}(I_\mu) \text{ if and only if } w\lambda = \mu$$

for some w in the Weyl group W of Σ. For any $\lambda \in \mathfrak{h}^*$, the sheaf $I_\lambda \mathcal{D}_{\mathfrak{h}}$ is a sheaf of two-sided ideals in $\mathcal{D}_{\mathfrak{h}}$; therefore $\mathcal{D}_\lambda = \mathcal{D}_{\mathfrak{h}}/I_\lambda \mathcal{D}_{\mathfrak{h}}$ is a sheaf of complex associative algebras on X. In the case when $\lambda = -\rho$, we have $I_{-\rho} = \mathfrak{h}\mathcal{U}(\mathfrak{h})$, hence $\mathcal{D}_{-\rho} = \mathcal{U}^\circ/\mathfrak{b}^\circ \mathcal{U}^\circ$, i.e., it is the sheaf of local differential operators on X. If $\lambda \in P(\Sigma)$, $\mathcal{D}_\lambda$ is the sheaf of differential operators on the invertible $\mathcal{O}_X$-module $\mathcal{O}(\lambda + \rho)$.

Let Y be a smooth complex algebraic variety. Denote by $\mathcal{O}_Y$ its structure sheaf. Let $\mathcal{D}_Y$ be the sheaf of local differential operators on Y. Denote by i_Y the natural homomorphism of the sheaf of rings $\mathcal{O}_Y$ into $\mathcal{D}_Y$. We can consider the category of pairs $(\mathcal{A}, i_\mathcal{A})$ where $\mathcal{A}$ is a sheaf of rings on Y and $i_\mathcal{A} : \mathcal{O}_Y \to \mathcal{A}$ a homomorphism of sheaves of rings. The morphisms are homomorphisms $\alpha :$ $\mathcal{A} \to \mathcal{B}$ of sheaves of algebras such that $\alpha \circ i_\mathcal{A} = i_\mathcal{B}$. A pair $(\mathcal{D}, i)$ is called a *twisted sheaf of differential operators* if Y has a cover by open sets U such that $(\mathcal{D}|U, i|U)$ is isomorphic to $(\mathcal{D}_U, i_U)$.

In general, the sheaves of algebras $\mathcal{D}_\lambda$, $\lambda \in \mathfrak{h}^*$, are twisted sheaves of differential operators on X.

Let θ be a Weyl group orbit in $\mathfrak{h}^*$ and $\lambda \in \theta$. Denote by $J_\theta = \gamma^{-1}(I_\lambda)$ the maximal ideal in $\mathcal{Z}(\mathfrak{g})$ determined by θ. We denote by χ_λ the homomorphism of $\mathcal{Z}(\mathfrak{g})$ into $\mathbb{C}$ with $\ker \chi_\lambda = J_\theta$ (as we remarked before, χ_λ depends only on the Weyl group orbit θ of λ). The elements of J_θ map into the zero section of $\mathcal{D}_\lambda$. Therefore, we have a canonical morphism of $\mathcal{U}_\theta = \mathcal{U}(\mathfrak{g})/J_\theta\mathcal{U}(\mathfrak{g})$ into $\Gamma(X, \mathcal{D}_\lambda)$.

3.2. THEOREM.

(i) *The morphism*

$$\mathcal{U}_\theta \to \Gamma(X, \mathcal{D}_\lambda)$$

is an isomorphism of algebras.

(ii) $H^i(X, \mathcal{D}_\lambda) = 0$ *for* $i > 0$.

PROOF. Let $\mathbb{C}_{\lambda+\rho}$ be a one-dimensional $\mathfrak{h}$-module defined by $\lambda + \rho$. Let

$$\cdots \to F^{-p} \to \cdots \to F^{-1} \to F^0 \to \mathbb{C}_{\lambda+\rho} \to 0$$

be a left free $\mathcal{U}(\mathfrak{h})$-module resolution of $\mathbb{C}_{\lambda+\rho}$. By tensoring with $\mathcal{D}_\mathfrak{h}$ over $\mathcal{U}(\mathfrak{h})$ we get

$$\cdots \to \mathcal{D}_\mathfrak{h} \otimes_{\mathcal{U}(\mathfrak{h})} F^{-p} \to \cdots \to \mathcal{D}_\mathfrak{h} \otimes_{\mathcal{U}(\mathfrak{h})} F^0 \to \mathcal{D}_\mathfrak{h} \otimes_{\mathcal{U}(\mathfrak{h})} \mathbb{C}_{\lambda+\rho} \to 0.$$

Since $\mathcal{D}_\mathfrak{h}$ is locally $\mathcal{U}(\mathfrak{h})$-free, this is an exact sequence. Therefore, by 3.1.(ii), it is a left resolution of $\mathcal{D}_\mathfrak{h} \otimes_{\mathcal{U}(\mathfrak{h})} \mathbb{C}_{\lambda+\rho} = \mathcal{D}_\lambda$ by $\Gamma(X, -)$-acyclic sheaves. This implies first that all higher cohomologies of $\mathcal{D}_\lambda$ vanish. Also, it gives, using 3.1.(i), the exact sequence

$$\cdots \to \mathcal{U}(\mathfrak{g}) \otimes_{\mathcal{Z}(\mathfrak{g})} F^{-p} \to \cdots \to \mathcal{U}(\mathfrak{g}) \otimes_{\mathcal{Z}(\mathfrak{g})} F^0 \to \Gamma(X, \mathcal{D}_\lambda) \to 0,$$

which yields $\mathcal{U}_\theta = \mathcal{U}(\mathfrak{g}) \otimes_{\mathcal{Z}(\mathfrak{g})} \mathbb{C}_{\lambda+\rho} = \Gamma(X, \mathcal{D}_\lambda)$. $\square$

Therefore, the twisted sheaves of differential operators $\mathcal{D}_\lambda$ on X can be viewed as "sheafified" versions of the quotients $\mathcal{U}_\theta$ of the enveloping algebra $\mathcal{U}(\mathfrak{g})$. This allows us to "localize" the modules over $\mathcal{U}_\theta$.

First, denote by $\mathcal{M}(\mathcal{U}_\theta)$ the category of $\mathcal{U}_\theta$-modules. Also, let $\mathcal{M}_{qc}(\mathcal{D}_\lambda)$ be the category of quasi-coherent $\mathcal{D}_\lambda$-modules on X. If $\mathcal{V}$ is a quasi-coherent $\mathcal{D}_\lambda$-module, its global sections (and higher cohomology groups) are modules over $\Gamma(X, \mathcal{D}_\lambda) = \mathcal{U}_\theta$. Therefore, we can consider the functors:

$$H^p(X, -) : \mathcal{M}_{qc}(\mathcal{D}_\lambda) \to \mathcal{M}(\mathcal{U}_\theta)$$

for $p \in \mathbb{Z}_+$.

The next two results can be viewed as a vast generalization of the Borel-Weil theorem. In idea, their proof is very similar to our proof of the Borel-Weil theorem. It is also based on the theorems of Serre on cohomology of invertible $\mathcal{O}$-modules on projective varieties, and a splitting argument for the action of $\mathcal{Z}(\mathfrak{g})$ [1].

The first result corresponds to 2.1.(i). We say that $\lambda \in \mathfrak{h}^*$ is *antidominant* if $\alpha\check{}(\lambda)$ is not a positive integer for any $\alpha \in \Sigma^+$. This generalizes the notion of antidominance for weights in $P(\Sigma)$ introduced in §2.

3.3. VANISHING THEOREM. *Let $\lambda \in \mathfrak{h}^*$ be antidominant. Let $\mathcal{V}$ be a quasi-coherent $\mathcal{D}_\lambda$-module on the flag variety X. Then the cohomology groups $H^i(X, \mathcal{V})$ vanish for $i > 0$.*

In particular, the functor

$$\Gamma : \mathcal{M}_{qc}(\mathcal{D}_\lambda) \to \mathcal{M}(\mathcal{U}_\theta)$$

is exact. The second result corresponds to 2.1.(ii).

3.4. NONVANISHING THEOREM. *Let $\lambda \in \mathfrak{h}^*$ be regular and antidominant and $\mathcal{V} \in \mathcal{M}_{qc}(\mathcal{D}_\lambda)$ such that $\Gamma(X, \mathcal{V}) = 0$. Then $\mathcal{V} = 0$.*

This has the following consequence:

3.5. COROLLARY. *Let $\lambda \in \mathfrak{h}^*$ be antidominant and regular. Then any $\mathcal{V} \in \mathcal{M}_{qc}(\mathcal{D}_\lambda)$ is generated by its global sections.*

PROOF. Denote by $\mathcal{W}$ the $\mathcal{D}_\lambda$-submodule of $\mathcal{V}$ generated by all global sections. Then, we have an exact sequence

$$0 \to \Gamma(X, \mathcal{W}) \to \Gamma(X, \mathcal{V}) \to \Gamma(X, \mathcal{V}/\mathcal{W}) \to 0,$$

of $\mathcal{U}_\theta$-modules, and therefore $\Gamma(X, \mathcal{V}/\mathcal{W}) = 0$. Hence, $\mathcal{V}/\mathcal{W} = 0$, and $\mathcal{V}$ is generated by its global sections. $\square$

Let $\lambda \in \mathfrak{h}^*$ and let θ be the corresponding Weyl group orbit. Then we can define a right exact covariant functor Δ_λ from $\mathcal{M}(\mathcal{U}_\theta)$ into $\mathcal{M}_{qc}(\mathcal{D}_\lambda)$ by

$$\Delta_\lambda(V) = \mathcal{D}_\lambda \otimes_{\mathcal{U}_\theta} V$$

for any $V \in \mathcal{M}(\mathcal{U}_\theta)$. It is called the *localization functor*. Since

$$\Gamma(X, \mathcal{W}) = \mathrm{Hom}_{\mathcal{D}_\lambda}(\mathcal{D}_\lambda, \mathcal{W})$$

for any $\mathcal{W} \in \mathcal{M}_{qc}(\mathcal{D}_\lambda)$, it follows that Δ_λ is a left adjoint functor to the functor of global sections Γ, i.e.,

$$\mathrm{Hom}_{\mathcal{D}_\lambda}(\Delta_\lambda(V), \mathcal{W}) = \mathrm{Hom}_{\mathcal{U}_\theta}(V, \Gamma(X, \mathcal{W})),$$

for any $V \in \mathcal{M}(\mathcal{U}_\theta)$ and $\mathcal{W} \in \mathcal{M}_{qc}(\mathcal{D}_\lambda)$. In particular, there exists a functorial morphism φ from the identity functor into $\Gamma \circ \Delta_\lambda$. For any $V \in \mathcal{M}(\mathcal{U}_\theta)$, it is given by the natural morphism $\varphi_V : V \to \Gamma(X, \Delta_\lambda(V))$.

3.6. LEMMA. *Let $\lambda \in \mathfrak{h}^*$ be antidominant. Then the natural map φ_V of V into $\Gamma(X, \Delta_\lambda(V))$ is an isomorphism of $\mathfrak{g}$-modules.*

PROOF. If $V = \mathcal{U}_\theta$ this follows from 3.2. Also, by 3.3, we know that Γ is exact in this situation. This implies that $\Gamma \circ \Delta_\lambda$ is a right exact functor. Let

$$(\mathcal{U}_\theta)^{(J)} \to (\mathcal{U}_\theta)^{(I)} \to V \to 0$$

be an exact sequence of $\mathfrak{g}$-modules. Then we have the commutative diagram

$$
\begin{array}{ccccccc}
(\mathcal{U}_\theta)^{(J)} & \longrightarrow & (\mathcal{U}_\theta)^{(I)} & \longrightarrow & V & \longrightarrow & 0 \\
\downarrow & & \downarrow & & \downarrow & & \\
\Gamma(X, \Delta_\lambda(\mathcal{U}_\theta))^{(J)} & \longrightarrow & \Gamma(X, \Delta_\lambda(\mathcal{U}_\theta))^{(I)} & \longrightarrow & \Gamma(X, \Delta_\lambda(V)) & \longrightarrow & 0
\end{array}
$$

with exact rows, and the first two vertical arrows are isomorphisms. This implies that the third one is also an isomorphism. $\square$

On the other hand, the adjointness gives also a functorial morphism ψ from $\Delta_\lambda \circ \Gamma$ into the identity functor. For any $\mathcal{V} \in \mathcal{M}_{qc}(\mathcal{D}_\lambda)$, it is given by the natural morphism $\psi_\mathcal{V}$ of $\Delta_\lambda(\Gamma(X, \mathcal{V})) = \mathcal{D}_\lambda \otimes_{\mathcal{U}_\theta} \Gamma(X, \mathcal{V})$ into $\mathcal{V}$. Assume that λ is also regular. Then, by 3.5, $\psi_\mathcal{V}$ is an epimorphism. Let $\mathcal{K}$ be the kernel of $\psi_\mathcal{V}$. Then we have the exact sequence of quasi-coherent $\mathcal{D}_\lambda$-modules

$$0 \to \mathcal{K} \to \Delta_\lambda(\Gamma(X, \mathcal{V})) \to \mathcal{V} \to 0$$

and, by applying Γ and using 3.3, we get the exact sequence

$$0 \to \Gamma(X, \mathcal{K}) \to \Gamma(X, \Delta_\lambda(\Gamma(X, \mathcal{V}))) \to \Gamma(X, \mathcal{V}) \to 0.$$

By 3.6, we see that $\Gamma(X, \mathcal{K}) = 0$. By 3.4, $\mathcal{K} = 0$ and $\psi_\mathcal{V}$ is an isomorphism. This implies the following result, which is known as the *Beilinson-Bernstein equivalence of categories.*

3.7. THEOREM (BEILINSON-BERNSTEIN). *Let $\lambda \in \mathfrak{h}^*$ be antidominant and regular. Then the functor Δ_λ from $\mathcal{M}(\mathcal{U}_\theta)$ into $\mathcal{M}_{qc}(\mathcal{D}_\lambda)$ is an equivalence of categories. Its inverse is Γ.*

3.8. REMARK. In general, if we assume only that λ is antidominant, we denote by $\mathcal{QM}_{qc}(\mathcal{D}_\lambda)$ the quotient category of $\mathcal{M}_{qc}(\mathcal{D}_\lambda)$ with respect to the subcategory of all quasi-coherent $\mathcal{D}_\lambda$-modules with no global sections. Clearly, Γ induces an exact functor from $\mathcal{QM}_{qc}(\mathcal{D}_\lambda)$ into $\mathcal{M}(\mathcal{U}_\theta)$ which we denote also by Γ. Then we have an equivalence of categories

$$\mathcal{QM}_{qc}(\mathcal{D}_\lambda) \xrightarrow{\Gamma} \mathcal{M}(\mathcal{U}_\theta).$$

The equivalence of categories allows one to transfer problems about $\mathcal{U}_\theta$-modules into problems about $\mathcal{D}_\lambda$-modules. The latter problems can be attacked by "local" methods. To make this approach useful we need to introduce a "sheafified" version of Harish-Chandra modules.[1] A *Harish-Chandra sheaf* is

   (i) a coherent $\mathcal{D}_\lambda$-module $\mathcal{V}$

   (ii) with an algebraic action of K;

   (iii) the actions of $\mathcal{D}_\lambda$ and K on $\mathcal{V}$ are compatible, i.e.,

      (a) the action of $\mathfrak{k}$ as a subalgebra of $\mathfrak{g} \subset \mathcal{U}_\theta = \Gamma(X, \mathcal{D}_\lambda)$ agrees with the differential of the action of K;

      (b) the action $\mathcal{D}_\lambda \otimes_{\mathcal{O}_X} \mathcal{V} \to \mathcal{V}$ is K-equivariant.

Morphisms of Harish-Chandra sheaves are K-equivariant $\mathcal{D}_\lambda$-module morphisms. Harish-Chandra sheaves form an abelian category denoted by $\mathcal{M}_{coh}(\mathcal{D}_\lambda, K)$. Because of completely formal reasons, the equivalence of categories has the following consequence, which is a K-equivariant version of 3.7.

   3.9. THEOREM. *Let $\lambda \in \mathfrak{h}^*$ be antidominant and regular. Then the functor Δ_λ from $\mathcal{M}(\mathcal{U}_\theta, K)$ into $\mathcal{M}_{coh}(\mathcal{D}_\lambda, K)$ is an equivalence of categories. Its inverse is Γ.*

Therefore, by 3.9 and its analogue in the singular case, the classification of all irreducible Harish-Chandra modules is equivalent to the following two problems:

   (a) the classification of all irreducible Harish-Chandra sheaves;

   (b) determination of all irreducible Harish-Chandra sheaves $\mathcal{V}$ with $\Gamma(X, \mathcal{V}) \neq 0$ for antidominant $\lambda \in \mathfrak{h}^*$.

In next sections we shall explain how to solve these two problems.

   3.10. REMARK. Although the setting of 3.7 is adequate for the formulation of our results, the proofs require a more general setup. The difference between 3.7 and the general setup is analogous to the difference between the Borel-Weil theorem and its generalization, the Borel-Weil-Bott theorem. To explain this we have to use the language of derived categories.

   Let $D^b(\mathcal{D}_\lambda)$ be the bounded derived category of the category of quasi-coherent $\mathcal{D}_\lambda$-modules. Let $D^b(\mathcal{U}_\theta)$ be the bounded derived category of the category of $\mathcal{U}_\theta$-modules. Then, we have the following result:

   3.11. THEOREM. *For a regular λ, the derived functors $R\Gamma : D^b(\mathcal{D}_\lambda) \to D^b(\mathcal{U}_\theta)$ and $L\Delta_\lambda : D^b(\mathcal{U}_\theta) \to D^b(\mathcal{D}_\lambda)$ are mutually inverse equivalences of categories.*

4. Algebraic $\mathcal{D}$-modules

In this section we review some basic notions and results from the algebraic theory of $\mathcal{D}$-modules. They will allow us to study the structure of Harish-Chandra sheaves. Interested readers can find details in [**4**].

Let X be a smooth algebraic variety and $\mathcal{D}$ a twisted sheaf of differential operators on X. Then the opposite sheaf of rings $\mathcal{D}^{\mathrm{opp}}$ is again a twisted sheaf

---

[1]This requires some technical machinery beyond the scope of this paper, so we shall be rather vague in this definition.

of differential operators on X. We can therefore view left $\mathcal{D}$-modules as right $\mathcal{D}^{\text{opp}}$-modules and vice versa. Formally, the category $\mathcal{M}_{qc}^L(\mathcal{D})$ of quasi-coherent left $\mathcal{D}$-modules on X is isomorphic to the category $\mathcal{M}_{qc}^R(\mathcal{D}^{\text{opp}})$ of quasi-coherent right $\mathcal{D}^{\text{opp}}$-modules on X. Hence one can freely use right and left modules depending on the particular situation.

For a category $\mathcal{M}_{qc}(\mathcal{D})$ of $\mathcal{D}$-modules we denote by $\mathcal{M}_{coh}(\mathcal{D})$ the corresponding subcategory of coherent $\mathcal{D}$-modules.

The sheaf of algebras $\mathcal{D}$ has a natural filtration $(\mathcal{D}_p; p \in \mathbb{Z})$ by the degree. If we take a sufficiently small open set U in X such that $\mathcal{D}|_U \cong \mathcal{D}_U$, this filtration agrees with the standard degree filtration on $\mathcal{D}_U$. If we denote by π the canonical projection of the cotangent bundle $T^*(X)$ onto X, we have $\operatorname{Gr} \mathcal{D} = \pi_*(\mathcal{O}_{T^*(X)})$.

For any coherent $\mathcal{D}$-module $\mathcal{V}$ we can construct a *good filtration* $\operatorname{F} \mathcal{V}$ of $\mathcal{V}$ as a $\mathcal{D}$-module:

(a) The filtration $\operatorname{F} \mathcal{V}$ is increasing, exhaustive and $\operatorname{F}_p \mathcal{V} = 0$ for "very negative" $p \in \mathbb{Z}$;

(b) $\operatorname{F}_p \mathcal{V}$ are coherent $\mathcal{O}_X$-modules;

(c) $\mathcal{D}_p \operatorname{F}_q \mathcal{V} = \operatorname{F}_{p+q} \mathcal{V}$ for large $q \in \mathbb{Z}$ and all $p \in \mathbb{Z}_+$.

The annihilator of $\operatorname{Gr} \mathcal{V}$ is a sheaf of ideals in $\pi_*(\mathcal{O}_{T^*(X)})$. Therefore, we can attach to it its zero set in $T^*(X)$. This variety is called the *characteristic variety* $\operatorname{Char}(\mathcal{V})$ of $\mathcal{V}$. One can show that it is independent of the choice of the good filtration of $\mathcal{V}$.

A subvariety Z of $T^*(X)$ is called *conical* if $(x, \omega) \in Z$, with $x \in X$ and $\omega \in T_x^*(X)$, implies $(x, \lambda\omega) \in Z$ for all $\lambda \in \mathbb{C}$.

4.1. LEMMA. *Let $\mathcal{V}$ be a coherent $\mathcal{D}$-module on X. Then*

(i) *The characteristic variety $\operatorname{Char}(\mathcal{V})$ is conical.*

(ii) $\pi(\operatorname{Char}(\mathcal{V})) = \operatorname{supp}(\mathcal{V})$.

The characteristic variety of a coherent $\mathcal{D}$-module cannot be "too small". More precisely, we have the following result.

4.2. THEOREM. *Let $\mathcal{V}$ be a nonzero coherent $\mathcal{D}$-module on X. Then*

$$\dim \operatorname{Char}(\mathcal{V}) \geq \dim X.$$

If $\dim \operatorname{Char}(\mathcal{V}) = \dim X$ or $\mathcal{V} = 0$, we say that $\mathcal{V}$ is a *holonomic* $\mathcal{D}$-module. Holonomic modules form an abelian subcategory of $\mathcal{M}_{coh}(\mathcal{D})$. Any holonomic $\mathcal{D}$-module is of finite length.

Modules in $\mathcal{M}_{coh}(\mathcal{D})$ which are coherent as $\mathcal{O}_X$-modules are called *connections*. Connections are locally free as $\mathcal{O}_X$-modules. Therefore, the support of a connection τ is a union of connected components of X. If $\operatorname{supp}(\tau) = X$, its characteristic variety is the zero section of $T^*(X)$; in particular τ is holonomic. On the other hand, a coherent $\mathcal{D}$-module with characteristic variety equal to the zero section of $T^*(X)$ is a connection supported on X.

Assume that $\mathcal{V}$ is a holonomic module with support equal to X. Since the characteristic variety of a holonomic module $\mathcal{V}$ is conical, and has the same dimension as X, there exists an open and dense subset U in X such that the characteristic variety of $\mathcal{V}|_U$ is the zero section of $T^*(U)$. Therefore, $\mathcal{V}|_U$ is a connection.

Now we define several functors between various categories of $\mathcal{D}$-modules.

Let $\mathcal{V}$ be a quasi-coherent $\mathcal{O}_X$-module. An endomorphism D of the sheaf of linear spaces $\mathcal{V}$ is called a *differential endomorphism* of $\mathcal{V}$ of degree $\leq n$, $n \in \mathbb{Z}_+$, if we have

$$[\ldots [[D, f_0], f_1], \ldots, f_n] = 0$$

for any $(n+1)$-tuple $(f_0, f_1, \ldots, f_n)$ of regular functions on any open set U in X.

First, let $\mathcal{L}$ be an invertible $\mathcal{O}_X$-module on X. Then $\mathcal{L} \otimes_{\mathcal{O}_X} \mathcal{D}$ has a natural structure of a right $\mathcal{D}$-module by right multiplication in the second factor. Let $\mathcal{D}^{\mathcal{L}}$ be the sheaf of differential endomorphisms of the $\mathcal{O}_X$-module $\mathcal{L} \otimes_{\mathcal{O}_X} \mathcal{D}$ (for the $\mathcal{O}_X$-module structure given by the left multiplication) which commute with the right $\mathcal{D}$-module structure. Then $\mathcal{D}^{\mathcal{L}}$ is a twisted sheaf of differential operators on X. We can define the *twist* functor from $\mathcal{M}^L_{qc}(\mathcal{D})$ into $\mathcal{M}^L_{qc}(\mathcal{D}^{\mathcal{L}})$ by

$$\mathcal{V} \longmapsto (\mathcal{L} \otimes_{\mathcal{O}_X} \mathcal{D}) \otimes_{\mathcal{D}} \mathcal{V}$$

for $\mathcal{V}$ in $\mathcal{M}^L_{qc}(\mathcal{D})$. As an $\mathcal{O}_X$-module,

$$(\mathcal{L} \otimes_{\mathcal{O}_X} \mathcal{D}) \otimes_{\mathcal{D}} \mathcal{V} = \mathcal{L} \otimes_{\mathcal{O}_X} \mathcal{V}.$$

The operation of twist is visibly an equivalence of categories. It preserves coherence of $\mathcal{D}$-modules and their characteristic varieties. Therefore, the twist preserves holonomicity.

Let $f : Y \to X$ be a morphism of smooth algebraic varieties. Put

$$\mathcal{D}_{Y \to X} = f^*(\mathcal{D}) = \mathcal{O}_Y \otimes_{f^{-1}\mathcal{O}_X} f^{-1}\mathcal{D}.$$

Then $\mathcal{D}_{Y \to X}$ is a right $f^{-1}\mathcal{D}$-module for the right multiplication in the second factor. Denote by $\mathcal{D}^f$ the sheaf of differential endomorphisms of the $\mathcal{O}_Y$-module $\mathcal{D}_{Y \to X}$ which are also $f^{-1}\mathcal{D}$-module endomorphisms. Then $\mathcal{D}^f$ is a twisted sheaf of differential operators on Y.

Let $\mathcal{V}$ be in $\mathcal{M}^L_{qc}(\mathcal{D})$. Put

$$f^+(\mathcal{V}) = \mathcal{D}_{Y \to X} \otimes_{f^{-1}\mathcal{D}} f^{-1}\mathcal{V}.$$

Then $f^+(\mathcal{V})$ is the *inverse image* of $\mathcal{V}$ (in the category of $\mathcal{D}$-modules), and f^+ is a right exact covariant functor from $\mathcal{M}^L_{qc}(\mathcal{D})$ into $\mathcal{M}^L_{qc}(\mathcal{D}^f)$. Considered as an $\mathcal{O}_Y$-module,

$$f^+(\mathcal{V}) = \mathcal{O}_Y \otimes_{f^{-1}\mathcal{O}_X} f^{-1}\mathcal{V} = f^*(\mathcal{V}),$$

where $f^*(\mathcal{V})$ is the inverse image in the category of $\mathcal{O}$-modules. The left derived functors $L^p f^+ : \mathcal{M}^L_{qc}(\mathcal{D}) \to M^L_{qc}(\mathcal{D}^f)$ of f^+ have analogous properties. One can show that derived inverse images preserve holonomicity.

Let Y be a smooth subvariety of X and $\mathcal{D}$ a twisted sheaf of differential operators on X. Then $\mathcal{D}^i$ is a twisted sheaf of differential operators on Y and $L^p i^+ : \mathcal{M}_{qc}^L(\mathcal{D}) \to \mathcal{M}_{qc}^L(\mathcal{D}^i)$ vanish for $p < -\operatorname{codim} Y$. Therefore, $i^! = L^{-\operatorname{codim} Y} i^+$ is a left exact functor.

To define the direct image functors for $\mathcal{D}$-modules one has to use derived categories. In addition, it is simpler to define them for right $\mathcal{D}$-modules. Let $D^b(\mathcal{M}_{qc}^R(\mathcal{D}^f))$ be the bounded derived category of quasi-coherent right $\mathcal{D}^f$-modules. Then we define

$$Rf_+(\mathcal{V}^{\cdot}) = Rf_*(\mathcal{V}^{\cdot} \overset{L}{\otimes}_{\mathcal{D}^f} \mathcal{D}_{Y \to X})$$

for any complex $\mathcal{V}^{\cdot}$ in $D^b(\mathcal{M}_{qc}^R(\mathcal{D}^f))$ (here we denote by Rf_* and $\overset{L}{\otimes}$ the derived functors of direct image f_* and tensor product). Let $\mathcal{V}^{\cdot}$ be the complex in $D^b(\mathcal{M}_{qc}^R(\mathcal{D}^f))$ which is zero in all degrees except 0, where it is equal to a quasi-coherent right $\mathcal{D}^f$-module $\mathcal{V}$. Then we put

$$R^p f_+(\mathcal{V}) = H^p(Rf_+(\mathcal{V}^{\cdot})) \text{ for } p \in \mathbb{Z},$$

i.e., we get a family $R^p f_+$, $p \in \mathbb{Z}$, of functors from $\mathcal{M}_{qc}^R(\mathcal{D}^f)$ into $\mathcal{M}_{qc}^R(\mathcal{D})$. We call $R^p f_+$ the pth *direct image* functor. Direct image functors also preserve holonomicity.

If $i : Y \to X$ is an immersion, $\mathcal{D}_{Y \to X}$ is a locally free $\mathcal{D}^i$-module. This implies that

$$R^p i_+(\mathcal{V}) = R^p i_*(\mathcal{V} \otimes_{\mathcal{D}^i} \mathcal{D}_{Y \to X})$$

for $\mathcal{V}$ in $\mathcal{M}_{qc}^R(\mathcal{D}^i)$. Therefore, $i_+ = R^0 i_+$ is left exact and $R^p i_+$ are its right derived functors. In addition, if Y is a closed in X, the functor $i_+ : \mathcal{M}_{qc}^R(\mathcal{D}^i) \to \mathcal{M}_{qc}^R(\mathcal{D})$ is exact.

Let $i : Y \to X$ be a closed immersion. The support of $i_+(\mathcal{V})$ is equal to the support of $\mathcal{V}$ considered as a subset of $Y \subset X$.

4.3. THEOREM (KASHIWARA'S EQUIVALENCE OF CATEGORIES). *Let $i : Y \to X$ be a closed immersion. Then the direct image functor i_+ is an equivalence of $\mathcal{M}_{qc}^R(\mathcal{D}^i)$ with the full subcategory of $\mathcal{M}_{qc}^R(\mathcal{D})$ consisting of modules with support in Y.*

This equivalence preserves coherence and holonomicity.

The inverse functor is given by $i^!$ (up to a twist caused by our use of right $\mathcal{D}$-modules in the discussion of i_+).

5. K-orbits in the flag variety

In this section we study K-orbits in the flag variety X in more detail. As before, let σ be the Cartan involution of $\mathfrak{g}$ such that $\mathfrak{k}$ is its fixed point set.

We first establish that the number of K-orbits in X is finite.

5.1. PROPOSITION. *The group K acts on X with finitely many orbits.*

To prove this result we can assume that $G = \mathrm{Int}(\mathfrak{g})$. Also, by abuse of notation, denote by σ the involution of G with differential equal to the Cartan involution σ. The key step in the proof is the following lemma. First, define an action of G on $X \times X$ by

$$g(x, y) = (gx, \sigma(g)y)$$

for any $g \in G$, $x, y \in X$.

5.2. LEMMA. *The group G acts on $X \times X$ with finitely many orbits.*

PROOF. We fix a point $v \in X$. Let B_v be the Borel subgroup of G corresponding to v, and put $B = \sigma(B_v)$. Every G-orbit in $X \times X$ intersects $X \times \{v\}$. Let $u \in X$. Then the intersection of the G-orbit Q through (u, v) with $X \times \{v\}$ is equal to $Bu \times \{v\}$. By the Bruhat decomposition, this implies the finiteness of the number of G-orbits in $X \times X$. □

Now we show that 5.1 is a consequence of 5.2. Let Δ be the diagonal in $X \times X$. By 5.2, the orbit stratification of $X \times X$ induces a stratification of Δ by finitely many irreducible subvarieties which are the irreducible components of the intersections of the G-orbits with Δ. These strata are K-invariant, and therefore unions of K-orbits. Let V be one of these subvarieties, $(x, x) \in V$ and Q the K-orbit of (x, x). If we let $\mathfrak{b}_x$ denote the Borel subalgebra of $\mathfrak{g}$ corresponding to x, the tangent space $T_x(X)$ of X at x can be identified with $\mathfrak{g}/\mathfrak{b}_x$. Let p_x be the projection of $\mathfrak{g}$ onto $\mathfrak{g}/\mathfrak{b}_x$. The tangent space $T_{(x,x)}(X \times X)$ to $X \times X$ at (x, x) can be identified with $\mathfrak{g}/\mathfrak{b}_x \times \mathfrak{g}/\mathfrak{b}_x$. If the orbit map $f : G \to X \times X$ is defined by $f(g) = g(x, x)$, its differential at the identity in G is the linear map $\xi \to (p_x(\xi), p_x(\sigma(\xi)))$ of $\mathfrak{g}$ into $\mathfrak{g}/\mathfrak{b}_x \times \mathfrak{g}/\mathfrak{b}_x$. Then the tangent space to V at (x, x) is contained in the intersection of the image of this differential with the diagonal in the tangent space $T_{(x,x)}(X \times X)$, i.e.

$$T_{(x,x)}(V) \subset \{(p_x(\xi), p_x(\xi)) \mid \xi \in \mathfrak{g} \text{ such that } p_x(\xi) = p_x(\sigma(\xi))\}$$
$$= \{(p_x(\xi), p_x(\xi)) \mid \xi \in \mathfrak{k}\} = T_{(x,x)}(Q).$$

Consequently the tangent space to V at (x, x) agrees with the tangent space to Q, and Q is open in V. By the irreducibility of V, this implies that V is a K-orbit, and therefore our stratification of the diagonal Δ is the stratification induced via the diagonal map by the K-orbit stratification of X. Hence, 5.1 follows.

5.3. LEMMA. *Let $\mathfrak{b}$ be a Borel subalgebra of $\mathfrak{g}$, $\mathfrak{n} = [\mathfrak{b}, \mathfrak{b}]$ and N the connected subgroup of G determined by $\mathfrak{n}$. Then:*

(i) *$\mathfrak{b}$ contains a σ-stable Cartan subalgebra $\mathfrak{c}$.*

(ii) *any two such Cartan subalgebras are $K \cap N$-conjugate.*

PROOF. Clearly, $\sigma(\mathfrak{b})$ is another Borel subalgebra of $\mathfrak{g}$. Therefore, $\mathfrak{b} \cap \sigma(\mathfrak{b})$ contains a Cartan subalgebra $\mathfrak{d}$ of $\mathfrak{g}$. Now, $\sigma(\mathfrak{d})$ is also a Cartan subalgebra of $\mathfrak{g}$ and both $\mathfrak{d}$ and $\sigma(\mathfrak{d})$ are Cartan subalgebras of $\mathfrak{b} \cap \sigma(\mathfrak{b})$. Hence, they are conjugate by $n = \exp(\xi)$ with $\xi \in [\mathfrak{b} \cap \sigma(\mathfrak{b}), \mathfrak{b} \cap \sigma(\mathfrak{b})] \subset \mathfrak{n} \cap \sigma(\mathfrak{n})$. By applying σ to $\sigma(\mathfrak{d}) = \operatorname{Ad}(n)\mathfrak{d}$, we get $\mathfrak{d} = \operatorname{Ad}(\sigma(n))\sigma(\mathfrak{d})$. It follows that

$$\mathfrak{d} = \operatorname{Ad}(\sigma(n)) \operatorname{Ad}(n)\mathfrak{d} = \operatorname{Ad}(\sigma(n)n)\mathfrak{d}.$$

This implies that the element $\sigma(n)n \in N \cap \sigma(N)$ normalizes $\mathfrak{d}$. Hence, it is equal to 1, i.e. $\sigma(n) = n^{-1}$. Then

$$\exp(\sigma(\xi)) = \sigma(n) = n^{-1} = \exp(-\xi).$$

Since the exponential map on $\mathfrak{n} \cap \sigma(\mathfrak{n})$ is injective, we conclude that $\sigma(\xi) = -\xi$. Hence, the element

$$n^{\frac{1}{2}} = \exp\left(\tfrac{1}{2}\xi\right)$$

satisfies

$$\sigma(n^{\frac{1}{2}}) = \sigma\left(\exp\left(\tfrac{1}{2}\xi\right)\right) = \exp\left(\sigma\left(\tfrac{1}{2}\xi\right)\right) = \exp\left(-\tfrac{1}{2}\xi\right) = (n^{\frac{1}{2}})^{-1}.$$

Put $\mathfrak{c} = \operatorname{Ad}(n^{\frac{1}{2}})\mathfrak{d}$. Then $\mathfrak{c} \subset \mathfrak{b}$ and

$$\sigma(\mathfrak{c}) = \sigma(\operatorname{Ad}(n^{\frac{1}{2}})\mathfrak{d}) = \operatorname{Ad}(\sigma(n^{\frac{1}{2}}))\sigma(\mathfrak{d}) = \operatorname{Ad}((n^{\frac{1}{2}})^{-1}) \operatorname{Ad}(n)\mathfrak{d} = \operatorname{Ad}(n^{\frac{1}{2}})\mathfrak{d} = \mathfrak{c}$$

and $\mathfrak{c}$ is σ-stable. This proves (i).

(ii) Assume that $\mathfrak{c}$ and $\mathfrak{c}'$ are σ-stable Cartan subalgebras of $\mathfrak{g}$ and $\mathfrak{c} \subset \mathfrak{b}$, $\mathfrak{c}' \subset \mathfrak{b}$. Then, as before, there exists $n \in N \cap \sigma(N)$ such that $\mathfrak{c}' = \operatorname{Ad}(n)\mathfrak{c}$. Therefore, by applying σ we get $\mathfrak{c}' = \operatorname{Ad}(\sigma(n))\mathfrak{c}$, and

$$\operatorname{Ad}(n^{-1}\sigma(n))\mathfrak{c} = \mathfrak{c}.$$

As before, we conclude that $n^{-1}\sigma(n) = 1$, i.e. $\sigma(n) = n$. If $n = \exp(\xi)$, $\xi \in \mathfrak{n}$, we get $\sigma(\xi) = \xi$ and $\xi \in \mathfrak{k} \cap \mathfrak{n}$. Hence, $n \in K \cap N$. $\square$

Let $\mathfrak{c}$ be a σ-stable Cartan subalgebra in $\mathfrak{g}$ and $k \in K$. Then $\operatorname{Ad}(k)(\mathfrak{c})$ is also a σ-stable Cartan subalgebra. Therefore, K acts on the set of all σ-stable Cartan subalgebras.

The preceding result implies that to every Borel subalgebra $\mathfrak{b}$ we can attach a K-conjugacy class of σ-stable Cartan subalgebras, i.e., we have a natural map from the flag variety X onto the set of K-conjugacy classes of σ-stable Cartan subalgebras. Clearly, this map is constant on K-orbits, hence to each K-orbit in X we attach a unique K-conjugacy class of σ-stable Cartan subalgebras. Since the set of K-orbits in X is finite by 5.1, this immediately implies the following classical result.

5.4. LEMMA. *The set of K-conjugacy classes of σ-stable Cartan subalgebras in $\mathfrak{g}$ is finite.*

Let Q be a K-orbit in X, x a point of Q, and $\mathfrak{c}$ a σ-stable Cartan subalgebra contained in $\mathfrak{b}_x$. Then σ induces an involution on the root system R in $\mathfrak{c}^*$. Let R^+ be the set of positive roots determined by $\mathfrak{b}_x$. The specialization map from the Cartan triple $(\mathfrak{h}^*, \Sigma, \Sigma^+)$ into the triple $(\mathfrak{c}^*, R, R^+)$ allows us to pull back σ

to an involution of Σ. From the construction, one sees that this involution on Σ depends only on the orbit Q, so we denote it by σ_Q. Let $\mathfrak{h} = \mathfrak{t}_Q \oplus \mathfrak{a}_Q$ be the decomposition of $\mathfrak{h}$ into σ_Q-eigenspaces for the eigenvalue 1 and -1. Under the specialization map this corresponds to the decomposition $\mathfrak{c} = \mathfrak{t} \oplus \mathfrak{a}$ of $\mathfrak{c}$ into σ-eigenspaces for the eigenvalue 1 and -1. We call $\mathfrak{t}$ the *toroidal part* and $\mathfrak{a}$ the *split part* of $\mathfrak{c}$. The difference $\dim \mathfrak{t} - \dim \mathfrak{a}$ is called the *signature* of $\mathfrak{c}$. Clearly, it is constant on a K-conjugacy class of σ-stable Cartan subalgebras.

We say that a σ-stable Cartan subalgebra is *maximally toroidal* (resp. *maximally split*) if its signature is maximal (resp. minimal) among all σ-stable Cartan subalgebras in $\mathfrak{g}$. It is well-known that all maximally toroidal σ-stable Cartan subalgebras and all maximally split σ-stable Cartan subalgebras are K-conjugate.

A root $\alpha \in \Sigma$ is called Q-*imaginary* if $\sigma_Q \alpha = \alpha$, Q-*real* if $\sigma_Q \alpha = -\alpha$ and Q-*complex* otherwise. This division depends on the orbit Q, hence we have

$$\Sigma_{Q,I} = Q\text{-imaginary roots},$$
$$\Sigma_{Q,\mathbb{R}} = Q\text{-real roots},$$
$$\Sigma_{Q,\mathbb{C}} = Q\text{-complex roots}.$$

Via specialization, these roots correspond to imaginary, real and complex roots in the root system R in $\mathfrak{c}^*$.

Put

$$D_+(Q) = \{\alpha \in \Sigma^+ \mid \sigma_Q \alpha \in \Sigma^+, \ \sigma_Q \alpha \neq \alpha\};$$

then $D_+(Q)$ is σ_Q-invariant and consists of Q-complex roots. Each σ_Q-orbit in $D_+(Q)$ consists of two roots, hence $d(Q) = \operatorname{Card} D_+(Q)$ is even. The complement of the set $D_+(Q)$ in the set of all positive Q-complex roots is

$$D_-(Q) = \{\alpha \in \Sigma^+ \mid -\sigma_Q \alpha \in \Sigma^+, \sigma_Q \alpha \neq -\alpha\}.$$

In addition, for an imaginary $\alpha \in R$, $\sigma \alpha = \alpha$ and the root subspace $\mathfrak{g}_\alpha$ is σ-invariant. Therefore, σ acts on it either as 1 or as -1. In the first case $\mathfrak{g}_\alpha \subset \mathfrak{k}$ and α is a *compact imaginary* root, in the second case $\mathfrak{g}_\alpha \not\subset \mathfrak{k}$ and α is a *noncompact imaginary* root. We denote by R_{CI} and R_{NI} the sets of compact, resp. noncompact, imaginary roots in R. Also, we denote the corresponding sets of roots in Σ by $\Sigma_{Q,CI}$ and $\Sigma_{Q,NI}$.

5.5. LEMMA.

(i) *The Lie algebra $\mathfrak{k}$ is the direct sum of $\mathfrak{t}$, the root subspaces $\mathfrak{g}_\alpha$ for compact imaginary roots α, and the σ-eigenspaces of $\mathfrak{g}_\alpha \oplus \mathfrak{g}_{\sigma\alpha}$ for the eigenvalue 1 for real and complex roots α.*

(ii) *The Lie algebra $\mathfrak{k} \cap \mathfrak{b}_x$ is spanned by $\mathfrak{t}$, $\mathfrak{g}_\alpha$ for positive compact imaginary roots α, and the σ-eigenspaces of $\mathfrak{g}_\alpha \oplus \mathfrak{g}_{\sigma\alpha}$ for the eigenvalue 1 for complex roots $\alpha \in R^+$ with $\sigma\alpha \in R^+$.*

5.6. LEMMA. *Let Q be a K-orbit in X. Then*

$$\dim Q = \tfrac{1}{2}(\operatorname{Card}\Sigma_{Q,CI} + \operatorname{Card}\Sigma_{Q,\mathbb{R}} + \operatorname{Card}\Sigma_{Q,\mathbb{C}} - d(Q)).$$

PROOF. The tangent space to Q at $\mathfrak{b}_x$ can be identified with $\mathfrak{k}/(\mathfrak{k}\cap\mathfrak{b}_x)$. By 5.5,

$$\dim Q = \dim\mathfrak{k} - \dim(\mathfrak{k}\cap\mathfrak{b}_x)$$
$$= \operatorname{Card}\Sigma_{Q,CI} + \tfrac{1}{2}(\operatorname{Card}\Sigma_{Q,\mathbb{R}} + \operatorname{Card}\Sigma_{Q,\mathbb{C}}) - \tfrac{1}{2}\operatorname{Card}\Sigma_{Q,CI} - \tfrac{1}{2}d(Q). \quad \square$$

By 5.6, since $D_+(Q)$ consists of at most half of all Q-complex roots, the dimension of K-orbits attached to $\mathfrak{c}$ lies between

$$\tfrac{1}{2}(\operatorname{Card}\Sigma_{Q,CI} + \operatorname{Card}\Sigma_{Q,\mathbb{R}} + \tfrac{1}{2}\operatorname{Card}\Sigma_{Q,\mathbb{C}})$$

and

$$\tfrac{1}{2}(\operatorname{Card}\Sigma_{Q,CI} + \operatorname{Card}\Sigma_{Q,\mathbb{R}} + \operatorname{Card}\Sigma_{Q,\mathbb{C}}).$$

The first, minimal, value corresponds to the orbits we call *Zuckerman orbits* attached to $\mathfrak{c}$. The second, maximal, value is attained on the K-orbits we call *Langlands orbits* attached to $\mathfrak{c}$. It can be shown that both types of orbits exist for any σ-stable Cartan subalgebra $\mathfrak{c}$. They clearly depend only on the K-conjugacy class of $\mathfrak{c}$.

Since X is connected, it has a unique open K-orbit. Its dimension is obviously $\tfrac{1}{2}\operatorname{Card}\Sigma$, hence by the preceding formulas, it corresponds to the Langlands orbit attached to the conjugacy class of σ-stable Cartan subalgebras with no noncompact imaginary roots. This immediately implies the following remark.

5.7. COROLLARY. *The open K-orbit in X is the Langlands orbit attached to the conjugacy class of maximally split σ-stable Cartan subalgebras in $\mathfrak{g}$.*

On the other hand, we have the following characterization of closed K-orbits in X.

5.8. LEMMA. *A K-orbit in the flag variety X is closed if and only if it consists of σ-stable Borel subalgebras.*

PROOF. Consider the action of G on $X \times X$ from 5.2. Let $(x, x) \in \Delta$. If B_x is the Borel subgroup which stabilizes $x \in X$, the stabilizer of (x, x) equals $B_x \cap \sigma(B_x)$. Therefore, if the Lie algebra $\mathfrak{b}_x$ of B_x is σ-stable, the stabilizer of (x, x) is B_x, and the G-orbit of (x, x) is closed. Let C be the connected component containing (x, x) of the intersection of this orbit with the diagonal Δ. Then C is closed. Via the correspondence set up in the proof of 5.1, C corresponds to the K-orbit of x under the diagonal imbedding of X in $X \times X$.

Let Q be a closed K-orbit, and $x \in Q$. Then the stabilizer of x in K is a solvable parabolic subgroup, i.e., it is a Borel subgroup of K. Therefore, by 5.5,

$$\dim Q = \tfrac{1}{2}(\dim\mathfrak{k} - \dim\mathfrak{t}) = \tfrac{1}{2}(\operatorname{Card}\Sigma_{Q,CI} + \tfrac{1}{2}(\operatorname{Card}\Sigma_{Q,\mathbb{C}} + \operatorname{Card}\Sigma_{Q,\mathbb{R}})).$$

Comparing this with 5.6, we get

$$\operatorname{Card}\Sigma_{Q,\mathbb{R}} + \operatorname{Card}\Sigma_{Q,\mathbb{C}} = 2d(Q).$$

Since $D_+(Q)$ consists of at most half of all Q-complex roots, we see that there are no Q-real roots, and all positive Q-complex root lie in $D_+(Q)$. This implies that all Borel subalgebras $\mathfrak{b}_x$, $x \in Q$, are σ-stable. $\square$

5.9. COROLLARY. *The closed K-orbits in X are the Zuckerman orbits attached to the conjugacy class of maximally toroidal Cartan subalgebras in $\mathfrak{g}$.*

5.10. THE K-ORBITS FOR $\mathrm{SL}(2,\mathbb{R})$. The simplest example corresponds to the group $\mathrm{SL}(2,\mathbb{R})$. For simplicity of the notation, we shall discuss the group $\mathrm{SU}(1,1)$ isomorphic to it. In this case $\mathfrak{g} = \mathfrak{sl}(2,\mathbb{C})$. We can identify the flag variety X of $\mathfrak{g}$ with the one-dimensional projective space $\mathbb{P}^1$. If we denote by $[x_0, x_1]$ the projective coordinates of $x \in \mathbb{P}^1$, the corresponding Borel subalgebra $\mathfrak{b}_x$ is the Lie subalgebra of $\mathfrak{sl}(2,\mathbb{C})$ which leaves the line x invariant. The Cartan involution σ is given by $\sigma(T) = JTJ$, $T \in \mathfrak{g}$, where

$$J = \begin{pmatrix} -1 & 0 \\ 0 & 1 \end{pmatrix}.$$

Then $\mathfrak{k}$ is the subalgebra of diagonal matrices in $\mathfrak{g}$, and K is the torus of diagonal matrices in $\mathrm{SL}(2,\mathbb{C})$ which stabilizes $0 = [1,0]$ and $\infty = [0,1]$. Hence, the K-orbits in $X = \mathbb{P}^1$ are $\{0\}$, $\{\infty\}$ and $\mathbb{C}^*$. There are two K-conjugacy classes of σ-stable Cartan subalgebras in $\mathfrak{g}$, the class of toroidal Cartan subalgebras and the class of split Cartan subalgebras. The K-orbits $\{0\}$, $\{\infty\}$ correspond to the toroidal class, the open K-orbit $\mathbb{C}^*$ corresponds to the split class.

5.11. THE K-ORBITS FOR $G_0 = \mathrm{SU}(2,1)$. This is a more interesting example. In this case, $\mathfrak{g} = \mathfrak{sl}(3,\mathbb{C})$. Let

$$J = \begin{pmatrix} -1 & 0 & 0 \\ 0 & -1 & 0 \\ 0 & 0 & 1 \end{pmatrix}.$$

The Cartan involution σ on $\mathfrak{g}$ is given $\sigma(T) = JTJ$, $T \in \mathfrak{g}$. The subalgebra $\mathfrak{k}$ consists of matrices

$$\begin{pmatrix} A & \begin{matrix} 0 \\ 0 \end{matrix} \\ 0 \quad 0 & -\operatorname{tr}A \end{pmatrix},$$

where A is an arbitrary 2×2 matrix. In addition, $K = \{A \in \mathrm{SL}(3,\mathbb{C}) \mid \sigma(A) = A\}$ consists of matrices

$$\begin{pmatrix} B & \begin{matrix} 0 \\ 0 \end{matrix} \\ 0 \quad 0 & (\det B)^{-1} \end{pmatrix},$$

where B is an arbitrary regular 2×2 matrix. There exist two K-conjugacy classes of σ-stable Cartan subalgebras. The conjugacy class of toroidal Cartan subalgebras is represented by the Cartan subalgebra of the diagonal matrices in

$\mathfrak{g}$. The conjugacy class of maximally split Cartan subalgebras is represented by the Cartan subalgebra of all matrices of the form

$$\begin{pmatrix} a & 0 & b \\ 0 & -2a & 0 \\ b & 0 & a \end{pmatrix}$$

where $a, b \in \mathbb{C}$ are arbitrary. The Cartan involution acts on this Cartan subalgebra by

$$\sigma \begin{pmatrix} a & 0 & b \\ 0 & -2a & 0 \\ b & 0 & a \end{pmatrix} = \begin{pmatrix} a & 0 & -b \\ 0 & -2a & 0 \\ -b & 0 & a \end{pmatrix}.$$

All roots attached to a toroidal Cartan subalgebra are imaginary. A pair of roots is compact imaginary and the remaining ones are noncompact imaginary. Hence, by 5.6 and 5.8, all K-orbits are one-dimensional and closed. Since the normalizer of such Cartan subalgebra in K induces the reflection with respect to the compact imaginary roots, the number of these K-orbits is equal to three. One of these, which we denote by C_0, corresponds to a set of simple roots consisting of two noncompact imaginary roots. The other two, C_+ and C_-, correspond to sets of simple roots containing one compact imaginary root and one noncompact imaginary root. The latter two are the "holomorphic" and "antiholomorphic" K-orbits.

If we consider a maximally split Cartan subalgebra, one pair of roots is real and the other roots are complex.

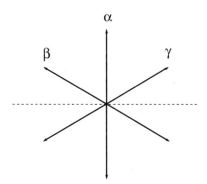

In the above figure, σ is the reflection with respect to the dotted line, the roots $\alpha, -\alpha$ are real, and the other roots are complex. By 5.6, we see that the K-orbits attached to the class of this Cartan subalgebra can have dimension equal to either 3 or 2. Since $J \in K$, the action of the Cartan involution on this Cartan subalgebra is given by an element of K, i.e., the sets of positive roots conjugate by σ determine the same orbit. Since the flag variety is three-dimensional, the open K-orbit O corresponds to the set of positive roots consisting of α, β and γ. The remaining two two-dimensional K-orbits, Q_+ and Q_-, correspond to the sets of positive roots α, β and $-\gamma$ and α, $-\beta$ and γ respectively.

Therefore, we have the following picture of the K-orbit structure in X.

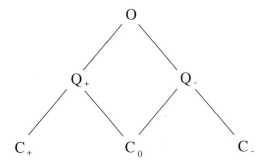

The top three K-orbits are attached to the K-conjugacy class of maximally split Cartan subalgebras, the bottom three are the closed K-orbits attached to the K-conjugacy class of toroidal Cartan subalgebras. The boundary of one K-orbit is equal to the union of all K-orbits below it connected to it by lines.

6. Standard Harish-Chandra sheaves

Now we shall apply the results from the algebraic theory of $\mathcal{D}$-modules we discussed in §4 to the study of Harish-Chandra sheaves. First we prove the following basic result.

6.1. THEOREM. *Harish-Chandra sheaves are holonomic $\mathcal{D}_\lambda$-modules. In particular, they are of finite length.*

This result is based on an analysis of characteristic varieties of Harish-Chandra modules. We start with the following observation.

6.2. LEMMA. *Any Harish-Chandra sheaf $\mathcal{V}$ has a good filtration $\mathrm{F}\mathcal{V}$ consisting of K-homogeneous coherent $\mathcal{O}_X$-modules.*

PROOF. By shifting with $\mathcal{O}(\mu)$ for sufficiently negative $\mu \in P(\Sigma)$ we can assume that λ is antidominant and regular. In this case, by the equivalence of categories, $\mathcal{V} = \mathcal{D}_\lambda \otimes_{\mathcal{U}_\theta} V$, where $V = \Gamma(X, \mathcal{V})$. Since V is an algebraic K-module and a finitely generated $\mathcal{U}_\theta$-module, there is a finite-dimensional K-invariant subspace U which generates V as a $\mathcal{U}_\theta$-module. Then $F_p\mathcal{D}_\lambda \otimes_\mathbb{C} U$, $p \in \mathbb{Z}_+$, are K-homogeneous coherent $\mathcal{O}_X$-modules. Since the natural map of $F_p\mathcal{D}_\lambda \otimes_\mathbb{C} U$ into $\mathcal{V}$ is K-equivariant, the image $\mathrm{F}_p\mathcal{V}$ is a K-homogeneous coherent $\mathcal{O}_X$-submodule of $\mathcal{V}$ for arbitrary $p \in \mathbb{Z}_+$.

We claim that $\mathrm{F}\mathcal{V}$ is a good filtration of the $\mathcal{D}_\lambda$-module $\mathcal{V}$. Clearly, this is a $\mathcal{D}_\lambda$-module filtration of $\mathcal{V}$ by K-homogeneous coherent $\mathcal{O}_X$-modules. Since $\mathcal{V}$ is generated by its global sections, to show that it is exhaustive it is enough to show that any global section v of $\mathcal{V}$ lies in $F_p\mathcal{V}$ for sufficiently large p. Since V is generated by U as a $\mathcal{U}_\theta$-module, there are $T_i \in \mathcal{U}_\theta$, $u_i \in U$, $1 \leq i \leq m$, such that $v = \sum_{i=1}^m T_i u_i$. On the other hand, there exists $p \in \mathbb{Z}_+$ such that T_i,

$1 \leq i \leq m$, are global sections of $F_p \mathcal{D}_\lambda$. This implies that $v \in F_p \mathcal{V}$. Finally, by the construction of $F \mathcal{V}$, it is evident that $F_p \mathcal{D}_\lambda F_q \mathcal{V} = F_{p+q} \mathcal{V}$ for all $p, q \in \mathbb{Z}_+$, i.e., $F \mathcal{V}$ is a good filtration. $\square$

We also need some notation. Let Y be a smooth algebraic variety and Z a smooth subvariety of Y. Then we define a smooth subvariety $N_Z(Y)$ of $T^*(Y)$ as the variety of all points $(z, \omega) \in T^*(Y)$ where $z \in Z$ and $\omega \in T_z^*(Y)$ is a linear form vanishing on $T_z(Z) \subset T_z(Y)$. We call $N_Z(Y)$ the *conormal variety* of Z in Y. The dimension of the conormal variety $N_Z(Y)$ of Z in Y is equal to $\dim Y$. To see this, we remark that the dimension of the space of all linear forms in $T_z^*(Y)$ vanishing on $T_z(Z)$ is equal to $\dim T_z(Y) - \dim T_z(Z) = \dim Y - \dim_z Z$. Hence, $\dim_z N_Z(Y) = \dim Y$.

Let $\lambda \in \mathfrak{h}^*$. Then, as we remarked before, $\mathrm{Gr}\, \mathcal{D}_\lambda = \pi_*(\mathcal{O}_{T^*(X)})$, where $\pi : T^*(X) \to X$ is the natural projection. Let $\xi \in \mathfrak{g}$. Then ξ determines a global section of $\mathcal{D}_\lambda$ of order ≤ 1, i.e. a global section of $F_1 \mathcal{D}_\lambda$. The symbol of this section is a global section of $\pi_*(\mathcal{O}_{T^*(X)})$ independent of λ. Let $x \in X$. Then the differential at $1 \in G$ of the orbit map $f_x : G \to X$, given by $f_x(g) = gx$, $g \in G$, maps the Lie algebra $\mathfrak{g}$ onto the tangent space $T_x(X)$ at x. The kernel of this map is $\mathfrak{b}_x$, i.e. the differential $T_1(f_x)$ of f_x at 1 identifies $\mathfrak{g}/\mathfrak{b}_x$ with $T_x(X)$. The symbol of the section determined by ξ is given by the function $(x, \omega) \longmapsto \omega(T_1(f_x)(\xi))$ for $x \in X$ and $\omega \in T_x^*(X)$.

Denote by $\mathcal{I}_K$ the ideal in the $\mathcal{O}_X$-algebra $\pi_*(\mathcal{O}_{T^*(X)})$ generated by the symbols of sections attached to elements of $\mathfrak{k}$. Let $\mathcal{N}_K$ be the set of zeros of this ideal in $T^*(X)$.

6.3. LEMMA. *The variety $\mathcal{N}_K$ is the union of the conormal varieties $N_Q(X)$ for all K-orbits Q in X. Its dimension is equal to $\dim X$.*

PROOF. Let $x \in X$ and denote by Q the K-orbit through x. Then,

$$\mathcal{N}_K \cap T_x^*(X) = \{\omega \in T_x^*(X) \mid \omega \text{ vanishes on } T_1(f_x)(\mathfrak{k})\}$$
$$= \{\omega \in T_x^*(X) \mid \omega \text{ vanishes on } T_x(Q)\} = N_Q(X) \cap T_x^*(X),$$

i.e. $\mathcal{N}_K$ is the union of all $N_Q(X)$.

For any K-orbit Q in X, its conormal variety $N_Q(X)$ has dimension equal to $\dim X$. Since the number of K-orbits in X is finite, $\mathcal{N}_K$ is a finite union of subvarieties of dimension $\dim X$. $\square$

Therefore, 6.1 is an immediate consequence of the following result.

6.4. PROPOSITION. *Let $\mathcal{V}$ be a Harish-Chandra sheaf. Then the characteristic variety $\mathrm{Char}(\mathcal{V})$ of $\mathcal{V}$ is a closed subvariety of $\mathcal{N}_K$.*

PROOF. By 6.2, $\mathcal{V}$ has a good filtration $F \mathcal{V}$ consisting of K-homogeneous coherent $\mathcal{O}_X$-modules. Therefore, the global sections of $\mathcal{D}_\lambda$ corresponding to $\mathfrak{k}$ map $F_p \mathcal{V}$ into itself for $p \in \mathbb{Z}$. Hence, their symbols annihilate $\mathrm{Gr}\, \mathcal{V}$ and $\mathcal{I}_K$ is contained in the annihilator of $\mathrm{Gr}\, \mathcal{V}$ in $\pi_*(\mathcal{O}_{T^*(X)})$. This implies that the characteristic variety $\mathrm{Char}(\mathcal{V})$ is a closed subvariety of $\mathcal{N}_K$. $\square$

Now we want to describe all irreducible Harish-Chandra sheaves. We start with the following remark.

6.5. LEMMA. *Let $\mathcal{V}$ be an irreducible Harish-Chandra sheaf. Then its support* $\text{supp}(\mathcal{V})$ *is the closure of a K-orbit Q in X.*

PROOF. Since K is connected, the Harish-Chandra sheaf $\mathcal{V}$ is irreducible if and only if it is irreducible as a $\mathcal{D}_\lambda$-module. To see this we may assume, by twisting with $\mathcal{O}(\mu)$ for sufficiently negative μ, that λ is antidominant and regular. In this case the statement follows from the equivalence of categories and the analogous statement for Harish-Chandra modules (which is evident).

Therefore, we know that $\text{supp}(\mathcal{V})$ is an irreducible closed subvariety of X. Since it must also be K-invariant, it is a union of K-orbits. The finiteness of K-orbits implies that there exists an orbit Q in $\text{supp}(\mathcal{V})$ such that $\dim Q = \dim \text{supp}(\mathcal{V})$. Therefore, $\bar{Q}$ is a closed irreducible subset of $\text{supp}(\mathcal{V})$ and $\dim \bar{Q} = \dim \text{supp}(\mathcal{V})$. This implies that $\bar{Q} = \text{supp}(\mathcal{V})$. $\square$

Let $\mathcal{V}$ be an irreducible Harish-Chandra sheaf and Q the K-orbit in X such that $\text{supp}(\mathcal{V}) = \bar{Q}$. Let $X' = X - \partial Q$. Then X' is an open subvariety of X and Q is a closed subvariety of X'. The restriction $\mathcal{V}|_{X'}$ of $\mathcal{V}$ to X' is again irreducible. Let $i : Q \to X$, $i' : Q \to X'$ and $j : X' \to X$ be the natural immersions. Hence, $i = j \circ i'$. Then $\mathcal{V}|_{X'}$ is an irreducible module supported in Q. Since Q is a smooth closed subvariety of X', by Kashiwara's equivalence of categories, $i'_+(\tau) = \mathcal{V}|_{X'}$ for $\tau = i^!(\mathcal{V})$. Also, τ is an irreducible $(\mathcal{D}_\lambda^i, K)$-module. Since $\mathcal{V}$ is holonomic by 6.1, τ is a holonomic $\mathcal{D}_\lambda^i$-module with the support equal to Q. This implies that there exists an open dense subset U in Q such that $\tau|_U$ is a connection. Since K acts transitively on Q, τ must be a K-homogeneous connection on Q.

Therefore, to each irreducible Harish-Chandra sheaf we attach a pair (Q, τ) consisting of a K-orbit Q and an irreducible K-homogeneous connection τ on Q such that:

(i) $\text{supp}(\mathcal{V}) = \bar{Q}$;
(ii) $i^!(\mathcal{V}) = \tau$.

We call the pair (Q, τ) *the standard data* attached to $\mathcal{V}$.

Let Q be a K-orbit in X and τ an irreducible K-homogeneous connection on Q in $\mathcal{M}_{coh}(\mathcal{D}_\lambda^i, K)$. Then, $\mathcal{I}(Q, \tau) = i_+(\tau)$ is a $(\mathcal{D}_\lambda, K)$-module. Moreover, it is holonomic and therefore coherent. Hence, $\mathcal{I}(Q, \tau)$ is a Harish-Chandra sheaf. We call it *the standard Harish-Chandra sheaf* attached to (Q, τ).

6.6. LEMMA. *Let Q be a K-orbit in X and τ an irreducible K-homogeneous connection on Q. Then the standard Harish-Chandra sheaf $\mathcal{I}(Q, \tau)$ contains a unique irreducible Harish-Chandra subsheaf.*

PROOF. Clearly,

$$\mathcal{I}(Q, \tau) = i_+(\tau) = j_+(i'_+(\tau)) = j_.(i'_+(\tau)),$$

where $j.$ is the sheaf direct image functor. Therefore, $\mathcal{I}(Q,\tau)$ contains no sections supported in ∂Q. Hence, any nonzero $\mathcal{D}_\lambda$-submodule $\mathcal{U}$ of $\mathcal{I}(Q,\tau)$ has a nonzero restriction to X'. By Kashiwara's equivalence of categories, $i'_+(\tau)$ is an irreducible $\mathcal{D}_\lambda|_{X'}$-module. Hence, $\mathcal{U}|_{X'} = \mathcal{I}(Q,\tau)|_{X'}$. Therefore, for any two nonzero $\mathcal{D}_\lambda$-submodules $\mathcal{U}$ and $\mathcal{U}'$ of $\mathcal{I}(Q,\tau)$, $\mathcal{U} \cap \mathcal{U}' \neq 0$. Since $\mathcal{I}(Q,\tau)$ is of finite length, it has a minimal $\mathcal{D}_\lambda$-submodule and by the preceding remark this module is unique. By its uniqueness it must be K-equivariant, therefore it is a Harish-Chandra subsheaf. $\square$

We denote by $\mathcal{L}(Q,\tau)$ the unique irreducible Harish-Chandra subsheaf of $\mathcal{I}(Q,\tau)$. The following result gives a classification of irreducible Harish-Chandra sheaves.

6.7. THEOREM (BEILINSON-BERNSTEIN).

 (i) An irreducible Harish-Chandra sheaf $\mathcal{V}$ with the standard data (Q,τ) is isomorphic to $\mathcal{L}(Q,\tau)$.
 (ii) Let Q and Q' be K-orbits in X, and τ and τ' irreducible K-homogeneous connections on Q and Q' respectively. Then $\mathcal{L}(Q,\tau) \cong \mathcal{L}(Q',\tau')$ if and only if $Q = Q'$ and $\tau \cong \tau'$.

PROOF. (i) Let $\mathcal{V}$ be an irreducible Harish-Chandra sheaf and (Q,τ) the corresponding standard data. Then, as we remarked before, $\mathcal{V}|X' = (i')_+(\tau)$. By the universal property of $j.$, there exists a nontrivial morphism of $\mathcal{V}$ into $\mathcal{I}(Q,\tau) = j.(i'_+(\tau))$ which extends this isomorphism. Since $\mathcal{V}$ is irreducible, the kernel of this morphism must be zero. Clearly, by 6.6, its image is equal to $\mathcal{L}(Q,\tau)$.

(ii) Since $\bar{Q} = \operatorname{supp} \mathcal{L}(Q,\tau)$, it is evident that $\mathcal{L}(Q,\tau) \cong \mathcal{L}(Q',\tau')$ implies $Q = Q'$. The rest follows from the formula $\tau = i^!(\mathcal{L}(Q,\tau))$. $\square$

From the construction it is evident that the quotient of the standard module $\mathcal{I}(Q,\tau)$ by the irreducible submodule $\mathcal{L}(Q,\tau)$ is supported in the boundary ∂Q of Q. In particular, if Q is closed, $\mathcal{I}(Q,\tau)$ is irreducible.

Let Q be a K-orbit and τ an irreducible K-homogeneous connection on Q in $\mathcal{M}_{coh}(\mathcal{D}^i_\lambda, K)$. Let $x \in Q$ and $T_x(\tau)$ be the geometric fibre of τ at x. Then $T_x(\tau)$ is finite dimensional, and the stabilizer S_x of x in K acts irreducibly in $T_x(\tau)$. The connection τ is completely determined by the representation ω of S_x in $T_x(\tau)$. Let $\mathfrak{c}$ be a σ-stable Cartan subalgebra in $\mathfrak{b}_x$. The Lie algebra $\mathfrak{s}_x = \mathfrak{k} \cap \mathfrak{b}_x$ of S_x is the semidirect product of the toroidal part $\mathfrak{t}$ of $\mathfrak{c}$ with the nilpotent radical $\mathfrak{u}_x = \mathfrak{k} \cap \mathfrak{n}_x$ of $\mathfrak{s}_x$. Let U_x be the unipotent subgroup of K corresponding to $\mathfrak{u}_x$; it is the unipotent radical of S_x. Let T be the Levi factor of S_x with Lie algebra $\mathfrak{t}$. Then S_x is the semidirect product of T with U_x. The representation ω is trivial on U_x, hence it can be viewed as a representation of the group T. The differential of the representation ω, considered as a representation of $\mathfrak{t}$, is a direct sum of a finite number of copies of the one dimensional representation defined by the restriction of the specialization of $\lambda + \rho$ to $\mathfrak{t}$. Therefore, we say that τ is compatible with $\lambda + \rho$.

If the group G_0 is linear, T is contained in a complex torus in the complexification of G_0, hence it is abelian. Therefore, in this case, ω is one-dimensional. Hence, if S_x is connected, it is completely determined by $\lambda + \rho$. Otherwise, Q can admit several K-homogeneous connections compatible with the same $\lambda + \rho$, as we can see from the following basic example.

6.8. STANDARD HARISH-CHANDRA SHEAVES FOR $\mathrm{SL}(2, \mathbb{R})$. Now we discuss the structure of standard Harish-Chandra sheaves for $\mathrm{SL}(2, \mathbb{R})$ (the more general situation of finite covers of $\mathrm{SL}(2, \mathbb{R})$ is discussed in [**13**]). In this case, as we discussed in 5.10, K has three orbits in $X = \mathbb{P}^1$, namely $\{0\}$, $\{\infty\}$ and $\mathbb{C}^*$.

The standard $\mathcal{D}_\lambda$-modules corresponding to the orbits $\{0\}$ and $\{\infty\}$ exist if and only if λ is a weight in $P(\Sigma)$. Since these orbits are closed, these standard modules are irreducible.

Therefore, it remains to study the standard modules attached to the open orbit $\mathbb{C}^*$. First we want to construct suitable trivializations of $\mathcal{D}_\lambda$ on the open cover of $\mathbb{P}^1$ consisting of $\mathbb{P}^1 - \{0\}$ and $\mathbb{P}^1 - \{\infty\}$. We denote by $\alpha \in \mathfrak{h}^*$ the positive root of $\mathfrak{g}$ and put $\rho = \frac{1}{2}\alpha$ and $t = \alpha^{\check{}}(\lambda)$, where $\alpha^{\check{}}$ is the dual root of α.

Let $\{E, F, H\}$ denote the standard basis of $\mathfrak{sl}(2, \mathbb{C})$:

$$
E = \begin{pmatrix} 0 & 1 \\ 0 & 0 \end{pmatrix} \quad F = \begin{pmatrix} 0 & 0 \\ 1 & 0 \end{pmatrix} \quad H = \begin{pmatrix} 1 & 0 \\ 0 & -1 \end{pmatrix}.
$$

They satisfy the commutation relations

$$
[H, E] = 2E \quad [H, F] = -2F \quad [E, F] = H.
$$

Also, H spans the Lie algebra $\mathfrak{k}$. Moreover, if we specialize at 0, H corresponds to the dual root $\alpha^{\check{}}$, but if we specialize at ∞, H corresponds to the negative of $\alpha^{\check{}}$.

First we discuss $\mathbb{P}^1 - \{\infty\}$. On this set we define the coordinate z by $z([1, x_1]) = x_1$. In this way one identifies $\mathbb{P}^1 - \{\infty\}$ with the complex plane $\mathbb{C}$. After a short calculation we get

$$
E = -z^2\partial - (t+1)z, \quad F = \partial, \quad H = 2z\partial + (t+1)
$$

in this coordinate system. Analogously, on $\mathbb{P}^1 - \{0\}$ with the natural coordinate $\zeta([x_0, 1]) = x_0$, we have

$$
E = \partial, \quad F = -\zeta^2\partial - (t+1)\zeta, \quad H = -2\zeta\partial - (t+1).
$$

On $\mathbb{C}^*$ these two coordinate systems are related by the inversion $\zeta = \frac{1}{z}$. This implies that $\partial_\zeta = -z^2\partial_z$, i. e., on $\mathbb{C}^*$ the second trivialization gives

$$
E = -z^2\partial, \quad F = \partial - \frac{1+t}{z} \quad H = 2z\partial - (t+1).
$$

It follows that the first and the second trivialization on $\mathbb{C}^*$ are related by the automorphism of $\mathcal{D}_{\mathbb{C}^*}$ induced by

$$
\partial \longrightarrow \partial - \frac{1+t}{z} = z^{1+t}\,\partial\,z^{-(1+t)}.
$$

Now we want to analyze the standard Harish-Chandra sheaves attached to the open K-orbit $\mathbb{C}^*$. If we identify K with another copy of $\mathbb{C}^*$, the stabilizer in K of any point in the orbit $\mathbb{C}^*$ is the group $M = \{\pm 1\}$. Let η_0 be the trivial representation of M and η_1 the identity representation of M. Denote by τ_k the irreducible K-equivariant connection on $\mathbb{C}^*$ corresponding to the representation η_k of M, and by $\mathcal{I}(\mathbb{C}^*, \tau_k)$ the corresponding standard Harish-Chandra sheaf in $\mathcal{M}_{coh}(\mathcal{D}_\lambda, K)$. To analyze these $\mathcal{D}_\lambda$-modules it is convenient to introduce a trivialization of $\mathcal{D}_\lambda$ on $\mathbb{C}^* = \mathbb{P}^1 - \{0, \infty\}$ such that H corresponds to the differential operator $2z\partial$ on the orbit $\mathbb{C}^*$ and $t \in K \cong \mathbb{C}^*$ acts on it by multiplication by t^2. We obtain this trivialization by restricting the original z-trivialization to $\mathbb{C}^*$ and twisting it by the automorphism

$$\partial \longrightarrow \partial - \frac{1+t}{2z} = z^{\frac{1+t}{2}} \partial z^{-\frac{1+t}{2}}.$$

This gives a trivialization of $\mathcal{D}_\lambda|_{\mathbb{C}^*}$ which satisfies

$$E = -z^2\partial - \frac{1+t}{2}z, \quad F = \partial - \frac{1+t}{2z}, \quad H = 2z\partial.$$

The global sections of τ_k on $\mathbb{C}^*$ form the linear space spanned by functions $z^{p+\frac{k}{2}}$, $p \in \mathbb{Z}$. To analyze irreducibility of the standard $\mathcal{D}_\lambda$-module $\mathcal{I}(\mathbb{C}^*, \tau_k)$ we have to study its behavior at 0 and ∞. By the preceding discussion, if we use the z-trivialization of $\mathcal{D}_\lambda$ on $\mathbb{C}^*$, $\mathcal{I}(\mathbb{C}^*, \tau_k)|\mathbb{P}^1 - \{\infty\}$ looks like the $\mathcal{D}_\mathbb{C}$-module which is the direct image of the $\mathcal{D}_{\mathbb{C}^*}$-module generated by $z^{\frac{k-t-1}{2}}$. This module is clearly reducible if and only if it contains functions regular at the origin, i.e., if and only if $\frac{k-t-1}{2}$ is an integer. Analogously, $\mathcal{I}(\mathbb{C}^*, \tau_k)|\mathbb{P}^1 - \{0\}$ is reducible if and only if $\frac{k+t+1}{2}$ is an integer. Therefore, $\mathcal{I}(\mathbb{C}^*, \tau_k)$ is irreducible if and only if $t + k$ is an odd integer.

We can summarize this as the *parity condition*: The following conditions are equivalent:

    (i) $\alpha^\vee(\lambda) + k \notin 2\mathbb{Z} + 1$;
    (ii) the standard module $\mathcal{I}(\mathbb{C}^*, \tau_k)$ is irreducible.

Therefore, if λ is not a weight, the standard Harish-Chandra sheaves $\mathcal{I}(\mathbb{C}^*, \tau_k)$, $k = 0, 1$, are irreducible. If λ is a weight, $\alpha^\vee(\lambda)$ is an integer, and depending on its parity, one of the standard Harish-Chandra sheaves $\mathcal{I}(\mathbb{C}^*, \tau_0)$ and $\mathcal{I}(\mathbb{C}^*, \tau_1)$ is reducible while the other one is irreducible. Assume that $\mathcal{I}(\mathbb{C}^*, \tau_k)$ is reducible. Then it contains the module $\mathcal{O}(\lambda + \rho)$ as the unique irreducible submodule and the quotient by this submodule is the direct sum of standard Harish-Chandra sheaves at $\{0\}$ and $\{\infty\}$.

Under the equivalence of categories, this describes basic results on classification of irreducible Harish-Chandra modules for $\mathrm{SL}(2, \mathbb{R})$. If $\mathrm{Re}\,\alpha^\vee(\lambda) \leq 0$ and $\lambda \neq 0$, the global sections of the standard Harish-Chandra sheaves at $\{0\}$ and $\{\infty\}$ represent the discrete series representations (holomorphic and anti-holomorphic series correspond to the opposite orbits). The global sections of the standard Harish-Chandra sheaves attached to the open orbit are the principal series representations. They are reducible if $\alpha^\vee(\lambda)$ is an integer and k is of the appropriate parity. In this case, they have irreducible finite-dimensional submodules, and their quotients by these submodules are direct sums of holomorphic and antiholomorphic discrete series. If $\lambda = 0$, the global sections of the irreducible standard Harish-Chandra sheaves attached to $\{0\}$ and $\{\infty\}$ are the limits of discrete series, the space of global sections of the irreducible standard Harish-Chandra sheaf attached to the open orbit is the irreducible principal series representation and the space of global sections of the reducible standard Harish-Chandra sheaf attached to the open orbit is the reducible principal series representation which splits into the sum of two limits of discrete series. The latter phenomenon is caused by the vanishing of global sections of $\mathcal{O}(\rho)$.

To handle the analogous phenomena in general, we have to formulate an analogous parity condition. We restrict ourselves to the case of linear group G_0 (the general case is discussed in [13]). In this case we can assume that K is a subgroup of the complexification G of G_0. Let α be a Q-real root. Denote by $\mathfrak{s}_\alpha$ the three-dimensional simple algebra generated by the root subspaces corresponding to α and $-\alpha$. Let S_α be the connected subgroup of G with Lie algebra $\mathfrak{s}_\alpha$; it is isomorphic either to $\mathrm{SL}(2,\mathbb{C})$ or to $\mathrm{PSL}(2,\mathbb{C})$. Denote by H_α the element of $\mathfrak{s}_\alpha \cap \mathfrak{c}$ such that $\alpha(H_\alpha) = 2$. Then $m_\alpha = \exp(\pi i H_\alpha) \in G$ satisfies $m_\alpha^2 = 1$. Moreover, $\sigma(m_\alpha) = \exp(-\pi i H_\alpha) = m_\alpha^{-1} = m_\alpha$. Clearly, $m_\alpha = 1$ if $S_\alpha \cong \mathrm{PSL}(2,\mathbb{C})$, and $m_\alpha \neq 1$ if $S_\alpha \cong \mathrm{SL}(2,\mathbb{C})$. In the latter case m_α corresponds to the negative of the identity matrix in $\mathrm{SL}(2,\mathbb{C})$. In both cases, m_α lies in T.

The set $D_-(Q)$ is the union of $-\sigma_Q$-orbits consisting of pairs $\{\beta, -\sigma_Q\beta\}$. Let A be a set of representatives of $-\sigma_Q$-orbits in $D_-(Q)$. Then, for an arbitrary Q-real root α, the number

$$\delta_Q(m_\alpha) = \prod_{\beta \in A} e^\beta(m_\alpha)$$

is independent of the choice of A and equal to ± 1.

Following B. Speh and D. Vogan [16][2], we say that τ satisfies the SL_2-*parity condition with respect to the Q-real root* α if the number $e^{i\pi\alpha^\vee(\lambda)}$ is not equal to $-\delta_Q(m_\alpha)\omega(m_\alpha)$. Clearly, this condition specializes to the condition (i) in 6.8.

The relation of the SL_2-parity condition with irreducibility of the standard modules can be seen from the following result. First, let

$$\Sigma_\lambda = \{\alpha \in \Sigma \mid \alpha^\vee(\lambda) \in \mathbb{Z}\}$$

be the root subsystem of Σ consisting of all roots integral with respect to λ. The following result is established in [8]. We formulate it in the case of linear group G_0, where it corresponds to the result of Speh and Vogan [16]. The discussion of the general situation can be found in [13].

---

[2]In fact, they consider the reducibility condition, while ours is the irreducibility condition.

6.9. THEOREM. *Let Q be a K-orbit in X, $\lambda \in \mathfrak{h}^*$, and τ an irreducible K-homogeneous connection on Q compatible with $\lambda + \rho$. Then the following conditions are equivalent:*

(i) *$D_-(Q) \cap \Sigma_\lambda = \emptyset$, and τ satisfies the SL_2-parity condition with respect to every Q-real root in Σ; and*

(ii) *the standard $\mathcal{D}_\lambda$-module $\mathcal{I}(Q, \tau)$ is irreducible.*

6.10. STANDARD HARISH-CHANDRA SHEAVES FOR $\mathrm{SU}(2,1)$. Consider again the case of $G_0 = \mathrm{SU}(2,1)$. In this case, the stabilizers in K of any point $x \in X$ are connected, so each K-orbit admits at most one irreducible K-homogeneous connection compatible with $\lambda + \rho$ for a given $\lambda \in \mathfrak{h}^*$. Therefore, we can denote the corresponding standard Harish-Chandra sheaf by $\mathcal{I}(Q, \lambda)$. If Q is any of the closed K-orbits, these standard Harish-Chandra sheaves exist if and only if $\lambda \in P(\Sigma)$. If Q is a nonclosed K-orbit, these standard Harish-Chandra sheaves exist if and only if $\lambda + \sigma_Q \lambda \in P(\Sigma)$.

Clearly, the standard Harish-Chandra sheaves attached to the closed orbits are always irreducible. By analyzing 6.9, we see that the standard Harish-Chandra sheaves for the other orbits are reducible if and only if λ is a weight. If Q is the open orbit O, the standard Harish-Chandra sheaf $\mathcal{I}(Q, \lambda)$ attached to $\lambda \in P(\Sigma)$ contains the homogeneous invertible $\mathcal{O}_X$-module $\mathcal{O}(\lambda + \rho)$ as its unique irreducible submodule, the standard Harish-Chandra sheaf $\mathcal{I}(C_0, \lambda)$ is its unique irreducible quotient, and the direct sum $\mathcal{L}(Q_+, \lambda) \oplus \mathcal{L}(Q_-, \lambda)$ is in the "middle" of the composition series. The standard Harish-Chandra sheaves $\mathcal{I}(Q_+, \lambda)$ and $\mathcal{I}(Q_-, \lambda)$ have unique irreducible submodules $\mathcal{L}(Q_+, \lambda)$ and $\mathcal{L}(Q_-, \lambda)$ respectively, and the quotients are

$$\mathcal{I}(Q_+, \lambda)/\mathcal{L}(Q_+, \lambda) = \mathcal{I}(C_+, \lambda) \oplus \mathcal{I}(C_0, \lambda)$$
$$\text{and } \mathcal{I}(Q_-, \lambda)/\mathcal{L}(Q_-, \lambda) = \mathcal{I}(C_-, \lambda) \oplus \mathcal{I}(C_0, \lambda).$$

7. Geometric classification of irreducible Harish-Chandra modules

In the preceding section we described the classification of all irreducible Harish-Chandra sheaves. Now, we use this classification to classify irreducible Harish-Chandra modules.

First, it is useful to use a more restrictive condition than antidominance. We say that $\lambda \in \mathfrak{h}^*$ is *strongly antidominant* if $\mathrm{Re}\, \alpha^\vee(\lambda) \leq 0$ for any $\alpha \in \Sigma^+$. Clearly, a strongly antidominant λ is antidominant.

Let V be an irreducible Harish-Chandra module. We can view V as an irreducible object in the category $\mathcal{M}(\mathcal{U}_\theta, K)$. We fix a strongly antidominant $\lambda \in \theta$. Then, as we remarked in §3, there exists a unique irreducible $\mathcal{D}_\lambda$-module $\mathcal{V}$ such that $\Gamma(X, \mathcal{V}) = V$. Since this $\mathcal{D}_\lambda$-module must be a Harish-Chandra sheaf, it is of the form $\mathcal{L}(Q, \tau)$ for some K-orbit Q in X and an irreducible K-homogeneous connection τ on Q compatible with $\lambda + \rho$. Hence, there is a unique pair (Q, τ) such that $\Gamma(X, \mathcal{L}(Q, \tau)) = V$. If λ is regular in addition,

this correspondence gives a parametrization of equivalence classes of irreducible
Harish-Chandra modules by all pairs (Q, τ). On the other hand, if λ is not regular, some of the pairs (Q, τ) correspond to irreducible Harish-Chandra sheaves
$\mathcal{L}(Q, \tau)$ with $\Gamma(X, \mathcal{L}(Q, \tau)) = 0$. Therefore, to give a precise formulation of this
classification of irreducible Harish-Chandra modules, we have to determine a nec-
essary and sufficient condition for nonvanishing of global sections of irreducible
Harish-Chandra sheaves $\mathcal{L}(Q, \tau)$.

For any root $\alpha \in \Sigma$ we have $\alpha^{\vee}(\lambda + \sigma_Q \lambda) \in \mathbb{R}$. In particular, if α is Q-
imaginary, $\alpha^{\vee}(\lambda)$ is real.

Let $\lambda \in \mathfrak{h}^*$ be strongly antidominant. Let

$$\Sigma_0 = \{\alpha \in \Sigma \mid \operatorname{Re} \alpha^{\vee}(\lambda) = 0\}.$$

Let Π be the basis in Σ corresponding to Σ^+. Put $\Sigma_0^+ = \Sigma_0 \cap \Sigma^+$ and $\Pi_0 = \Pi \cap \Sigma_0$.
Since λ is strongly antidominant, Π_0 is the basis of the root system Σ_0 determined
by the set of positive roots Σ_0^+.

Let $\Sigma_1 = \Sigma_0 \cap \sigma_Q(\Sigma_0)$; equivalently, Σ_1 is the largest root subsystem of Σ_0
invariant under σ_Q. Let

$$\Sigma_2 = \{\alpha \in \Sigma_1 \mid \alpha^{\vee}(\lambda) = 0\}.$$

This set is also σ_Q-invariant. Let $\Sigma_2^+ = \Sigma_2 \cap \Sigma^+$, and denote by Π_2 the corresponding basis of the root system Σ_2. Clearly, $\Pi_0 \cap \Sigma_2 \subset \Pi_2$, but this inclusion
is strict in general.

The next theorem gives the simple necessary and sufficient condition for
$\Gamma(X, \mathcal{L}(Q, \tau)) \neq 0$, that was alluded to before. In effect, this completes the
classification of irreducible Harish-Chandra modules. The proof can be found in
[8].

7.1. THEOREM. *Let $\lambda \in \mathfrak{h}^*$ be strongly antidominant. Let Q be a K-orbit in
X and τ an irreducible K-homogeneous connection on Q compatible with $\lambda + \rho$.
Then the following conditions are equivalent:*

(i) $\Gamma(X, \mathcal{L}(Q, \tau)) \neq 0$;

(ii) *the following conditions hold:*

   (a) *the set Π_2 contains no compact Q-imaginary roots;*

   (b) *for any positive Q-complex root α with $\alpha^{\vee}(\lambda) = 0$, the root $\sigma_Q \alpha$ is
   also positive;*

   (c) *for any Q-real α with $\alpha^{\vee}(\lambda) = 0$, τ must satisfy the SL_2-parity
   condition with respect to α.*

The proof of this result is based on the use of the intertwining functors I_w
for w in the subgroup W_0 of the Weyl group W generated by reflections with
respect to roots in Σ_0 [2], [13]. The vanishing of $\Gamma(X, \mathcal{L}(Q, \tau))$ is equivalent
with $I_w(\mathcal{L}(Q, \tau)) = 0$ for some $w \in W_0$. Let $\alpha \in \Pi_0$ and s_α the corresponding
reflection. Then, essentially by an $\mathrm{SL}(2, \mathbb{C})$-calculation, $I_{s_\alpha}(\mathcal{L}(Q, \tau)) = 0$ if and

only if a condition in (ii) fails for α, i.e., $\alpha\check{}(\lambda) = 0$ and α is either a compact Q-imaginary root, or a Q-complex root with $-\sigma_Q\alpha \in \Sigma^+$, or a Q-real root and the SL_2-parity condition for τ fails for α. Otherwise, either $\alpha\check{}(\lambda) = 0$ and $\mathcal{L}(Q,\tau)$ is a quotient of $I_{s_\alpha}(\mathcal{L}(Q,\tau))$, or $\alpha\check{}(\lambda) \neq 0$ and $I_{s_\alpha}(\mathcal{L}(Q,\tau)) = \mathcal{L}(Q',\tau')$ for some K-orbit Q' and irreducible K-homogeneous connection τ' on Q' compatible with $s_\alpha\lambda + \rho$ and $\Gamma(X, \mathcal{L}(Q,\tau)) = \Gamma(X, \mathcal{L}(Q',\tau'))$. Since intertwining functors satisfy the product formula

$$I_{w'w''} = I_{w'}I_{w''} \text{ for } w', w'' \in W \text{ such that } \ell(w'w'') = \ell(w') + \ell(w''),$$

by induction in the length of $w \in W_0$, one checks that (i) holds if and only if (ii) holds for all roots in Σ_0.

In general, there are several strongly antidominant λ in θ, and an irreducible Harish-Chandra module V correspond to different standard data (Q, τ). Still, all such K-orbits Q correspond to the same K-conjugacy class of σ-stable Cartan subalgebras [8].

8. Geometric classification versus Langlands classification

At the first glance it is not clear how the "geometric" classification in §7 relates to the other classification schemes. To see its relation to the Langlands classification, it is critical to understand the asymptotic behavior of the matrix coefficients of the irreducible Harish-Chandra modules $\Gamma(X, \mathcal{L}(Q, \tau))$. Although the asymptotic behavior of the matrix coefficients is an "analytic" invariant, its connection with the $\mathfrak{n}$-homology of Harish-Chandra modules studied by Casselman and the author in [6], [12] shows that it also has a simple, completely algebraic, interpretation. Together with the connection of the $\mathfrak{n}$-homology of $\Gamma(X, \mathcal{L}(Q, \tau))$ with the derived geometric fibres of $\mathcal{L}(Q, \tau)$ (see, for example, [9]), this establishes a precise relationship between the standard data and the asymptotics of $\Gamma(X, \mathcal{L}(Q, \tau))$ [8].

To formulate some important consequences of this relationship, for $\lambda \in \mathfrak{h}^*$ and a K-orbit Q, we introduce the following invariant:

$$\lambda_Q = \frac{1}{2}(\lambda - \sigma_Q\lambda).$$

8.1. THEOREM. *Let $\lambda \in \mathfrak{h}^*$ be strongly antidominant, Q a K-orbit in X and τ an irreducible K-homogeneous connection on Q compatible with $\lambda + \rho$ such that $V = \Gamma(X, \mathcal{L}(Q, \tau)) \neq 0$. Then:*

(i) *V is tempered if and only if $\operatorname{Re}\lambda_Q = 0$;*
(ii) *V is square-integrable if and only if $\sigma_Q = 1$ and λ is regular.*

If $\operatorname{Re}\lambda_Q = 0$, then $\operatorname{Re}\alpha\check{}(\lambda) = \operatorname{Re}(\sigma_Q\alpha)\check{}(\lambda)$. Hence, if α is Q-real, $\operatorname{Re}\alpha\check{}(\lambda) = 0$ and α is in the subset Σ_1 introduced in the preceding section. If α is in $D_-(Q)$, $\alpha, -\sigma_Q\alpha \in \Sigma^+$ and, since λ is strongly dominant, we conclude that $\operatorname{Re}\alpha\check{}(\lambda) = \operatorname{Re}(\sigma_Q\alpha)\check{}(\lambda) = 0$, i.e., α is also in Σ_1. It follows that all roots in $D_-(Q)$ and all Q-real roots are in Σ_1.

Hence, 8.1, 7.1 and 6.9 have the following consequence which was first proved by Ivan Mirković [**15**].

8.2. THEOREM. *Let* $\lambda \in \mathfrak{h}^*$ *be strongly antidominant. Let* Q *be a* K-*orbit in* X *and* τ *an irreducible* K-*homogeneous connection on* Q. *Assume that* $\mathrm{Re}\,\lambda_Q = 0$. *Then* $\Gamma(X, \mathcal{L}(Q, \tau)) \neq 0$ *implies that* $\mathcal{I}(Q, \tau)$ *is irreducible, i.e.,* $\mathcal{L}(Q, \tau) = \mathcal{I}(Q, \tau)$.

Thus 8.2 explains the simplicity of the classification of tempered irreducible Harish-Chandra modules: every tempered irreducible Harish-Chandra module is the space of global sections of an irreducible standard Harish-Chandra sheaf.

The analysis becomes especially simple in the case of square-integrable irreducible Harish-Chandra modules. By 8.1.(ii) they exist if and only if rank $\mathfrak{g} =$ rank K – this is a classical result of Harish-Chandra. If this condition is satisfied, the Weyl group orbit θ must in addition be regular and real. Since it is real, θ contains a unique strongly antidominant λ. This λ is regular and $\Gamma(X, \mathcal{L}(Q, \tau))$ is square-integrable if and only if $\sigma_Q = 1$. Therefore, all Borel subalgebras in Q are σ-stable. By 5.8, the K-orbit Q is necessarily closed. The stabilizer in K of a point in Q is a Cartan subgroup of K. Hence, an irreducible K-homogeneous connection τ compatible with $\lambda + \rho$ exists on the K-orbit Q if and only if $\lambda + \rho$ specializes to a character of this Cartan subgroup. If G_0 is linear, this means that λ is a weight in $P(\Sigma)$. The connection $\tau = \tau_{Q,\lambda}$ is completely determined by $\lambda + \rho$. Hence, the map $Q \to \Gamma(X, \mathcal{I}(Q, \tau_{Q,\lambda}))$ is a bijection between closed K-orbits in X and equivalence classes of irreducible square-integrable Harish-Chandra modules with infinitesimal character determined by θ.

By definition, the *discrete series* is the set of equivalence classes of irreducible square-integrable Harish-Chandra modules.

If we drop the regularity assumption on λ, for a closed K-orbit Q in X and an irreducible K-homogeneous connection τ compatible with $\lambda + \rho$, $\Gamma(X, \mathcal{I}(Q, \tau)) \neq 0$ if and only if there exists no compact Q-imaginary root $\alpha \in \Pi$ such that $\alpha^{\check{}}(\lambda) = 0$. These representations are tempered irreducible Harish-Chandra modules. They constitute the *limits of discrete series*.

Using the duality theorem of [**7**], one shows that the space of global sections of a standard Harish-Chandra sheaf is a standard Harish-Chandra module, as is explained in [**18**]. In particular, irreducible tempered representations are irreducible unitary principal series representations induced from limits of discrete series [**10**]. More precisely, if $\Gamma(X, \mathcal{I}(Q, \tau))$ is not a limit of discrete series, we have $\mathfrak{a}_Q \neq \{0\}$. Then $\mathfrak{a}_Q$ determines a parabolic subgroup in G_0. The standard data (Q, τ) determine, by "restriction", the standard data of a limit of discrete series representation of its Levi factor. The module $\Gamma(X, \mathcal{I}(Q, \tau))$ is the irreducible unitary principal series representation induced from the limits of discrete series representation attached to these "restricted" data. If the standard Harish-Chandra sheaf $\mathcal{I}(Q, \tau)$ with $\mathrm{Re}\,\lambda_Q = 0$ is reducible, its space of global sections represents a reducible unitary principal series representation induced from a limits of discrete series representation. These reducible standard Harish-Chandra sheaves can be analyzed in more detail. This leads to a $\mathcal{D}$-module theoretic explanation of the results of Knapp and Zuckerman on the reducibility of unitary principal series representations [**10**]. This analysis has been done by Ivan Mirković in [**15**].

It remains to discuss nontempered irreducible Harish-Chandra modules, i.e., the Langlands representations. In this case $\mathrm{Re}\,\lambda_Q \neq 0$ and it defines a nonzero linear form on $\mathfrak{a}_Q$. This form determines a parabolic subgroup of G_0 such that the roots of its Levi factor are orthogonal to the specialization of $\mathrm{Re}\,\lambda_Q$. The "restriction" of the standard data (Q, τ) to this Levi factor determines tempered standard data. The module $\Gamma(X, \mathcal{L}(Q, \tau))$ is equal to the unique irreducible submodule of the principal series representation $\Gamma(X, \mathcal{I}(Q, \tau))$ corresponding to this parabolic subgroup, and induced from the tempered representation of the Levi factor attached to the "restricted" standard data. By definition, this unique irreducible submodule is a Langlands representation. A detailed analysis of this construction leads to a completely algebraic proof of the Langlands classification [**8**].

In the following we analyze in detail the case of $\mathrm{SU}(2, 1)$. In this case the K-orbit structure and the structure of standard Harish-Chandra sheaves are rather simple. Still, all situations from 7.1.(ii) appear there.

8.3. Discrete series of $\mathrm{SU}(2, 1)$. If G_0 is $\mathrm{SU}(2, 1)$, we see that the discrete series are attached to all regular weights λ in the negative chamber. Therefore, we have the following picture:

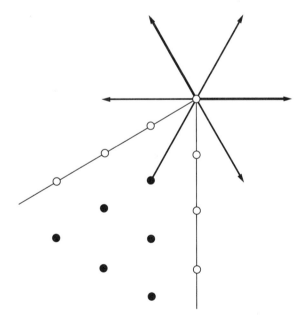

The black dots correspond to weights λ to which a discrete series representation is attached for a particular orbit. If the orbit in question is C_0, these are the "non-holomorphic" discrete series and the white dots in the walls correspond to the limits of discrete series. If the orbit is either C_+ or C_-, these are either "holomorphic" or "anti-holomorphic" discrete series. Since one of the simple roots is compact imaginary in these cases, the standard Harish-Chandra sheaves corresponding to the white dots in the wall orthogonal to this root have no global sections. The white dots in the other wall are again the limits of discrete series.

8.4. TEMPERED REPRESENTATIONS OF SU(2, 1). Except the discrete series and the limits of discrete series we already discussed, the other irreducible Harish-Chandra modules are attached to the open orbit O and the two-dimensional orbits Q_+ and Q_-. The picture for the open orbit is:

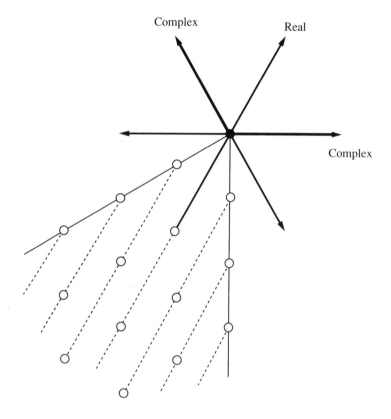

As we discussed in 6.10, the standard Harish-Chandra sheaves $\mathcal{I}(O, \lambda)$ on the open orbit O exist (in the negative chamber) only for $\operatorname{Re} \lambda$ on the dotted lines. As we remarked, $\mathcal{I}(O, \lambda)$ are reducible if and only if λ is a weight (i.e. one of the dots in the picture). At these points, $\mathcal{I}(O, \lambda)$ have the invertible $\mathcal{O}_X$-modules $\mathcal{O}(\lambda + \rho)$ as the unique irreducible submodules, i.e., $\mathcal{L}(O, \lambda) = \mathcal{O}(\lambda + \rho)$. The length of these standard Harish-Chandra sheaves is equal to 4. Their composition series consist of the irreducible Harish-Chandra sheaves attached to K-orbits O, Q_+, Q_- and C_0. The standard Harish-Chandra sheaf corresponding to C_0 is the unique irreducible quotient of $\mathcal{I}(O,\lambda)$ and $\mathcal{I}(O,\lambda)/\mathcal{O}(\lambda+\rho)$ contains the direct sum of $\mathcal{L}(Q_+,\lambda)$ and $\mathcal{L}(Q_-,\lambda)$ as a submodule.

The only tempered modules can be obtained for $\operatorname{Re}\lambda_O = 0$, which in this situation corresponds to $\operatorname{Re}\lambda = 0$. Since $\lambda = 0$ corresponds to the invertible $\mathcal{O}_X$-module $\mathcal{O}(\rho)$ with no cohomology, we see that the only irreducible tempered Harish-Chandra modules in this case correspond to $\operatorname{Re}\lambda = 0$, $\lambda \neq 0$. These representations are irreducible unitary spherical principal series.

It remains to study the case of tempered Harish-Chandra modules attached to the orbits Q_+ and Q_-. The picture in these cases is:

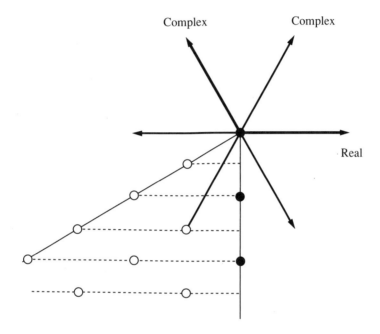

Assume that we are looking at the picture for Q_+. Again, the standard Harish-Chandra sheaves on this orbit (in the negative chamber) exist only for $\operatorname{Re}\lambda$ on the dotted lines. The standard Harish-Chandra sheaves $\mathcal{I}(Q_+,\lambda)$ are reducible if and only if λ is a weight (i.e., one of the dots in the picture). In this case $\mathcal{I}(Q_+,\lambda)$ has length 3, and the quotient $\mathcal{I}(Q_+,\lambda)/\mathcal{L}(Q_+,\lambda)$ is the direct sum of the standard modules on C_0 and C_+. The temperedness condition is satisfied for $\operatorname{Re}\lambda$ in the wall corresponding to the real root. The corresponding standard Harish-Chandra sheaves are irreducible, except in the case of λ being one of the black dots. Their global sections are various irreducible unitary principal series. The standard modules at the black dots correspond to the reducible unitary principal series. Since in these cases the global sections of $\mathcal{L}(Q_+,\lambda)$ vanish, the global sections of $\mathcal{I}(Q_+,\lambda)$, for $\lambda \neq 0$, are direct sums of "non-holomorphic" and "holomorphic" limits of discrete series representations. If $\lambda = 0$, the "holomorphic" limit of discrete series also "disappears," hence the spherical unitary principal series representation is irreducible and equal to the "non-holomorphic" limit of discrete series.

8.5. LANGLANDS REPRESENTATIONS OF SU(2,1). As we already remarked, Langlands representations are attached only to non-closed K-orbits. They are either irreducible non-unitary principal series representations, or unique irreducible submodules of reducible principal series. In the latter case, for the open K-orbit, the Langlands representations are irreducible finite-dimensional representations by the Borel-Weil theorem.

Putting all of this information together we can describe the structure of principal series representations for SU(2,1). Let P be the minimal parabolic subgroup of SU(2,1) and $P = MAN$ its Langlands decomposition. Then the group MA is a connected maximally split Cartan subgroup in G_0. Therefore the principal series representations are parametrized by pairs (δ, μ) where δ is a representation of the circle group M and μ is a linear form on the complexified Lie algebra $\mathfrak{a}$ of A. The Lie algebra $\mathfrak{a}$ is spanned by the dual root $\alpha^{\check{}}$ of a real root α. Because of the duality between principal series, it is enough to describe their structure for $\operatorname{Re}\alpha^{\check{}}(\mu) \leq 0$. These parameters correspond to the dotted lines in our next figure.

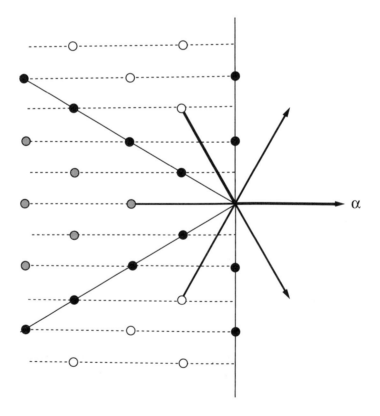

This picture is a "union" of the pictures for the orbits O, Q_+ and Q_-. A detailed explanation of this phenomenon in general can be found in [**18**]. The principal series are generically irreducible. The length of their composition series is two at the black dots, three at the white dots and four at the gray dots. The representations on the intersection of the dotted lines with the vertical wall are unitary principal series. They are either irreducible or sums of two limits of the discrete series, one "holomorphic" and one "non-holomorphic". The spherical unitary principal series at the origin is actually equal to a limit of "non-holomorphic" discrete series. The white dots correspond to the representations which contain infinite dimensional Langlands representations as unique irreducible submodules and direct sums of two discrete series, one "holomorphic" and one "non-holomorphic", as quotients. At two non-vertical walls the composition series consists of an infinite dimensional Langlands representation as a submodule and a limit of "non-holomorphic" discrete series as a quotient (since the limit of "holomorphic" discrete series "vanishes" in these walls). The gray dots correspond to representations which contain finite-dimensional representations as unique irreducible submodules, the "non-holomorphic" discrete series as unique irreducible quotients, and the direct sums of the infinite dimensional Langlands representations attached to Q_+ and Q_- in the "middle" of the composition series.

REFERENCES

1. A. Beilinson, J. Bernstein, *Localisation de $\mathfrak{g}$-modules*, C. R. Acad. Sci. Paris, Ser. I **292** (1981), 15–18.

2. A. Beilinson, J. Bernstein, *A generalization of Casselman's submodule theorem*, Representation Theory of Reductive Groups (Peter C. Trombi, ed.), Progress in Mathematics, vol. 40, Birkhäuser, Boston, 1983, pp. 35–52.

3. A. Borel, *Représentations linéaires et espaces homogénes kähleriens des groupes simples compacts*, (inédit, mars 1954), Armand Borel Œuvres Collected Papers, Vol. I, Springer Verlag, Berlin–Heidelberg–New York–Tokyo, 1983, pp. 392–396.

4. A. Borel et al., *Algebraic $\mathcal{D}$-modules*, Academic Press, Boston, 1987.

5. R. Bott, *Homogeneous vector bundles*, Annals of Math. **66** (1957), 203–248.

6. W. Casselman, D. Miličić, *Asymptotic behavior of matrix coefficients of admissible representations*, Duke Math. Journal **49** (1982), 869–930.

7. H. Hecht, D. Miličić, W. Schmid, J. A. Wolf, *Localization and standard modules for real semisimple Lie groups I: The duality theorem*, Inventiones Math. **90** (1987), 297–332.

8. H. Hecht, D. Miličić, W. Schmid, J. A. Wolf, *Localization and standard modules for real semisimple Lie groups II: Irreducibility, vanishing theorems and classification*, (in preparation).

9. H. Hecht, D. Miličić, *On the cohomological dimension of the localization functor*, Proc. Amer. Math. Soc. **108** (1990), 249-254.

10. A. W. Knapp, G. J. Zuckerman, *Classification of irreducible tempered representations of semisimple groups*, Part I, Annals of Math. **116** (1982), 389–455; Part II, Annals of Math. **116** (1982), 457–501.

11. R. P. Langlands, *On the classification of irreducible representations of real algebraic groups*, Representation theory and harmonic analysis on semisimple Lie groups, Math. Surveys No. 31, Amer. Math. Soc., Providence, R.I., 1989, pp. 101–170.

12. D. Miličić, *Asymptotic behavior of matrix coefficients of discrete series*, Duke Math. Journal **44** (1977), 59–88.

13. D. Miličić, *Intertwining functors and irreducibility of standard Harish-Chandra sheaves*, Harmonic Analysis on Reductive Groups (William Barker, Paul Sally, ed.), Progress in Mathematics, vol. 101, Birkhäuser, Boston, 1991, pp. 209–222.
14. D. Miličić, *Localization and representation theory of reductive Lie groups*, (mimeographed notes), to appear.
15. I. Mirković, *Classification of irreducible tempered representations of semisimple Lie groups*, Ph. D. Thesis, University of Utah, 1986.
16. B. Speh, D. Vogan, *Reducibility of generalized principal series representations*, Acta Math. **145** (1980), 227–299.
17. W. Schmid, *Recent developments in representation theory*, Proceedings of Arbeitstagung, Bonn 1984, Lecture Notes in Math., vol. 1111, Springer-Verlag, Berlin–Heidelberg–New York–Tokyo, 1985, pp. 135–153.
18. W. Schmid, *Construction and classification of irreducible Harish-Chandra modules*, Harmonic Analysis on Reductive Groups (William Barker, Paul Sally, ed.), Progress in Mathematics, vol. 101, Birkhäuser, Boston, 1991, pp. 235–275.
19. D. Vogan, *Representations of real reductive Lie groups*, Progress in Mathematics, vol. 15, Birkhäuser, Boston, 1981.

DEPARTMENT OF MATHEMATICS, UNIVERSITY OF UTAH, SALT LAKE CITY, UTAH 84112

E-mail address: milicic@math.utah.edu

Contemporary Mathematics
Volume **154**, 1993

Parabolic Invariant Theory and Geometry

TOBY N. BAILEY

1. Introduction

This is a report on some recent joint work with Michael Eastwood and Robin Graham [**BEG**], which is based on some results of Rod Gover [**Go**]. The problems we address concern the invariant theory of certain modules for parabolic subgroups of semisimple Lie groups and they have important geometrical applications. I will outline the way in which these problems arise geometrically (a more complete discussion can be found in [**Gr**] and [**BEG**]) and then sketch our methods for one particular simple example. I hope this will give an idea of what the subject is about and of the methods we use.

Problems of this sort were first considered by Fefferman [**F**]. He was studying the local invariants of C-R structures and he found a geometric way of writing down many such invariants in terms of an auxiliary indefinite Ricci-flat Kähler metric, which is defined only up to a certain "order". He reduced the question of whether, up to this "order", all C-R invariants are obtained in this way to the algebraic question of determining whether all invariants of a module for a parabolic subgroup of $SU(n + 1, 1)$ are obtained by a certain construction. Fefferman achieved partial results on this algebraic problem by means of some difficult arguments.

In [**BEG**] we give a complete answer (in the affirmative) to Fefferman's algebraic question. A particular application that Fefferman had in mind was to the asymptotic expansion of the Bergman kernel near the boundary of a pseudoconvex domain in $\mathbb{C}^n$—the connection being that the coefficients of the expansion are invariants of the C-R structure of the boundary, and thus a list of all such invariants provides information on the asymptotics. These results are summarized in [**BEG**].

1991 *Mathematics Subject Classification.* Primary 53A55; Secondary 53A30, 32C16.

A full account of this work will appear elsewhere.

We also give solutions to a similar problem which Fefferman posed (and almost completely solved) in [**F**], and to a related problem which has applications in conformal differential geometry. The methods we use for all these problems are very similar.

2. Invariant Theory of the Orthogonal Group

Let us recall a little of Weyl's invariant theory of the orthogonal groups [**W**]. Suppose we are given $\mathbb{R}^n$ equipped with a positive definite bilinear form g and corresponding volume form $\epsilon \in \Lambda^n \mathbb{R}^n$. Let $\mathcal{U}$ be a space whose elements consist of collections of tensors $(u^{(0)}, \dots, u^{(m)})$ on $\mathbb{R}^n$ with given ranks and, perhaps, specified symmetries. Since g gives an isomorphism of $\mathbb{R}^n$ with $\mathbb{R}^{n*}$, we lose no generality by considering only the case where $u^{(k)} \in \otimes^{l_k} \mathbb{R}^{n*}$ for some l_k. We will write such tensors with their components as subscripts. An invariant is a polynomial on $\mathcal{U}$ (i.e. a polynomial in the components of the $u^{(k)}$) which is invariant under the orthogonal group $\mathrm{SO}(g)$.

There are two simplifications to make, which apply to all our considerations of invariants henceforth. Firstly, any invariant can be written as a sum of invariants each of which is a homogeneous polynomial. Thus we lose no generality in considering henceforth only invariants that are homogeneous polynomials of some *degree* (which we will usually denote by d). Secondly, any invariant can be uniquely decomposed as a sum of an *even* and an *odd* invariant, where an invariant is even if it is unchanged under orientation reversal and odd if it changes sign. We consider the two cases separately.

Write $g^{-1} \in \otimes^2 \mathbb{R}^n$ for the inverse metric (with components g^{ab}). The generalization to spaces of tensors of what, in the case of vectors, is called the First Main Theorem of invariant theory says that every even invariant is a linear combination of complete contractions of the form

$$(1) \qquad \mathrm{contr}(g^{-1} \otimes \cdots \otimes g^{-1} \otimes u^{(k_1)} \otimes \cdots \otimes u^{(k_d)})$$

and every odd invariant is a linear combination of complete contractions of the form

$$\mathrm{contr}(\epsilon \otimes g^{-1} \otimes \cdots \otimes g^{-1} \otimes u^{(k_1)} \otimes \cdots \otimes u^{(k_d)}).$$

Here "contr" means some complete contraction of the tensor—i.e. each subscript index is paired with a superscript one and a trace taken over every such pair so as to produce a scalar. To be able to do this do this, the tensor in the parentheses must of course have the same number of subscript as superscript indices. An example of an even invariant of three tensors of ranks $1, 2$ and 3 is

$$g^{ab} g^{cd} g^{ef} g^{gh} (3 u_a^{(1)} u_{bc}^{(2)} u_{eh}^{(2)} u_{fdg}^{(3)} - 7 u_d^{(1)} u_{bf}^{(2)} u_{gh}^{(2)} u_{ace}^{(3)}).$$

The tensors of which we wish to form invariants will always have subscript indices, and the inverse metric and volume form with which we contract will always have superscript indices. We take this as a convention henceforth.

Note for later use that if all the $u^{(k)}$ are *symmetric* tensors, then a non-zero odd invariant necessarily has degree at least n, because if $d < n$, in each complete contraction the volume form must have two indices contracted into the same tensor.

Although one can write down complete contractions involving more than one volume form (and indeed they are invariants), it is unnecessary to include such things because the tensor product of two volume forms can be written in terms of g^{-1}.

3. Parabolic Geometry

We call a geometric structure *parabolic* if its flat model is a homogeneous space G/P, where $P \subset G$ is a parabolic subgroup of a semisimple Lie group, and G acts as the group of automorphisms of the geometry. Examples include conformal structures, C-R structures and projective structures. By contrast, Riemannian geometry is not parabolic—its flat model is Euclidean space $\mathbb{E}^n = E(n)/\mathrm{SO}(n)$, where $E(n)$ denotes the group of identity connected Euclidean motions.

Given a geometric structure, one is often interested in the problem of finding all "invariants" of some type. Two examples from Riemannian geometry will serve to illustrate the sort of things we have in mind. First, let us consider the flat model and ask what are all the differential invariants of functions on $\mathbb{E}^n$—where by a differential invariant we mean a scalar valued $E(n)$-equivariant differential operator. The answer to our question is well known—write g^{-1} for the inverse Euclidean metric, ϵ for the volume form and $\nabla^k f$ for the symmetric k-tensor which is the k-fold derivative of a function f (with the convention that $\nabla^0 f = f$). Then every such even invariant is a linear combination of complete contractions of the form

$$\mathrm{contr}(g^{-1} \otimes \cdots \otimes g^{-1} \otimes \nabla^{k_1} f \otimes \cdots \otimes \nabla^{k_d} f)$$

and every such odd invariant is a linear combination of complete contractions of the form

$$\mathrm{contr}(\epsilon \otimes g^{-1} \otimes \cdots \otimes g^{-1} \otimes \nabla^{k_1} f \otimes \cdots \otimes \nabla^{k_d} f).$$

An example of an even complete contraction (and hence of an even invariant) is

$$g^{ab} g^{cd} g^{ij} g^{kl} (\nabla_a \nabla_c \nabla_i \nabla_j f)(\nabla_b \nabla_k f)(\nabla_d \nabla_l f).$$

As our second example, consider the problem of writing down all local invariants of Riemannian structures. To be more precise, by a "local invariant" we mean a polynomial in the components g_{ij} of a Riemannian metric, its coordinate derivatives $\partial_a \ldots \partial_d g_{ij}$, and $(\det(g_{ij}))^{-1}$ at a point, which is invariant under change of coordinates. The answer to this problem is also well known. If we write g^{-1} for the inverse metric, ϵ for the volume form and $\nabla^k R$ for the k-fold covariant derivative of the completely covariant Riemann curvature tensor (i.e. we "lower an index" on the usual curvature tensor with g) then every even invariant is a linear combination of complete contractions of the form

$$\text{contr}(g^{-1} \otimes \cdots \otimes g^{-1} \otimes \nabla^{k_1} R \otimes \cdots \otimes \nabla^{k_d} R),$$

and every odd one is a linear combination of complete contractions of the form

$$\text{contr}(\epsilon \otimes g^{-1} \otimes \cdots \otimes g^{-1} \otimes \nabla^{k_1} R \otimes \cdots \otimes \nabla^{k_d} R).$$

Atiyah, Bott and Patodi give a proof of this in their paper on the heat equation and the index theorem [**ABP**]. They need the result because the coefficients in the asymptotic expansion of the heat kernel are necessarily such invariants. In proving both the above results, one first reduces the problem to that of listing the invariants of a collection of tensors on $\mathbb{R}^n$ under the orthogonal group (in the first case the set of all the $\nabla^k f$, in the second the set of all the $\nabla^k R$) and then uses the "First Main Theorem" of §2.

Let us turn now to parabolic geometries, where the corresponding problems are much harder because they reduce to the invariant theory of the parabolic P. We will discuss a parabolic analogue of the first example problem above in the main body of this article. An analogue of the second problem is that of listing all conformal invariants of a conformal structure. A conformal structure is a Riemannian metric known only up to multiplication by nowhere vanishing smooth functions. A *conformal invariant* is a local invariant I of Riemannian structures which under the "conformal rescaling" $g \mapsto \Omega^2 g$ simply transforms according to $I \mapsto \Omega^q I$ for some q. Another of the parabolic invariant theory problems that are solved in [**BEG**] leads to a list of all conformal invariants in odd dimension. The reduction of the geometrical problem to the algebraic was achieved by Fefferman and Graham [**FG2**] and uses their "ambient metric construction" [**FG1**]. The reason that we only achieve strong results in odd dimension is that, as with several problems of this sort, the relation between the "geometric" problem and the "algebraic" is obstructed at some finite order in the even case.

4. An Example Problem

The problem we will consider as an example is essentially the analogue for conformal geometry of the "differential invariants of a function" problem of the previous section. I will start by taking a geometric problem and translating it into an algebraic one. This is doing things backwards from a historical point of view, since this is a case of Fefferman's "model problem" and it was only given a geometric interpretation by Eastwood and Graham [**EG**].

Let

$$(2) \qquad\qquad X^I = \begin{pmatrix} X^0 \\ X^i \\ X^\infty \end{pmatrix}, \; i = 1, \ldots, n$$

be coordinates on $\mathbb{R}^{n+2}$ equipped with the indefinite quadratic form $\tilde{g}$ given in coordinates by

$$(3) \qquad g_{IJ}U^I U^J = g_{ij}U^i U^j + 2U^0 U^\infty,$$

where (g_{ij}) is the usual positive definite quadratic form on $\mathbb{R}^n$ and we use the summation convention throughout. Fix the point e_0 with coordinates

$$e_0^I = \begin{pmatrix} 1 \\ 0 \\ \vdots \\ 0 \end{pmatrix}$$

and define the parabolic subgroup P of $G = O_0(\tilde{g})$ (the subscript denotes identity connected component) by

$$(4) \qquad P = \{g \in G : ge_0 = \lambda e_0,\ \mathbb{R} \ni \lambda > 0\}$$

We will denote by σ_q the character given by λ^{-q}. The space $\mathcal{E}(l)$ of jets at e_0 of functions on $\mathbb{R}^{n+2}$ homogeneous of degree l has a natural P-module structure. This is slightly subtle—the usual action of an element g (as in (4)) on a jet at e_0 would give a jet at $\lambda^{-1}e_0$, but the homogeneity condition allows one to recover a jet at e_0 from this. There is a natural "evaluation at e_0" which is a P-equivariant map from $\mathcal{E}(l)$ to σ_l. Whenever we evaluate a jet at e_0, we mean evaluation in this equivariant sense.

An *invariant* of a P-module is a P-equivariant polynomial on the module taking values in a 1-dimensional representation. As remarked in §2, we consider only invariants homogeneous of some degree d as a polynomial. The homogeneous space G/P is the space of generators of the null-cone of $\tilde{g}$ and it can be identified with the sphere S^n with its usual flat conformal structure and G acting by conformal automorphisms. The "invariant" problem we wish to address is (roughly) that of finding all (not necessarily linear) G-equivariant differential operators from functions on S^n to sections of some G-homogeneous line bundle over S^n. These are of course determined by their form at the identity coset, and so our problem is equivalent to that of finding all invariants of the P-module of jets of functions at the reference point $[e_0] \in S^n$. (By "jets" we always mean infinite formal power series.) This space of jets splits as a direct sum $\mathbb{R} \oplus \mathcal{J}$ of P-modules, where $\mathbb{R}$ is the jets of constants and $\mathcal{J}$ is the space of jets that vanish to zeroth order at $[e_0]$. The precise problem we will address is to find all the invariants of $\mathcal{J}$. (Or to return to the original formulation, differential operators which agree on functions that differ only by a constant.)

One might ask why we are looking for invariants of $\mathcal{J}$ rather than of the whole space of jets. In fact, since the complement of $\mathcal{J}$ is 1-dimensional, if one has a list of invariants of $\mathcal{J}$ then one can form a list of invariants of the whole space of jets. This is a special feature of this particular problem however. In general, for the problems considered in [**BEG**], listing the invariants of the "whole space of jets" is an unsolved problem. Fortunately, it is often the easier problems analogous to finding invariants of $\mathcal{J}$ which arise in applications.

5. The Algebraic Problem

We need a different realization of $\mathcal{J}$ which gives us a way of writing down at least some invariants. In what follows, we will often speak of jets as though they are actual functions. This should cause no confusion.

As we remarked before, S^n is the space of generators of the null-cone Q of $\tilde{g}$. Thus $\mathcal{J}$ can be identified with the space of jets at $e_0 \in Q$ of functions on Q homogeneous of degree zero, modulo constants. Let us write Δ for the indefinite Laplacian

$$\Delta = \tilde{g}^{IJ} \frac{\partial^2}{\partial X^I \partial X^J}$$

on $\mathbb{R}^{n+2}$, where $\tilde{g}^{IJ}$ are the components of the inverse metric and we are using the summation convention. We will say that f is *harmonic* if $\Delta f = 0$. It is shown in [**EG**] that if n is odd then locally any function f on Q homogeneous of degree zero admits a unique harmonic extension off Q to all orders. Thus *for n odd, $\mathcal{J} \cong \mathcal{H}_0$* where

(5) $$\mathcal{H}_0 = \frac{\left\{ \begin{array}{l} \text{Jets at } e_0 \text{ on } \mathbb{R}^{n+2} \text{ of harmonic functions} \\ \text{homogeneous of degree zero} \end{array} \right\}}{\mathbb{R}}.$$

(The "subscript zero" on $\mathcal{H}_0$ is so that our notation agrees with [**BEG**], where the example we are considering is one of a family of problems.) That $\mathcal{H}_0$ is a P-module follows from the remarks about jets of homogeneous functions generally just below (4).

Our methods work perfectly well for $\mathcal{H}_0$, whether n is odd or even. Because unique harmonic extension of homogeneity zero functions off Q fails for n even however, the results give only limited information about invariants of $\mathcal{J}$ in that case. We see here a simple example of the type of obstructions that often arise in the relation between the geometric problems and the algebraic. Henceforth, we take our problem to be that of finding all invariants of $\mathcal{H}_0$.

Regarding $\mathbb{R}^{n+2}$ as a P-module by restriction, the line generated by e_0 is a submodule isomorphic to σ_{-1}. Equivalently, there is a preferred element of $\mathbb{R}^{n+2} \otimes \sigma_1$ which we will denote by e. We use this in the following algebraic description of elements of $\mathcal{H}_0$ as lists of tensors.

PROPOSITION 5.1.

$$\mathcal{H}_0 \cong \left\{ \begin{array}{l} (T^{(1)}, T^{(2)}, \dots) : T^{(k)} \in \bigodot_0^k \mathbb{R}^{n+2*} \otimes \sigma_{-k}, \\ T^{(k)} = -k\, e \lrcorner T^{(k+1)} \text{ for } k \geq 1, \; e \lrcorner T^{(1)} = 0 \end{array} \right\}$$

where "$\lrcorner$" denotes interior multiplication (i.e. contraction) and "$\bigodot_0$" denotes trace-free (with respect to $\tilde{g}$) symmetric tensor product. The P-action is (apart *from the σ_{-k} factors) just the restriction of the G-action on each tensor separately.*

The proof of the proposition is easy. If f is a jet, define the $T^{(k)}$ by

$$T^{(k)}_{A\ldots D} = \frac{\partial}{\partial X^A} \cdots \frac{\partial}{\partial X^D} f\bigg|_{e_0},$$

where the evaluation at e_0 is a generalization to jets of homogeneous tensors of the P-equivariant evaluation map for homogeneous functions mentioned below (4). That the tensors belong to the indicated spaces follows from the observation that differentiation with respect to the coordinates lowers homogeneity by one.

The "$\lrcorner$" relations now follow from Euler's equation for homogeneous functions and the $T^{(k)}$ are clearly symmetric. That they are also trace-free is equivalent to f's being harmonic. All the modules we deal with have this general form where elements are given by lists of G-tensors with P-invariant (but not G-invariant) "linking relations".

The great advantage of this realization of $\mathcal{H}_0$ is that there are many obvious invariants that one can write down. Anything that would be a G-invariant of a list of tensors (ignoring the linking relations) is a P-invariant of $\mathcal{H}_0$. For simplicity, until §7 we will consider only even invariants of $\mathcal{H}_0$.

DEFINITION 5.2. *An even Weyl invariant of $\mathcal{H}_0$ is a linear combination of complete contractions of the form*

$$\text{contr}(\tilde{g}^{-1} \otimes \cdots \otimes \tilde{g}^{-1} \otimes T^{(k_1)} \otimes \cdots \otimes T^{(k_d)})$$

all taking their values in the same σ_q.

The basic theorem in this case which provides a list of all even invariants is:

THEOREM 5.3. *Every even invariant of $\mathcal{H}_0$ is a Weyl invariant.*

6. An Outline of the Proof

A complete proof of Theorem 5.3 is contained in [**BEG**]. Here I just want to give an outline of the ideas. The central idea is this: Suppose that one has a linear combination C of partial contractions (same definition as complete contraction, except one is allowed to leave some indices uncontracted) constructed from the $T^{(k)}_{A\ldots D}$, $\tilde{g}^{AB}$ and e^I, which is trace-free and symmetric, and such that

(6) $$C^{AB\cdots D} = \underbrace{e^A e^B \cdots e^D}_{m} I$$

where I is some map $\mathcal{H}_0 \to \sigma_q$. Then I is an invariant. We call such invariants *Weak Weyl Invariants*. The notion includes that of Weyl invariant as the special case $m = 0$. In plain language, C has all components vanishing except $C^{0\cdots 0} = I$. Weak Weyl invariants are important because of:

PROPOSITION 6.1. *Every even P-invariant of $\mathcal{H}_0$ is a weak Weyl invariant.*

It is worth noting that this proposition does not in itself provide a list of the invariants, because there is no way of seeing exactly which linear combinations of partial contractions are weak Weyl invariants.

I will try and outline the proof of the Proposition. Let I be an even P-invariant of $\mathcal{H}_0$. The orthogonal group $\mathrm{SO}(n)$ sits in P as part of a Levi factor—it essentially acts in the normal way on the lower case indices (as defined by (2)). The invariant I is thus by restriction an even $\mathrm{SO}(n)$ invariant. Define tensors $u^{(p,q,r)}$ on $\mathbb{R}^n$ by

$$(7) \qquad u_{a\cdots d}^{(p,q,r)} = T_{\underbrace{0\cdots 0}_{p}\underbrace{a\cdots d}_{q}\underbrace{\infty\cdots\infty}_{r}}^{(k)}, \ p+q+r = k.$$

There are relations between them, but that is not really important. The classical invariant theory of §2 tells us that I can be written as a linear combination of complete contractions of the form

$$\mathrm{contr}(g^{-1} \otimes \cdots \otimes g^{-1} \otimes u^{(p_1,q_1,r_1)} \otimes \cdots \otimes u^{(p_d,q_d,r_d)}),$$

where g^{-1} denotes the inverse Euclidean metric on $\mathbb{R}^n$. Take this linear combination, and replace the u's with the corresponding components of the T's according to (7). The idea then is to replace lower case contractions with the $\mathbb{R}^n$ metric by upper case contractions with $\tilde{g}^{-1}$ using the explicit relation between these as given by (3). This generates many lower infinity and zero indices. Surplus lower zero indices can be eliminated using the linking relations, and that leaves just lower infinity indices. But these are equivalent to upper zero indices (given the non-diagonal form of $\tilde{g}$), which correspond exactly to the component we are trying to obtain.

The details of this process are somewhat long and not so important, but at the end one has a linear combination of partial contractions of the required form with the invariant living in the correct component. It remains to see that all the other components vanish—this follows from an elementary representation theory argument, the crucial facts being that (after complexification) the required component is the highest weight space of an irreducible $\mathfrak{g}_{\mathbb{C}}$-module (recall that C is trace-free symmetric), and that all the raising operators in $\mathfrak{g}_{\mathbb{C}}$ are contained in $\mathfrak{p}_{\mathbb{C}}$.

In all the problems we have considered, every invariant is a weak Weyl invariant by arguments of the above sort.

6.1. Low Degree Invariants. Our methods now differ for high and low degree. In low degree we have at least two distinct arguments, but they both eventually come down to "second main theorems" for classical invariant theory. Second main theorems for the orthogonal group concern the relations between the invariants described in §2. We will consider only even invariants. Fix a dimension n. It may happen that a particular linear combination of complete contractions of the form (1), while not zero when considered as a formal expression, is zero as a polynomial in the components of the tensors—we say it *vanishes on substitution in dimension n*. The results we need say essentially that this can only happen if the formal expression contains an antisymmetrization over $n+1$ indices. In particular, it can not happen at all for a linear combination of complete contractions of degree $\leq n$ in a collection of *symmetric* tensors, simply because there are not $n+1$ indices over which one can non-trivially antisymmetrize. These results need a careful statement and proof, which is given in the appendix of [**BEG**].

I will sketch the simplest way of applying these results to our problem. Let I be an invariant of degree $d < n$ and consider the expression (6) for I as a weak Weyl invariant. Suppose $m > 0$, so that all but one of the components of C vanish. We can (roughly speaking) construct from this an $SO(n)$ invariant of $d+1$ symmetric tensors on $\mathbb{R}^n$ (the tensors are essentially a subset of the $u^{(p,q,r)}$'s of (7) plus one auxiliary tensor) that vanishes on substitution in dimension n. From the above observations, it must vanish as a formal expression. One can deduce from this that the original weak Weyl invariant must vanish as a formal expression. Thus either the invariant is zero, or $m = 0$ and the weak Weyl invariant is already a Weyl invariant. This method proves Theorem 5.3 for $d < n$.

In [**BEG**] we use the above argument in the "conformal curvature case", but use the same "second main theorem" ideas in a different way for this example.

6.2. High degree case. Our methods for high degree are quite different. They are adaptations of the ideas of Gover [**Go**] and rely on a return to regarding $\mathcal{H}_0$ as a space of harmonic jets. Suppose that in the linear combination of partial contractions that constitute the left hand side of (6) we make the replacements

$$T^{(k)}_{A\cdots D} \rightsquigarrow \frac{\partial}{\partial X^A} \cdots \frac{\partial}{\partial X^D} f \text{ and } e^I \rightsquigarrow X^I$$

and regard $\tilde{g}^{-1}$ as defining a (constant) jet, then the left hand side of (6) defines a map $\widetilde{C}$ from $\mathcal{H}_0$ to the space of jets at e_0 of functions on Q of homogeneity $q+m$ taking values in $\bigodot_0^m \mathbb{R}^{n+2}$, and with the property that $\widetilde{C}|_{e_0} = C$. We claim that if one also substitutes $e^I \rightsquigarrow X^I$ on the right hand side, then there exists a map $\tilde{I}$ from $\mathcal{H}_0$ to the space of jets at e_0 of functions on Q of homogeneity q with $\tilde{I}|_{e_0} = I$ such that

$$(8) \qquad \widetilde{C}^{AB\cdots D} = \underbrace{X^A X^B \cdots X^D}_{m} \tilde{I}.$$

The proof of this claim is an application of Frobenius reciprocity, and so it relies on the fact that $\mathcal{H}_0$ is actually a $(\mathfrak{g}, P)$-module.

Having obtained (8), the final step is to hit both sides with a "differentiation operator" D. The operator D is defined on functions or tensor fields on Q homogeneous of degree $s \neq 1 - n/2$ to be the result of taking the first order harmonic extension off Q, differentiating with respect to the coordinates, and restricting back to Q. It is not hard to get an explicit formula for this operator— it is given on a function f of homogeneity s by

$$D_I f = \left(\partial_I f - \frac{\tilde{g}_{IJ} X_J \Delta f}{(n + 2s - 2)} \right) \Bigg|_Q ,$$

where to make sense of the right-hand side, one must choose an extension of f off Q. It is easy to check that the the expression is independent of that choice. The crucial property possessed by D is that for any homogeneous function or tensor field f on Q,

(9) $$D_I(X^I f) = \text{constant} \times f.$$

Provided firstly that the homogeneities of the jets to which we need to apply D are never equal to $1 - n/2$, and secondly that the constant in (9) (which also depends on homogeneity) is non-zero, then one sees after some thought that

$$I = D_A \ldots D_D C^{A \cdots D} \big|_{e_0}$$

realises I as a Weyl invariant. It turns out that this works precisely for I having degree $\geq n$, which covers all the cases not included in the low degree arguments, and so the proof of Theorem 5.3 is complete.

7. Odd Invariants

We will now sketch the odd case. Let $\tilde{\epsilon}$ denote the volume form on $\mathbb{R}^{n+2}$, and write $\tilde{\epsilon}_0$ for the $(n+1)$-form $e \lrcorner \tilde{\epsilon}$. An odd Weyl invariant is a linear combination of complete contractions, each of which is of the form

$$\text{contr}(\tilde{\epsilon} \otimes \tilde{g}^{-1} \otimes \cdots \otimes \tilde{g}^{-1} \otimes T^{(k_1)} \otimes \cdots \otimes T^{(k_d)})$$

or

$$\text{contr}(\tilde{\epsilon}_0 \otimes \tilde{g}^{-1} \otimes \cdots \otimes \tilde{g}^{-1} \otimes T^{(k_1)} \otimes \cdots \otimes T^{(k_d)}).$$

Given that we are now allowing e to appear in our Weyl invariants, one might ask why we did not do so before. The answer is that in even invariants, or in odd invariants except only when it appears contracted into the volume form, it can be eliminated using the linking relations. It follows from the fact that both $\tilde{\epsilon}$ and $\tilde{\epsilon}_0$ are antisymmetric that a non-zero odd Weyl invariant necessarily has degree $\geq n + 1$, because if two indices of the same $T^{(k)}$ are contracted into the volume form, the result necessarily vanishes (remember the $T^{(k)}$ are symmetric).

On the other hand, an odd P-invariant is also an odd $SO(n)$ invariant of the $u^{(p,q,r)}$ (just as for the even case—see below Proposition 6.1). Since the $u^{(p,q,r)}$ are symmetric, the argument at the end of §2 applies, and a non-zero odd invariant must have degree at least n.

A similar argument to the even case shows that every odd invariant is a weak Weyl invariant (with the definition suitably modified so that a volume form appears in each partial contraction). There is no "low degree" case for odd invariants. Our high degree methods work much as before for odd invariants to give:

THEOREM 7.1. *There are no non-zero odd invariants of $\mathcal{H}_0$ of degree $< n$. Those of degree $\geq n+1$ are Weyl invariants, and those of degree exactly n are not.*

We refer to invariants which are not Weyl as *exceptional* (or sometimes as "vile"). A question does arise as to whether one can list the exceptional invariants. Gover has just done this for a related problem associated with projective structures, and we have recently extended this [**BG**] to cover all the exceptional invariants that arise in [**BEG**]. It turns out that all exceptional invariants are linear combinations of *basic exceptional invariants*, of which there are but a finite number. We give a quite explicit formula for these involving the D operator. This then completes the work of [**BEG**] in that we now have a means of listing all the invariants of the modules considered there.

8. Final Remarks

There is one difference between the results for low and high degrees that is worth noting. If an even P-invariant is a polynomial in the components of the tensors $T^{(k)}$ for $1 \leq k \leq K$ for some K, then if it is of low degree (which means $\leq n$, if one uses the "other argument" alluded to at the end of §6.1), then it can be written as a Weyl invariant depending only on $T^{(k)}$ for $1 \leq k \leq K$. On the other hand, in the high degree case (and also for odd invariants of degree $> n$), one generally needs to use $T^{(k)}$ with $k > K$ in order to write it as a Weyl invariant. (These are generated in our methods by the differentiation operator D.)

Our methods are somewhat *ad hoc* and we have only applied them to a few algebraic problems that happen to be of geometric interest. It is tempting to ask what general results there might be. In particular, the only exceptional invariants in the problem we considered are necessarily exceptional by elementary arguments using the classical invariant theory of the Levi factor, and the same is true for all the problems considered in [**BEG**]. It would be interesting to know whether this is the case for some class of similar modules. We know of no counterexamples in any of the problems we have considered (although we do know of problems where our methods fail).

From the geometrical perspective, the outstanding problem is to obtain some better results in the cases such as even dimensional conformal invariants and invariants of C-R structures, where the relationship between the geometric and algebraic problems (as currently posed) is obstructed. This would seem to require some new ideas!

REFERENCES

[ABP] M. Atiyah, R. Bott, and V. K. Patodi, *On the heat equation and the index theorem*, Inventiones Math. **19** (1973), 279–330.

[BEG]   T. N. Bailey, M. G. Eastwood, and & C. R. Graham, *Invariant theory for conformal and C-R geometry*, Annals of Math. (to appear).

[BG]   T. N. Bailey and R. Gover, *Exceptional invariants in the parabolic invariant theory of conformal geometry*, preprint, 1992.

[EG]   M. G. Eastwood and C. R. Graham, *Invariants of conformal densities*, Duke Math. Jour. **63** (1991), 633–671.

[F]   C. Fefferman *Parabolic invariant theory in complex analysis*, Adv. in Math. **31** (1979), 131–262.

[FG1]   C. Fefferman and C. R. Graham, *Conformal invariants*, Élie Cartan et les Mathématiques d'Aujourdui, Astérisque (1985), 95–116.

[FG2]   _____ , in preparation.

[Go]   R. Gover, *Invariants on projective space*, preprint, 1991.

[Gr]   C. R. Graham, *Invariant theory of parabolic geometries*, Complex Geometry, Proc. 1990 Osaka Conference, Marcel Dekker Lecture Notes vol. 143, 1993, pp. 53–66.

[W]   H. Weyl, *The classical groups*, Princeton University Press, 1939.

DEPARTMENT OF MATHEMATICS, UNIVERSITY OF EDINBURGH, JAMES CLERK MAXWELL BUILDING, THE KING'S BUILDINGS, MAYFIELD ROAD, EDINBURGH EH9 3JZ, SCOTLAND
    E-mail address: tnb@mathematics.edinburgh.ac.uk

Contemporary Mathematics
Volume **154**, 1993

Kaehler Structures on $K_{\mathbb{C}}/N$

MENG-KIAT CHUAH AND VICTOR GUILLEMIN

1. Introduction

Let K be a compact semi-simple Lie group, let $G = K_{\mathbb{C}}$ and let KAN be the Iwasawa decomposition of G. We will denote by T the centralizer of A in K (which, in this case, is a Cartan subgroup of K). Let $X = G/N$. Since G and N are complex groups, X is a complex manifold. We will denote by $\mathcal{O}$ the ring of holomorphic functions on X and by ρ the natural representation of K on $\mathcal{O}$. By the Bott-Borel-Weil theorem, ρ has the following property:

(*)   Every irreducible representation of K occurs exactly once as a subrepresentation of ρ.

Indeed one can deduce this from the usual version of Bott-Borel-Weil as follows: Since TA normalizes N, there is a right action of TA on X which commutes with the left action of K. Since TA is a complex group, its irreducible representations are in 1-1 correspondence with holomorphic characters, $\chi : TA \to \mathbb{C}^*$. Fix such a character and consider the subspace, $\mathcal{O}_\chi$, of $\mathcal{O}$ consisting of those holomorphic functions which transform under the action of TA according to χ. $\mathcal{O}_\chi$ can be regarded as the space of sections of the holomorphic line bundle over G/B induced by χ (where B is the Borel subgroup TAN); so, by Bott-Borel-Weil, the representation of K on $\mathcal{O}_\chi$ is either irreducible or zero. Moreover, for a given irreducible representation, there is exactly one χ for which the representation of K on $\mathcal{O}_\chi$ is this given representation.

A unitary representation of the group K with the property (*) is called a *model*. This terminology is due to Gelfand and Zelevinski, who give in [**GZ**] several ingenious constructions of models for the classical compact groups. The example we've just discussed is not a model in this sense, however, since the representation of K on $\mathcal{O}$ is not a unitary representation but just a representation

1991 *Mathematics Subject Classification.* Primary 53C55.

The second author was supported in part by NSF Grant DMS 89 0771.

This paper is in final form and no version of it will be submitted for publication elsewhere.

in the abstract algebraic sense. A couple of years ago, A.S. Schwarz suggested that it might be possible to remedy this defect as follows: Equip X with a $K \times T$-invariant Kaehler form, ω, and apply to (X, ω) the standard machinery of geometric quantization. By this means, the action of $K \times T$ gets converted into a unitary representation which is multiplicity-free by Bott-Borel-Weil, and therefore has a chance of being a model (given the right choice of ω. A result of this nature is true for the Virasoro group. See [**LNS**].) We will show below that this proposal doesn't quite work, but that it is possible to choose ω so that the corresponding representation contains all the non-degenerate irreducible representations of K.

In §2 we will give a complete classification of $K \times T$-invariant Kaehler structures on X. We will show that every Kaehler structure is defined by a potential function, F; i.e. is of the form

$$(1.1) \qquad\qquad \omega_F = \sqrt{-1}\partial\bar{\partial}F,$$

where F is K-invariant.

By Iwasawa, $X = KA$ so every such function is just a function on A. Let $\mathfrak{a}$ be the Lie algebra of A. The exponential map takes $\mathfrak{a}$ bijectively onto A so we can identify $\mathfrak{a}$ with A. Moreover as subalgebras of the complex Lie algebra, $\mathfrak{g}$, $\mathfrak{a} = \sqrt{-1}\mathfrak{t}$, so by choosing a basis of lattice vectors in $\mathfrak{t}$, we get identifications

$$\mathfrak{a} \cong \mathfrak{t} \cong \mathfrak{t}^* \cong \mathbb{R}^n.$$

Hence we can think of the function, F, above as being a function on $\mathbb{R}^n$. Recall that T is a Cartan subgroup of K. Let $\mathfrak{t}^*_+$ be the positive Weyl chamber in $\mathfrak{t}^*$. Our first main result is the following:

THEOREM I *The two-form* (1.1) *is a Kaehler form if and only if:*
   (a) *F is a strictly convex function, and*
   (b) *the image of the mapping*

$$(1.2) \qquad\qquad \frac{1}{2}\frac{\partial F}{\partial x} : \mathbb{R}^n \longrightarrow \mathbb{R}^{n*} = \mathfrak{t}^*$$

is entirely contained in Int $\mathfrak{t}^*_+$.

This will be proved in §2. The map (1.2) has the following intrinsic interpretation: From (1.1) it is easy to see that the action of T on X is Hamiltonian. Let $J : X \longrightarrow \mathfrak{t}^*$ be its moment map. J is K-invariant, so if we regard X as the product $K \times A$, it is determined entirely by its restriction to A. We will show in §2 that this restriction is the mapping (1.2). Thus, from Theorem I we conclude

THEOREM II *The image of J is contained in the interior of the positive Weyl chamber.*

Let ω_F be the form (1.1) and let $\beta = -\sqrt{-1}\partial F$. Then $\omega = d\beta$; so, in particular, $[\omega] = 0$. Let $(\mathbb{L}, \nabla)$ be a line bundle-connection pair with $curv(\nabla) = \omega$. Since $[\omega] = 0$, $\mathbb{L}$ is trivial as a line bundle. However, the Hermitian structure on $\mathbb{L}$ is not the trivial Hermitian structure. In §3 we will describe the structure of $\mathbb{L}$ as a Hermitian bundle. More explicitly we will show that there exists a non-vanishing holomorphic section, s, of $\mathbb{L}$ with the property

$$(1.3) \qquad -\sqrt{-1}\frac{\nabla s}{s} = \beta;$$

where β is, as above, the one-form, $-\sqrt{-1}\partial F$.

In fact s is uniquely determined by (1.3) up to a non-zero scalar multiple and is $K \times T$-invariant. The Hermitian structure on $\mathbb{L}$ will then be determined by the identity:

$$(1.4) \qquad \langle s, s \rangle = e^{-F}.$$

Via the identification, $X = K \times A$, the Haar measure on $K \times A$ gets identified with a measure on X which is $K \times TA$-invariant and which we will denote by μ. We will denote by $\mathcal{H}_F$ the space of all holomorphic sections of $\mathbb{L}$ for which the integral

$$(1.5) \qquad \int_X \langle s, s \rangle \mu$$

is finite.

$\mathcal{H}_F$ is a Hilbert space, and there is a natural representation of $K \times T$ on it which we will denote by ρ_F. In §4 we will prove the following result (which is the main result of this paper):

THEOREM III *Let λ be an integer lattice point in $\mathfrak{t}^*_+$. Then the irreducible representation of K with maximal weight, λ, occurs as a subrepresentation of ρ_F if and only if λ is in the image of the moment map, J.*

Remarks:

(1) If the representation with maximal weight, λ, occurs as a subrepresentation of ρ_F, then by Bott-Borel-Weil, it occurs exactly with multiplicity one.

(2) By Theorem II, the image of J is entirely contained in the interior of $\mathfrak{t}^*_+$, so this theorem says that no representation whose maximal weight lies in a wall of the positive Weyl chamber can occur as a subrepresentation of ρ_F. In particular, ρ_F can never be a model. In §5 we will study the Kaehler form, ω_F, for some special choices of F. Of particular interest is the choice

$$(1.6) \qquad F(x) = \sum e^{\lambda(x)}$$

where λ sums over a set of fundamental dominant weights of K. This function is convex and the range of (1.2) is the interior of the positive Weyl chamber. Thus by Theorem III the representation, ρ_F, contains as a subrepresentation all the irreducible representations of K whose maximal weights are in the interior of the positive Weyl chamber; i.e. all the *non-degenerate* irreducible representations of K.

The representations which are missing from this model are the representations whose maximal weights lie on walls of the positive Weyl chamber. We will describe elsewhere how to construct "models" for these representations by the same methods as above. (A brief outline of how to do this is given in §5.)

Acknowledgements:

We would like to thank A.S. Schwarz for stimulating discussions about the $K_\mathbb{C}/N$ model and its generalizations to infinite dimensional groups. We would also like to thank Richard Melrose whose informative lectures on the theory of Reinhardt domain prompted us to begin thinking about the questions addressed here. Finally we owe a large debt of gratitude to Reyer Sjamaar and David Vogan for helping us to correct some errors in the first draft of this paper.

2. Kaehler structures on G/N

The goal of this chapter is to classify all $K \times T$-invariant Kaehler forms on $X = G/N$, and in particular to show that each such form is given by a global potential function.

Since K is a compact semi-simple Lie group,

$$H^1(X, \mathbb{R}) = H^1(KA, \mathbb{R}) = H^1(K, \mathbb{R}) = 0.$$

In particular, every $K \times T$ invariant Kaehler form is of the form

$$\omega = d\alpha,$$

where α is a $K \times T$-invariant real 1-form. Let

$$\alpha = \alpha^{10} + \alpha^{01}$$

be the Dolbeault decomposition of α, α^{10} and α^{01} being $K \times T$-invariant forms of type *(1,0)* and *(0,1)*. Since ω is a closed real form of type *(1,1)*,

(2.1) $\alpha^{10} = \overline{\alpha^{01}}, \quad \partial\alpha^{10} = \bar{\partial}\alpha^{01} = 0, \quad \text{and} \quad \omega = \bar{\partial}\alpha^{10} + \partial\alpha^{01}.$

Hence α^{01} defines a cohomology class in $H^{0,1}(X, \mathbb{C})$. In fact, we claim

LEMMA α^{01} is $\bar{\partial}$ exact.

Proof: By the Dolbeault theorem,

$$H^{0,1}(X, \mathbb{C}) = H^1(X, \mathcal{O}_X),$$

where $\mathcal{O}_X$ is the sheaf of holomorphic functions on X. The holomorphic functions on X that transform by $\lambda \in \mathfrak{t}^*$ under the right T-action can be identified with holomorphic sections of the homogeneous bundle

$$\mathbb{L}_\lambda = G \times_B \mathbb{C}_\lambda, \qquad (g.t, z) \sim (g, \chi(t)z)$$

over $B = TAN$, where $\chi(\exp x) = e^{2\pi i \lambda(x)}$. Hence

$$H^1(X, \mathcal{O}_X)_\lambda \cong H^1(G/B, \tilde{\mathbb{L}}_\lambda),$$

where $\tilde{\mathbb{L}}_\lambda$ is the sheaf of holomorphic sections of $\mathbb{L}_\lambda$. In particular, since α^{01} is right T-invariant, $[\alpha^{01}]$ is an element of $H^1(G/B, \tilde{\mathbb{L}}_0)$. To complete the proof, it suffices to show that $H^1(G/B, \tilde{\mathbb{L}}_0) = 0$. But

$$H^1(G/B, \tilde{\mathbb{L}}_0) = H^1(G/B, \mathcal{O}) = H^{0,1}(G/B, \mathbb{C})$$

and since G/B is compact,

$$dim\, H^{0,1}(G/B, \mathbb{C}) = \frac{\beta_1(G/B)}{2} = 0$$

since G/B is simply connected, which proves the lemma.

Therefore $\alpha^{01} = \bar{\partial} f$, for some K-invariant function f. Let

$$F = \sqrt{-1}(-f + \bar{f}).$$

It follows from (2.1) that $\omega = \sqrt{-1} \partial \bar{\partial} F$. Thus we have proved

PROPOSITION 2.1. *Every $K \times T$-invariant Kaehler form on X has a K-invariant potential function.*

Let $\omega_F = \sqrt{-1} \partial \bar{\partial} F$ be a $K \times T$-invariant Kaehler form on X, and W be the orbit of TA through the identity coset. Then the inclusion

$$\iota : W \hookrightarrow X$$

is a holomorphic imbedding. Since $\iota^* \omega_F$ is a T-invariant Kaehler form on W, $\iota^* F$ is a strictly plurisubharmonic T-invariant function on $W = TA$. By the identification

$$\iota^* F \in \mathcal{C}^\infty(A) = \mathcal{C}^\infty(\mathfrak{a}),$$

it is a strictly convex function on $\mathfrak{a}$ (see e.g. [La]). As a complex manifold, $W = TA = \mathbb{C}^n/\mathbb{Z}^n$. From now on we will make this identification, and let

$$z = x + iy = (x_1 + iy_1, ..., x_n + iy_n)$$

be a system of complex coordinates on W. Consider now the T-action on the Kaehler manifold $(\mathbb{C}^n/\mathbb{Z}^n, \iota^*\omega_F)$. Its infinitesimal generators are the vector fields, $\frac{\partial}{\partial x_i}$, $i = 1, ..., n$. Since

$$\iota(\frac{\partial}{\partial x_i})\omega_F = \frac{1}{2}\sum \frac{\partial^2 F}{\partial y_i \partial y_j}dy_j = \frac{1}{2}d\frac{\partial F}{\partial y_i}$$

the moment map associated with this T-action is the map

(2.2) $$J : W \longrightarrow \mathbb{R}^{n*}, \quad (x, y) \rightarrow \frac{1}{2}\frac{\partial F}{\partial y},$$

which is, up to a factor of 2, the Legendre transform associated with ι^*F. Since ι^*F is strictly convex, the image of J is an open convex subset of $\mathbb{R}^{n*}$.

Let

(2.3) $$\phi_F : X \longrightarrow \mathfrak{k}^*$$

be the moment map corresponding to the K action on X. Note that $\omega_F = -d\alpha$, where

$$\alpha = Im\bar{\partial}F.$$

We can set the moment map to be

$$\phi_F^\xi = \iota(\xi^\sharp)\alpha$$

for all $\xi \in \mathfrak{k}$, where $\xi^\sharp$ is the vector field on X corresponding to ξ. Note that the moment map is independent of the choice of F or α, since K is semi-simple (see [**GS**] Theorem 26.1). We will show below that its restriction to W is the map (2.2). (Since ϕ_F is K-equivariant this will determine it completely.) This simple result turns out to require a surprisingly cumbersome proof: Recall that $\mathfrak{g} = \mathfrak{k} + \sqrt{-1}\mathfrak{k}$. Given $\xi \in \mathfrak{k}$ let $\eta = \sqrt{-1}\xi$. We claim that

(2.4) $$\phi_F^\xi = \frac{1}{2}\eta^\sharp F.$$

Proof: By definition, $\alpha = Im\bar{\partial}F$. Since $Re\bar{\partial}F = \frac{1}{2}dF$, and F is K-invariant,

$$\sqrt{-1}\iota(\xi^\sharp)\alpha = \iota(\xi^\sharp)\bar{\partial}F.$$

The term on the right can be written as

$$\frac{1}{2}\iota(\xi^\sharp + \sqrt{-1}\eta^\sharp)\bar{\partial}F + \frac{1}{2}\iota(\xi^\sharp - \sqrt{-1}\eta^\sharp)\bar{\partial}F$$

and since the vector field figuring in the second expression is holomorphic, the second term above is zero. Moreover, in the first term we can make the substitution

$$\bar{\partial}F = dF - \partial F$$

and since the vector field figuring in the first term is anti-holomorphic, the whole sum reduces to

$$\frac{1}{2}\iota(\xi^\sharp + \sqrt{-1}\eta^\sharp)dF.$$

However, F is K-invariant, so $\iota(\xi^{\sharp})dF = 0$, and we end up with the identity

$$\sqrt{-1}\phi_F^{\xi} = \sqrt{-1}\iota(\xi^{\sharp})\alpha = \frac{1}{2}\sqrt{-1}\iota(\eta^{\sharp})dF;$$

hence, $\phi_F^{\xi} = \frac{1}{2}\eta^{\sharp}F$ as claimed.

Now let Δ_+ be a system of positive roots of the Lie algebra $\mathfrak{g}$ and let

(2.5) $$\xi_{\alpha}, \xi_{-\alpha}, h_{\alpha}; \quad \alpha \in \Delta_+$$

be a Weyl basis of $\mathfrak{g}$ over $\mathbb{C}$. (See [H1], page 421.) Let $\eta_{\pm\alpha} = \sqrt{-1}\xi_{\pm\alpha}$. Then the Lie algebra, $\mathfrak{n}$, of N is the linear span over $\mathbb{R}$ of the elements of $\mathfrak{g}$:

(2.6) $$\xi_{\alpha}, \eta_{\alpha}; \quad \alpha \in \Delta_+,$$

and $\mathfrak{k}$ is the linear span over $\mathbb{R}$ of

(2.7) $$h_{\alpha}, \zeta_{\alpha} = \xi_{\alpha} - \xi_{-\alpha}, \text{ and } \gamma_{\alpha} = \eta_{\alpha} + \eta_{-\alpha}; \quad \alpha \in \Delta_+.$$

Since F is K-invariant, $\zeta_{\alpha}^{\sharp}F = \gamma_{\alpha}^{\sharp}F = 0$ and hence

(2.8) $$\xi_{\alpha}^{\sharp}F = \xi_{-\alpha}^{\sharp}F \text{ and } \eta_{\alpha}^{\sharp}F = -\eta_{-\alpha}^{\sharp}F.$$

In particular, if $p \in W$, the stabilizer group of p in G is N, so the vector fields $\xi_{\alpha}^{\sharp}$ and $\eta_{\alpha}^{\sharp}$ vanish at p. Thus, by (2.4) and (2.8):

(2.9) $$\phi_F^{\zeta_{\alpha}} = \phi_F^{\gamma_{\alpha}} = 0$$

at p. Thus, denoting by $\mathfrak{t}^{\natural}$ the subspace of $\mathfrak{k}^*$ annihilated by the elements ζ_{α} and γ_{α}, $\alpha \in \Delta_+$, of $\mathfrak{k}$; we've proved:

PROPOSITION 2.2. *The moment map* (2.3) *maps W into $\mathfrak{t}^{\natural}$.*

A more intrinsic characterization of $\mathfrak{t}^{\natural}$ is given by:

$$\mathfrak{t}^{\natural} = \{\lambda \in \mathfrak{k}^*, \ Ad^*(a)\lambda = \lambda \text{ for all } a \in T\}.$$

Moreover, if $\imath$ is the inclusion map of $\mathfrak{t}$ into $\mathfrak{g}$ its transpose, restricted to $\mathfrak{t}^{\natural}$,

(2.10) $$\imath^* : \mathfrak{t}^{\natural} \longrightarrow \mathfrak{t}^*$$

is a bijection. However the restriction of $\imath^* \cdot \phi_F$ to W is just the moment map associated with the action of T on W; so we have proved:

PROPOSITION 2.3. *Via the identification* (2.10), *the restriction of ϕ_F to W is the mapping* (2.2).

We have seen that for ω_F to be a Kaehler form, condition (a) of Theorem I has to be satisfied. We will show next that ω_F is a Kaehler form if and only if conditions (a) and (b) of Theorem I are satisfied: To show that ω_F is positive definite at all points of X, it suffices, by K-invariance, to show that ω_F is positive definite at all points, p, of W. Let $\lambda = \phi_F(p)$, and let ξ and η be elements of $\mathfrak{k}$. Then, at the point p,

$$\omega_F(\xi^\sharp, \eta^\sharp) = \phi_F^{[\xi,\eta]} = \lambda([\xi,\eta]).$$

The subspace of T_pX spanned by the vectors, $\zeta_\alpha^\sharp$ and $\gamma_\alpha^\sharp$, is a symplectic orthocomplement to T_pW; so for ω_F to be positive definite on all of T_pX it suffices that it be positive definite on this space. The complex structure on this space is that defined by the mapping

$$(2.11) \qquad\qquad \gamma_\alpha^\sharp \longrightarrow \zeta_\alpha^\sharp \text{ and } \zeta_\alpha^\sharp \longrightarrow -\gamma_\alpha^\sharp,$$

for $\alpha \in \Delta_+$. Therefore, ω_F is positive definite on this space if and only if

$$(2.12) \qquad\qquad \omega_F(\gamma_\alpha^\sharp, \zeta_\alpha^\sharp) = \lambda([\gamma_\alpha, \zeta_\alpha]) > 0$$

for all $\alpha \in \Delta_+$. However, $[\gamma_\alpha, \zeta_\alpha] = h_\alpha$ so this reduces to the conditions

$$(2.13) \qquad\qquad \lambda(h_\alpha) > 0 \text{ for all } \alpha \in \Delta_+$$

which is equivalent to the assertion: $\lambda \in Int\ \mathfrak{t}_+^*$.

3. Prequantization

Since ω_F is exact, the action of $K \times T$ on X can be prequantized: There is a line bundle-connection pair $(\mathbb{L}, \nabla)$ such that

$$curv(\nabla) = \omega_F.$$

Moreover, $\mathbb{L}$ is equipped with a Hermitian inner product which is invariant under parallel translation, and there is a lifting to $\mathbb{L}$ of the $K \times T$ action on X which preserves ∇ and preserves this inner product. (See [**K**].) Since ω_F is exact, $\mathbb{L}$ is not very interesting as an abstract line bundle: it admits a global non-vanishing section. However, we will show below that the inner product on $\mathbb{L}$ is very interesting. First we will show that $\mathbb{L}$ possesses a global non-vanishing section which is holomorphic and $K \times T$-invariant. Let

$$(3.1) \qquad\qquad \beta = -\sqrt{-1}\partial F.$$

Note that $d\beta = \sqrt{-1}\partial\bar{\partial}F = \omega_F$. We will prove

PROPOSITION 3.1. *There exists a non-vanishing section, s, of $\mathbb{L}$ with the property*

(3.2)
$$\beta = \frac{1}{\sqrt{-1}} \frac{\nabla s}{s}.$$

This section is unique up to a non-zero scalar multiple, is holomorphic and is $K \times T$-invariant.

Proof : *Existence:* As we pointed out above $\mathbb{L}$ possesses a non-vanishing section, s_0. Let

$$\alpha = \frac{1}{\sqrt{-1}} \frac{\nabla s_0}{s_0}.$$

The definition of the curvature form of $\mathbb{L}$ requires that $d\alpha = \omega_F$, so $d(\beta - \alpha) = 0$. Since K is semi-simple

$$H^1(X, \mathbb{R}) = H^1(KA, \mathbb{R}) = H^1(K, \mathbb{R}) = 0,$$

so $\beta - \alpha$ is exact: there exists a complex-valued function, f, for which $\beta = \alpha + df$. Now let $s = (exp\sqrt{-1}f)s_0$. Then

$$\frac{1}{\sqrt{-1}} \frac{\nabla s}{s} = \frac{1}{\sqrt{-1}} \frac{\nabla s_0}{s_0} + df = \beta.$$

Uniqueness: Suppose s_1 and s_2 are non-vanishing sections of $\mathbb{L}$. Let $h = \frac{s_2}{s_1}$. Then

$$\frac{1}{\sqrt{-1}} \frac{\nabla s_2}{s_2} = \frac{1}{\sqrt{-1}} \frac{\nabla s_1}{s_1} + \frac{1}{\sqrt{-1}} d\log h$$

so if s_1 and s_2 both satisfy (3.2) h has to be constant.

By this uniqueness result it is clear that s has to be $K \times T$-invariant. Finally, if v is an anti-holomorphic vector field

$$\frac{1}{\sqrt{-1}} \frac{\nabla_v s}{s} = \iota(v)\beta = 0$$

since β is a form of type $(1, 0)$; so s is holomorphic. $\square$

Next we will compute the function, $\langle s, s \rangle$, and show that it is given by (1.4). Since $\langle s, s \rangle$ and F are both K-invariant, it suffices to show that (1.4) holds for the restrictions of $\langle s, s \rangle$ and F to W. To compute $\langle s, s \rangle$ on W we will, as in §2, make the identification

$$W = \mathbb{C}^n / \mathbb{Z}^n.$$

Let ι be the inclusion map of W into X. Then

$$(3.3) \qquad \imath^*\beta = -\sqrt{-1}\partial_z F = \frac{1}{2}\sum \frac{\partial F}{\partial y_i} dz_i.$$

Let $\nabla_i = \nabla_{\frac{\partial}{\partial y_i}}$. Then

$$\frac{\partial}{\partial y_i}\langle s, s\rangle = \langle \nabla_i s, s\rangle + \langle s, \nabla_i s\rangle.$$

However, by (3.3)

$$\frac{\nabla_i s}{s} = \sqrt{-1}(\beta, \frac{\partial}{\partial y_i}) = -\frac{1}{2}\frac{\partial F}{\partial y_i}$$

so

$$\frac{\partial}{\partial y_i}\log\langle s, s\rangle = -\frac{\partial F}{\partial y_i}$$

and hence, up to a non-zero constant multiple,

$$\langle s, s\rangle = e^{-F}$$

as claimed.

4. Quantization

Let us denote by $\mathcal{O}(\mathbb{L})$ the space of holomorphic sections of $\mathbb{L}$. There is a natural representation of $K \times T$ on $\mathcal{O}(\mathbb{L})$. Moreover, letting s be the section of $\mathbb{L}$ defined by (3.2) we can identify $\mathcal{O}(\mathbb{L})$ with $\mathcal{O}$ by the map

$$(4.1) \qquad \mathcal{O} \longrightarrow \mathcal{O}(\mathbb{L}), \qquad h \mapsto hs.$$

Since s is $K \times T$-invariant this map intertwines the representation of $K \times T$ on $\mathcal{O}$ with the representation of $K \times T$ on $\mathcal{O}(\mathbb{L})$. Let $\mathcal{H}_F$ be the space of holomorphic sections of $\mathbb{L}$ for which the L^2-norm (1.5) is finite, and let ρ_F be the representation of $K \times T$ on this space. Via (4.1) we can identify $\mathcal{H}_F$ with the subspace of $\mathcal{O}$ consisting of those holomorphic functions, h, for which the weighted L^2-norm:

$$(4.2) \qquad \int_X |h|^2 e^{-F} \mu$$

is finite. Let λ be an integer lattice point in $\mathfrak{t}^*_+$ and let χ be the character of T defined by

$$\chi(\exp t) = \exp\sqrt{-1}\lambda(t)$$

We will denote by V_λ the set of all holomorphic functions on X which transform under the right action of T according to χ. Since the representation of K on $\mathcal{O}$ commutes with the representation of T, this space is K-invariant. In fact, as we pointed out in §1, the gist of the Bott-Borel-Weil theorem is that the representation, ρ_λ, of K on V_λ is finite dimensional and irreducible.

THEOREM $\quad \rho_\lambda$ *is contained in* ρ_F *if and only if* λ *is in the image of the moment map* (1.2).

Proof: Recall that $X = KA$ and that the measure, μ, figuring in (4.2) is just $dk\, da$, dk and da being the Haar measures on K and A respectively. As in §2 let us identify A with $\mathbb{R}^n$ and da with the standard Lebesgue measure, dy. Given $k \in K$, let $T_k : \mathcal{O} \to \mathcal{O}$ be the operator

$$T_k h(p) = h(kp).$$

Let $h_1, ..., h_N$ be a basis of V_λ which is orthonormal with respect to the (unique) K-invariant inner product on V_λ. Then, for any $h \in V_\lambda$,

$$h = \sum c_i h_i$$

so

$$h(ky) = T_k h(y) = (\rho_\lambda(k^{-1})h)(y) = \sum c_i a_{ir}(k) h_r(y)$$

where $a_{ir}(k)$ is the ir^{th} matrix coefficient of the representation, ρ_λ, with respect to the basis above. Thus

$$\int |h(ky)|^2 dk = \sum c_i \overline{c_j} \Big(\int a_{ir}(k) \overline{a_{js}}(k) dk \Big) h_r(y) \overline{h_s}(y).$$

However, by Peter-Weyl the inner integral is equal to

$$\frac{1}{N} \delta_{ij} \delta_{rs},$$

(see [**C**], page 186) so the integral (4.2) reduces to the following integral over $\mathbb{R}^n$:

$$(4.3) \qquad \frac{1}{N} \|h\|^2 \int \sum |h_r(y)|^2 e^{-F(y)} dy$$

where $\|h\|$ is the norm of h with respect to the given inner product structure on V_λ. However, each of the functions, $h_r(y)$, transforms under the action of T according to the character, χ, and, therefore, being holomorphic, transforms under the action of TA according to the complexified character, $\chi_{\mathbb{C}}$. In particular, $|h_r(y)|^2$ is a constant multiple of $e^{2\lambda(y)}$, and hence if $h \neq 0$, (4.3) is a non-zero constant multiple of the integral

$$(4.4) \qquad \int e^{-F(y)+2\lambda(y)} dy.$$

By the theorem in $A5$ of the appendix this integral converges if and only if 2λ is in the image of the Legendre transform

$$\frac{\partial F}{\partial y} : \mathbb{R}^n \longrightarrow \mathbb{R}^{n*},$$

or in other words if and only if λ is in the image of the moment mapping (1.2).

5. Examples

Let $\lambda_1, ..., \lambda_K$ be elements of $(\mathbb{R}^n)^*$ whose linear span is all of $(\mathbb{R}^n)^*$ and consider the function

$$F(y) = \sum e^{\lambda_i(y)}.$$

We show in the appendix that this function is strictly convex and that the image of the Legendre transform

$$\frac{\partial F}{\partial y} : \mathbb{R}^n \longrightarrow (\mathbb{R}^n)^*$$

is the open convex cone,

$$C(\lambda_1, ..., \lambda_K) = \{s_1\lambda_1 + ... + s_K\lambda_K, \quad s_i > 0\}.$$

Let's identify $(\mathbb{R}^n)^*$ with $\mathfrak{t}^*$ and, as in §1, choose the λ_i's to be the elements of $\mathfrak{t}^*_+$. Then F satisfies the hypotheses of Theorem I and by Theorem III the irreducible representations of K that occurs as subrepresentations of ρ_F are exactly the representations whose maximal weights are contained in $C(\lambda_1, ..., \lambda_K)$. In particular, if we take the λ_i's to be the dominant fundamental weights (see [**H2**]) this cone is the positive Weyl chamber itself; so *every* irreducible representation whose maximal weight lies in the interior of the positive Weyl chamber occurs as a subrepresentation of ρ_F. We will investigate elsewhere the function

$$F_W(y) = \sum e^{\lambda_i(y)}, \quad \lambda_i \in \overline{W},$$

where the λ_i's are, as above, the dominant fundamental weights and W is a fixed wall of the positive Weyl chamber. This function fails, of course, to meet either of the criteria of Theorem I. However, the two-form

$$\omega_W = \sqrt{-1}\partial\bar{\partial}F_W$$

is of constant rank, and the reduction, X_W, of X by ω_W is a Kaehler manifold on which K acts in a Hamiltonian fashion. The quantization of X_W gives rise to representation, ρ_W, of K in which the irreducibles that occur are exactly those whose maximal weights lie on W.

Appendix

A1 Minkowski's lemma on support hyperplanes:

Let C be an open convex subset of $\mathbb{R}^n$ and let p be a point not in C. Then there exists a hyperplane, H, such that $p \in H$, and $C \cap H = \emptyset$. H is called a *support hyperplane* of C since C is entirely contained in one of the two components of $\mathbb{R}^n - H$.

A2 Let $F \in \mathcal{C}^\infty(\mathbb{R}^n)$ be a strictly convex function. Then the image of the Legendre transform:

(A1)
$$\frac{\partial F}{\partial x} : \mathbb{R}^n \longrightarrow (\mathbb{R}^n)^*$$

is an open convex set. We will denote it by C_F.

A3 A consequence of the Minkowski's lemma is the following

PROPOSITION *Suppose* $0 \notin C_F$. *Then there exists a vector, a, such that for all* $x \in \mathbb{R}^n$, $a \cdot \frac{\partial F}{\partial x} > 0$.

Let U_ϵ be the open set, $\{x \in \mathbb{R}^n, \quad F(x) < \epsilon\}$.

COROLLARY *If* $x \in U_\epsilon$, *all points,* $x - ta$, $t > 0$, *are in* U_ϵ.

This corollary implies that

(A2)
$$\mu_{Lebesgue}(U_\epsilon) = \infty.$$

A4 If $0 \in C_F$, there exist positive constants, ϵ and r, such that

(A3)
$$F(x) \geq \epsilon|x|$$

when $|x| \geq r$.

Proof: *Case 1:* $n = 1$, $x = t$. Since $\frac{d^2 F}{dt^2} > 0$, $\frac{dF}{dt}$ is strictly increasing. If $\frac{dF}{dt} = 0$ at $t = 0$, then for every $t_0 > 0$ there exists an ϵ such that $\frac{dF}{dt} > \epsilon$ when $t > t_0$. Thus $F(t) > \epsilon t + F(t_0)$ on the interval, $t > t_0$.

Case 2: n *arbitrary.* Let p_0 be the unique critical point of F. Let a be a unit vector and apply the previous result to $F(p_0 + ta)$. $\square$

A5 The convex function, F, is said to be *stable* if $0 \in C_F$ and *unstable* if $0 \notin C_F$.

THEOREM *The integral,*

(A4)
$$\int e^{-F(x)} dx$$

is convergent if F is stable and divergent if F is unstable.

Proof: If F is stable, (A4) converges by (A3); and if F is unstable, it diverges by (A2).

A6 Let Δ be a finite subset of $(\mathbb{R}^n)^*$ whose linear span is all of $(\mathbb{R}^n)^*$. Let C_Δ be the open convex cone generated by Δ. We will assume that

$$C_\Delta \neq (\mathbb{R}^n)^*.$$

Let F be the exponential sum

(A5) $$F(x) = \sum e^{\alpha(x)}, \qquad \alpha \in \Delta.$$

Then

(A6) $$\partial F = \sum e^{\alpha(x)} \alpha$$

and

(A7) $$\partial^2 F = \sum e^{\alpha(x)} \alpha \otimes \alpha.$$

It is clear, by (A7) that F is strictly convex. We will prove:

(A8) $$C_F = C_\Delta.$$

Proof: It is clear, by (A6), that $C_F \subseteq C_\Delta$; so we have to prove that $C_\Delta \subseteq C_F$. Let C_Δ^* be the open cone dual to C_Δ in $\mathbb{R}^n$. If $x \in C_\Delta^*$, $\alpha(x) > 0$ for all $\alpha \in \Delta$, so the limit, as t tends to $-\infty$, of $\partial F(tx)$ is zero by (A6). Thus zero is in the closure of C_F. Let Δ_ϵ be the extremal elements of Δ. The (n-1)-dimensional faces of C_F^* are in one-one correspondence with the elements of Δ_ϵ. Indeed, if $\alpha \in \Delta_\epsilon$, the subset of $\mathbb{R}^n$ defined by the inequalities:

(A9) $$\alpha(x) = 0, \beta(x) > 0, \ \beta \in \Delta, \beta \neq \alpha$$

is an (n-1)-dimensional face of C_F, and every (n-1)-dimensional face is of this form. Let x_0 be a point on the (n-1)-dimensional face corresponding to α. Then if x is an exterior point of C_F^* which is sufficiently close to x_0,

$$\alpha(x) < 0 \text{ and } \beta(x) > 0, \ \beta \in \Delta, \beta \neq \alpha.$$

Hence the curve

$$t \to \partial F(tx) = e^{\alpha(x)t} \alpha + \sum_{\beta \neq \alpha} e^{\beta(x)t} \beta$$

tends asymptotically to the ray through α as t tends to $-\infty$. Since C_F is convex the line segment joining $\partial F(tx)$ to the origin lies in C_F, so the ray through α lies in the closure of C_F. Since C_Δ is the convex hull of these rays, $C_\Delta \subseteq C_F$.

References

[C]   C. Chevalley, *Theory of Lie Groups*, Princeton University Press, 1946.

[GZ] I. M. Gelfand and A. Zelevinski, *Models of representations of classical groups and their hidden symmetries*, Funct. Anal. Appl. **18** (1984), 183–198.

[GS] V. Guillemin and S. Sternberg, *Symplectic Techniques in Physics*, Cambridge University Press, 1984.

[H1] S. Helgason, *Differential geometry, Lie groups, and Symmetric Spaces*, Academic Press, 1978.

[H2] _____, *Groups and Geometric Analysis*, Academic Press, 1984.

[K]   B. Kostant, *Quantization and unitary representations*, Lecture Notes in Math., vol. 170, Springer, 1970, pp. 87–208.

[LNS] HoSeong La, P. Nelson, A. S. Schwarz, *Virasoro Model Space*, Comm. Math. Phys. **134** (1990), 523–537.

[La] M. Lassalle, *Deux généralisations du théorème des trois cercles de Hadamard*, Math. Ann. **249** (1980), 17–26.

DEPARTMENT OF MATHEMATICS, MASSACHUSETTS INSTITUTE OF TECHNOLOGY, CAMBRIDGE, MASSACHUSETTS 02139, USA

E-mail address: mchuah@math.mit.edu

DEPARTMENT OF MATHEMATICS, MASSACHUSETTS INSTITUTE OF TECHNOLOGY, CAMBRIDGE, MASSACHUSETTS 02139, USA

E-mail address: vwg@math.mit.edu

Contemporary Mathematics
Volume **154**, 1993

Cousin Complexes and Resolutions of Representations

JOHN W. RICE

Introduction

Cousin complexes, invented by Grothendieck and introduced in [**5**], are complexes attached to a sheaf in terms of a chain of closed subspaces. A finite descending chain $M = Z_0 \supset Z_1 \supset \cdots \supset Z_n$ of closed subspaces of a topological space M is called a filtration of M. Filtrations arise, for example, from cell decompositions, by letting Z_k be the union of all cells of codimension $\geq k$. Just as an open cover of M associates to each sheaf its Cech complex, so a filtration associates to each sheaf its Cousin complex. For example, if the filtration arises from a triangulation, which is a special case of a cell decomposition, then the Cousin complex attached to the constant sheaf $\mathbb{R}$ is the usual simplicial complex of the triangulation. The Cousin complex attached to $\mathbb{R}$ via a general cell decomposition is called the cellular complex of the cell decomposition [**7**]. More recently it has been called the Witten complex on account of Witten's description and infinite dimensional generalisations of it in connection with Morse theory [**11**].

Just as the Cech complex of a sheaf computes its cohomology under certain conditions, so the Cousin complex of a sheaf computes its cohomology under certain conditions. For example when the sheaf is $\mathbb{R}$ and the filtration arises from a cell decomposition, the Cousin complex, which in this case is the cellular complex, computes the cohomology of $\mathbb{R}$, i.e. the cohomology of the manifold M itself. In analysis sheaf cohomology is most often realised as de Rham or Dolbeault cohomology, and the isomorphism between Cech and de Rham cohomology is made via a double complex, namely the Cech-de Rham complex, which always computes de Rham cohomology, and for a good cover also computes Cech cohomology [**2**]. However the exposition of Cousin complexes has been left at the level of general injective or flabby resolutions rather than anything de Rham or

1991 *Mathematics Subject Classification.* Primary 14F05, 22E47; Secondary 58A12, 55N30.
This paper is in final form and no version of it will be submitted for publication elsewhere.

Dolbeault, and their connection with sheaf cohomology has been made through the general machinery of spectral sequences. Although it may not advance the theory appreciably, for the sake of the vast army of analysts who have yet to make their peace with spectral sequences it seems worthwhile to give an account of Cousin complexes in de Rham terms, and to show that there is effectively a Cousin de Rham double complex which always computes de Rham cohomology, and which under suitable circumstances computes Cousin cohomology also.

A particularly satisfying consequence of the theory is that the identity between de Rham cohomology and the classical simplicial cohomology of a triangulation is established as directly as between de Rham and Cech cohomology. This also proves equality of the Euler characteristic of the de Rham complex and the Euler characteristic of the triangulation in the traditional sense of the alternating sum of the number of its cells in each dimension. Indeed, the Betti numbers of the simplicial complex are the same as those of the de Rham complex and so their Euler characteristics are the same. But it is well known and simple linear algebra that for any complex $\mathcal{C}^\bullet$ of finite dimensional vector spaces

$$\sum_{k=0}^{n}(-1)^k \dim H^k(\mathcal{C}) = \sum_{k=0}^{n}(-1)^k \dim \mathcal{C}^k.$$

For the simplicial complex the sum on the right is $n_0 - n_1 + n_2 - \ldots$, which is the alternating sum of the number of cells of the triangulation in each dimension, i.e. the classical Euler characteristic.

In representation theory filtrations often arise via the orbits of the action of a lie group on a manifold X. If G is a complex reductive lie group and B a Borel subgroup then the orbits of B on the flag manifold $X = G/B$ are called Schubert cells, and they form a cell decomposition. In particular, the Cousin complex of $\mathbb{R}$ based on the Schubert cell filtration computes the de Rham cohomology of X. However, because they are complex cells the Schubert cells have even real dimensions, and so the Cousin complex attached to $\mathbb{R}$ is zero in odd degrees. All of its differentials are therefore equal to zero, and the cohomology of this complex, and hence the de Rham cohomology of X, is just the Cousin complex itself. This particular advantage in computing the cohomology of a manifold from a cell decomposition which has cells in only even dimensions was first observed and exploited by Morse in the form of his Morse Lacunary Principle.

If $\mathcal{L}_\lambda$ is the line bundle associated with a dominant integral weight λ then the sheaf of holomorphic sections $H^0(X, \mathcal{L}_\lambda)$ is an irreducible representation of G. Kempf [6] noticed that the meromorphic version of the Cousin complex associated to $\mathcal{L}_\lambda$ by the Schubert cell filtration could be identified with the dual of a resolution of this irreducible representation invented by Bernstein Gelfand and Gelfand [1]. Their construction was combinatorial, using properties of the Weyl group to piece together inclusions of Verma modules with appropriate signs so as to make a resolution by direct sums of Verma modules, ending on the right with the canonical map of a Verma module onto its irreducible quotient.

They pointed out that the Weyl Character Formula was a consequence of this resolution. In Cousin-Dolbeault theory the Cousin complex for $\mathcal{L}_\lambda$ determined by the Schubert cell filtration is in each degree a direct sum of terms attached to Schubert cells, and one can show that the terms are the duals of Verma modules. As Kempf pointed out, the fact that the Cousin complex is a resolution of the irreducible representation associated with $\mathcal{L}_\lambda$ follows because by Bott's Theorem the Dolbeault complex is already such a resolution. The Weyl Character Formula follows by the equality of the Euler characteristics of the two complexes treated as characters of representations, and so reveals itself as a generalisation of the equality between the Euler characteristic of the de Rham complex of a manifold, and the Euler characteristic of a triangulation.

Cousin de Rham theory is based on local cohomology, and a de Rham exposition of local cohomology is sketched in sections 1 and 2. The Cousin de Rham double complex is an instance of a general piece of algebraic machinery described in section 3. It might aptly be called the double complex attached to a filtered complex. It is a lifting of the E_1 term of a spectral sequence to the level of complexes, which allows us to deal with only the beginning part of the theory of spectral sequences, effectively avoiding the derivation of exact couples and convergence. In order to treat Cousin Dolbeault theory in the same way as Cousin de Rham, we would need to consider Dolbeault complexes with hyperfunction coefficients. However, for the meromorphic version which produces the BGG resolution, currents or distribution coefficients are just the right thing. This is because the Malgrange Division Theorem [9] shows that currents are acyclic for algebraic local cohomology, which is the kind arising in the meromorphic Cousin complex, and which produces the BGG resolution [10].

The differentials of a Cousin complex occur as connecting homomorphisms, which in the meromorphic case amount to taking higher order residues or principal parts of meromorphic forms along subvarieties, as we illustrate in section 5. For this reason Cousin complexes in the meromorphic case have sometimes been referred to as residue complexes, and this homological interpretation of residues in part explains their role in 'Residues and Duality' [5]. The analytical description of Cousin complexes, sketched in sections 4 and 5, is joint work with M.K.Murray, and full details appear in [10].

1. Local cohomology

Cousin complexes are defined in terms of local cohomology, which can be regarded as cohomology restricted or localised to sections of sheaves supported on some closed set, such as a submanifold. However, if we attempt to enact this idea using the de Rham complex, taking the subcomplex of the de Rham complex consisting of forms supported on some closed submanifold, we will be disappointed to obtain only the zero complex. In order to obtain something worthwhile let us copy the theory of distributions by defining for any open set U,

$$\mathbb{D}^k(U) = \Omega_c^{n-k}(U)'$$

where E' means the algebraic dual of the vector space E. We will call the elements of $\mathbb{D}^k(U)$ the generalised forms of degree k on U, noting that by contrast with the theory of distributions, we ignore questions of topology completely. Restricting ourselves to the case where U is orientable, each actual k-form $\omega \in \Omega^k(U)$ defines a generalised form $\hat{\omega} \in \mathbb{D}^k(U)$ via integration

$$\hat{\omega}(\phi) = \int_U \omega_{\wedge}\phi.$$

The differential defined on each $T \in \mathbb{D}^k(U)$ by

$$dT(\phi) = (-1)^{n-k}T(d\phi)$$

makes $\mathbb{D}^\bullet(U)$ into a complex and

$$\Omega^\bullet(U) \xrightarrow{\wedge} \mathbb{D}^\bullet(U)$$

into a morphism of complexes, meaning that $\widehat{d\omega}(\phi) = d\hat{\omega}(\phi)$, or more explicitly that

$$\int_U d\omega \wedge \phi = (-1)^k \int_U \omega \wedge d\phi.$$

If $V \subset U$ then $\Omega_c^\bullet(V)$ can be regarded as the subspace of $\Omega_c^\bullet(U)$ obtained by extending compactly supported forms on V by zero. Hence the restriction map $\mathbb{D}^\bullet(U) \to \mathbb{D}^\bullet(V)$ defined by restriction of linear functions on $\Omega_c^\bullet(U)$ to $\Omega_c^\bullet(V)$ makes $\mathbb{D}^\bullet$ into a presheaf. Partition of unity arguments show that, for a cover $\{V_\alpha\}$ of U, compatible generalised forms $\sigma_\alpha \in \mathbb{D}^k(V_\alpha)$ patch together to a unique $\sigma \in \mathbb{D}^k(U)$, and so $\mathbb{D}^\bullet$ is actually a sheaf. Moreover, $\omega \to \hat{\omega}$ is compatible with restriction, so that $\Omega^\bullet \xrightarrow{\wedge} \mathbb{D}^\bullet$ is a morphism of complexes of sheaves. It allows us to regard the complex of generalised forms as an enlargement of the de Rham complex. Let us call $\mathbb{D}^\bullet$ the generalised de Rham complex.

If U is diffeomorphic to $\mathbb{R}^n$ then the Poincaré Lemma says that $\Omega^\bullet(U)$ is exact except in degree 0 where its cohomology is $\mathbb{R}$. On the other hand the compactly supported Poincaré Lemma says that $\Omega_c^\bullet(U)$ is exact except in degree n, where its cohomology is $\mathbb{R}$. By duality $\mathbb{D}^\bullet(U)$ has the same cohomology as $\Omega^\bullet(U)$, so that the compactly supported Poincaré Lemma can be regarded by duality as the Poincaré Lemma for the generalised de Rham complex. It follows that $\mathbb{D}^\bullet$ is every bit as good for computing cohomology as $\Omega^\bullet$. For example, the arguments using the Cech de Rham complex $\mathbb{C}^\bullet(\mathcal{U}, \Omega^\bullet)$ for a good cover $\mathcal{U}$, which prove that Cech cohomology is isomorphic to the cohomology of $\Omega^\bullet(M)$, will also work with the generalised Cech de Rham complex $\mathbb{C}^\bullet(\mathcal{U}, \mathbb{D}^\bullet)$ and show that Cech cohomology coincides with that of $\mathbb{D}^\bullet(M)$. In particular, $\Omega^\bullet(M)$ and $\mathbb{D}^\bullet(M)$ have isomorphic cohomology, and the arguments show that $\Omega^\bullet(M) \xrightarrow{\wedge} \mathbb{D}^\bullet(M)$ induces the isomorphism, which is a statement of Poincaré Duality.

Let us pause to consider these observations from an advanced standpoint. A morphism of complexes which induces an isomorphism between their cohomology groups is called a quasi-isomorphism. The Poincaré Lemmas imply that $\Omega^\bullet(U) \xrightarrow{\wedge} \mathbb{D}^\bullet(U)$ is a quasi-isomorphism if U is diffeomorphic to $\mathbb{R}^n$, and hence $\Omega^\bullet \xrightarrow{\wedge} \mathbb{D}^\bullet$ is a quasi-isomorphism of complexes of sheaves. The fact that the morphism induced between global sections $\Omega^\bullet(M) \xrightarrow{\wedge} \mathbb{D}^\bullet(M)$ is also a quasi-isomorphism, i.e. Poincaré Duality, can be deduced directly from a general result of homological algebra, namely the Acyclicity Theorem [8]. It applies because both the de Rham complex and generalised de Rham complex admit partitions of unity, i.e. are complexes of fine sheaves, and hence are acyclic for the global section functor. In fact $\mathbb{D}^\bullet$ is better than fine, it is actually injective, so that from the derived functor point of view we should take $\mathbb{D}^\bullet(M)$ as the definition of the cohomology of M. As a halfway proof of injectivity let us show that $\mathbb{D}^\bullet$ is flabby, meaning that for any closed set Z, $\mathbb{D}^\bullet(M) \to \mathbb{D}^\bullet(M\backslash Z)$ is onto, or that any generalised form on $M\backslash Z$ extends to one on M. This is true simply because any linear function on the vector subspace $\Omega_c(M\backslash Z) \subset \Omega_c(M)$ extends to one on $\Omega_c(M)$. Proof of injectivity is an argument of the same kind.

We define $\mathbb{D}_Z^\bullet(M)$ to be the kernel of the restriction map above, i.e. we define it by the short exact sequence

$$0 \to \mathbb{D}_Z^\bullet(M) \to \mathbb{D}^\bullet(M) \to \mathbb{D}^\bullet(M\backslash Z) \to 0$$

and say that the generalised forms in $\mathbb{D}_Z^\bullet(M)$ are supported in Z. An important example of a generalised form supported in an oriented submanifold Z of codimension k is the Thom form $\tau_Z \in \mathbb{D}_Z^k(M)$ defined by

$$\tau_Z(\psi) = \int_Z i^*\psi \qquad \psi \in \Omega_c^{n-k}(M), \qquad Z \xrightarrow{i} M.$$

If the support of ψ lies in the complement of Z then the integral is zero, showing that τ_Z lies in $\mathbb{D}_Z^\bullet(M)$. If Z is a point p then τ_p is just the δ-form at p.

The local cohomology $H_Z^\bullet(M)$ of M along Z is defined to be the cohomology of the complex $\mathbb{D}_Z^\bullet(M)$. In terms of general theory, since $\mathbb{D}^\bullet$ is an injective resolution of $\mathbb{R}$, these are the right derived functors $R^\bullet\Gamma_Z(\mathbb{R})$, where the functor Γ_Z is defined on sheaves by

$$0 \to \Gamma_Z(M, \mathcal{F}) \to \Gamma(M, \mathcal{F}) \to \Gamma(M\backslash Z, \mathcal{F}).$$

Note that

$$d\tau_Z(\psi) = (-1)^{k+1}\tau(d\psi) = (-1)^{k+1}\int_Z d\psi = 0$$

by Stokes Theorem. Hence τ_Z is closed, and its cohomology class in $H_Z^k(M)$ is called the Thom class of Z.

2. Localisation and the Thom isomorphism

The short exact sequence of complexes that defines $\mathbb{D}_Z^\bullet(M)$ produces a long exact sequence in cohomology, which will allow us to compute $H^\bullet(M)$ from a knowledge of $H^\bullet(M\backslash Z)$ and $H_Z^\bullet(M)$. This is feasible because of two principles for the calculation of $H_Z^\bullet(M)$, called Localisation and the Thom Isomorphism.

Localisation, also known as excision, says that the restriction map $\mathbb{D}_Z^\bullet(M) \to \mathbb{D}_Z^\bullet(U)$ is a quasi-isomorphism for $U \supset Z$ (and U is open). In fact, it is elementary to prove that the restriction map is an isomorphism of complexes in this case. The restriction map $\mathbb{D}^\bullet(M) \to \mathbb{D}^\bullet(M\backslash Z)$ factorises as a composition of restriction maps $\mathbb{D}^\bullet(M) \to \mathbb{D}^\bullet(U) \to \mathbb{D}^\bullet(M\backslash Z)$, allowing us to prove surjectivity, while multiplication of test forms by a cut off function, which is supported in U and has the value 1 in a neighbourhood of Z, allows us to prove injectivity.

For any orientable submanifold Z of codimension k we can define

$$\omega \wedge \tau_Z(\psi) = \int_Z \omega \wedge i^*\psi$$

for each $\psi \in \Omega_c^{n-k-p}(Z)$, so that $\omega \wedge \tau_Z \in \mathbb{D}^{k+p}(M)$. This defines a morphism of complexes

$$\Omega^\bullet(Z) \xrightarrow{\wedge\tau_Z} \mathbb{D}_Z^\bullet(M)[k]$$

where $A^\bullet[k]$ denotes $A^\bullet$ shifted down by k degrees and with $(-1)^k d$ for its differential. The Thom Ismorphism says that $\wedge\tau_Z$ is a quasi-isomorphism. In order to prove the Thom Isomorphism it suffices, according to Localisation, to replace M by a tubular neighbourhood $E \xrightarrow{\pi} Z$. For a tubular neighbourhood we can define a morphism $\mathbb{D}_Z^\bullet(E) \xrightarrow{\pi_*} \mathbb{D}^\bullet(Z)$ by

$$\pi_*\theta(\psi) = \theta(\rho\pi^*\psi)$$

where ρ is any tubular cut-off function. We will show that this induces the inverse to the Thom morphism, noting that $H^\bullet(Z)$ is the cohomology of both $\Omega^\bullet(Z)$ and $\mathbb{D}^\bullet(Z)$. Indeed, the diagram

$$\mathbb{D}_Z^\bullet(E)$$
$$\wedge\tau_Z \nearrow \qquad \searrow \pi_*$$
$$\Omega^\bullet(Z) \xrightarrow{\wedge} \mathbb{D}^\bullet(Z)$$

commutes, and since the bottom line induces the Poincaré duality isomorphism in cohomology, it follows that $\wedge\tau_Z$ is injective and π_* surjective on cohomology. If π_* can be proved injective on cohomology then both $\wedge\tau_Z$ and π_* induce isomorphisms on cohomology which are inverse to each other.

The injectivity of π_* on cohomology is proved using the general Poincaré Lemma. It says that for a vector bundle $E \xrightarrow{\pi} Z$, if $Z \xrightarrow{s} E$ is the zero section, then π^* and s^* induce inverse isomorphisms in cohomology. It does this by establishing a homotopy formula

$$\psi - \pi^* s^* \psi = d(F\psi) + F(d\psi)$$

where F is the homotopy map on forms [2]. In order to prove π_* injective on cohomology we must show that if $\theta \in \mathbb{D}_Z^k(E)$ is closed, and $\pi_*\theta$ is exact, then θ is exact also. By duality, in order that θ be closed it is necessary and sufficient that it vanish on all exact test forms ψ. In order that it be exact it is necessary and sufficient that it vanish on all closed test forms. Suppose therefore that $d\psi = 0$, and that ρ is a tubular cut-off function which is one in a neighbourhood of Z and of compact support along the fibres of $E \xrightarrow{\pi} Z$. Then we have

$$
\begin{aligned}
\theta(\psi) &= \theta(\rho\psi) \\
&= \theta(\rho\pi^* s^* \psi) + \theta(\rho d(F\psi)) \\
&= \pi_*\theta(s^*\psi) + \theta(d(\rho F\psi)) - \theta(d\rho \wedge F\psi).
\end{aligned}
$$

Since ψ is closed so is $s^*\psi$ and hence the first term is zero by the exactness of $\pi_*\theta$. The second term is zero because θ is closed, and the last is zero because $d\rho$ has support in the complement of Z. It follows that θ vanishes on any closed test form, and is therefore exact. This proves that π_* is injective on cohomology.

We conclude this section with some basic examples of the computation of local cohomology using the Thom Isomorphism.

Example 1

Suppose that Z is diffeomorphic to a k-cell, i.e. to $\mathbb{R}^k$. Then the quasi-isomorphism $\Omega^\bullet(Z) \Rightarrow \mathbb{D}_Z^\bullet(M\backslash\partial Z)[k]$ implies that that

$$
\begin{aligned}
H_Z^j(M\backslash\partial Z) &= \mathbb{R} & j = k \\
&= 0 & j \neq k.
\end{aligned}
$$

Example 2

Given an oriented triangulation of an oriented manifold M, let Z_k be the union of all faces of dimension $\geq k$. Let n_k be the number of faces of dimension k. Then $Z_k\backslash Z_{k+1}$ is a disjoint union of n_k cells, and

$$
\begin{aligned}
H_{Z_k\backslash Z_{k+1}}^j(M\backslash Z_{k+1}) &= R^{n_k} & j = k \\
&= 0 & j \neq k.
\end{aligned}
$$

3. The double complex of a filtered complex

A filtration of M

$$M = Z_0 \supset Z_1 \supset Z_2 \supset \ldots Z_n$$

produces the obvious filtration of $\mathbb{D}^\bullet(M)$ by supports

$$\mathbb{D}^\bullet(M) \supseteq \mathbb{D}_{Z_1}^\bullet(M) \supseteq \mathbb{D}_{Z_2}^\bullet(M) \supseteq \cdots .$$

Note that the map $\mathbb{D}_{Z_k}^\bullet(M) \to \mathbb{D}_{Z_k\backslash Z_{k+1}}^\bullet(M\backslash Z_{k+1})$ defined by restriction to the complement of Z_{k+1} is surjective, by the same argument which showed that

$\mathbb{D}^\bullet(M) \to \mathbb{D}^\bullet(U)$ is surjective for any open U. Its kernel consists of those $\theta \in \mathbb{D}^\bullet_{Z_k}(M)$ which vanish on $Z_k \backslash Z_{k+1}$ i.e. outside Z_{k+1}. It follows that we have an exact sequence

$$0 \to \mathbb{D}^\bullet_{Z_{k+1}}(M) \to \mathbb{D}^\bullet_{Z_k}(M) \to \mathbb{D}^\bullet_{Z_k \backslash Z_{k+1}}(M \backslash Z_{k+1}) \to 0.$$

Example 2 computes the cohomology of this last complex $H^\bullet_{Z_k \backslash Z_{k+1}}(M \backslash Z_{k+1})$, as $\mathbb{R}^{n_k}$ in degree k and 0 otherwise, in the case where $\{Z_k\}$ is a cell decomposition.

The connecting homomorphism of the exact sequence above, followed by the restriction map at the next level, gives homomorphisms

$$H^p_{Z_k \backslash Z_{k+1}}(M \backslash Z_{k+1}) \to H^{p+1}_{Z_{k+1} \backslash Z_{k+2}}(M \backslash Z_{k+2}).$$

Consider what would transpire if we could produce morphisms $\mathbb{D}^\bullet_{Z_k \backslash Z_{k+1}} \xrightarrow{\delta} \mathbb{D}^\bullet_{Z_{k+1} \backslash Z_{k+2}}[1]$ which not only induce these homomorphisms in cohomology, but chain together into an exact sequence of complexes

$$0 \to \mathbb{D}^\bullet(M) \to \mathbb{D}^\bullet(M \backslash Z_1) \xrightarrow{\delta} \mathbb{D}^\bullet_{Z_1 \backslash Z_2}(M \backslash Z_2)[1] \xrightarrow{\delta} \mathbb{D}^\bullet_{Z_2 \backslash Z_3}(M \backslash Z_3)[2] \xrightarrow{\delta} \cdots.$$

Then the δ's would constitute the differentials of a double complex whose terms are

$$\mathbb{C}s^{pq}(\{Z_k\}) = \mathbb{D}^p_{Z_q \backslash Z_{q+1}}(M \backslash Z_{q+1})$$

and whose rows are exact except at the left hand end, where the kernel is $\mathbb{D}^\bullet(M)$. By the zig-zig lemma [2] we would have a quasi-isomorphism

$$\mathbb{D}^\bullet(M) \Rightarrow \mathrm{Tot}^\bullet(\mathbb{C}s^{\bullet\bullet}(\{Z_k\})).$$

Hence, like the Cech de Rham complex, the double complex $\mathbb{C}s^{\bullet\bullet}(\{Z_k\})$ would always compute the de Rham cohomology.

Consider the case of a filtration arising from a cell decomposition. By Example 2, $H^k_{Z_k \backslash Z_{k+1}}(M \backslash Z_{k+1}) = \mathbb{R}^{n_k}$, but because we have shifted the complexes down one degree at a time, this vertical cohomology occurs in degree zero, i.e. in the bottom row, and is zero elsewhere. The map induced by δ in (vertical) cohomology gives a differential, and the cohomology of the complex occurring in the bottom row

$$0 \to \mathbb{R}^{n_0} \xrightarrow{\delta} \mathbb{R}^{n_1} \xrightarrow{\delta} \mathbb{R}^{n_2} \cdots \xrightarrow{\delta} \qquad\qquad [*]$$

also computes the cohomology of $\mathrm{Tot}^\bullet(\mathbb{C}s^{\bullet\bullet}(\{Z_k\}))$. It therefore computes the cohomology of $\mathbb{D}^\bullet(M)$, and of $\Omega^\bullet(M)$ also. This establishes the identification of the classical Euler characteristic defined in terms of triangulations, and the Euler characteristic of the de Rham complex, as explained in section 1, as well as showing that the cohomology of the flag manifold in any dimension is equal to the number of Schubert cells in that dimension.

Given these exciting consequences, it is a great sadness that it is not possible to invent the kind of differential required to produce a double complex from $\mathbb{D}^\bullet_{Z_k \backslash Z_{k+1}}(M \backslash Z_{k+1})$. However, nothing turns out to be lost, since all of these consequences are derived from the cohomology of these complexes, rather than from the complexes themselves. If we can find quasi-isomorphic covers

$$\tilde{\mathbb{D}}^\bullet_{Z_k \setminus Z_{k+1}} (M \setminus Z_{k+1}) \Rightarrow \mathbb{D}^\bullet_{Z_k \setminus Z_{k+1}} (M \setminus Z_{k+1})$$

which admit such differentials, then we shall be every bit as well off as if we could have produced differentials between the complexes themselves.

Following [**10**], let us explain how to produce such a lifted double complex in the general setting of a filtered complex

$$F^\bullet = F^{\bullet 0} \supset F^{\bullet 1} \supset F^{\bullet 2} \supset F^{\bullet 3} \supset \cdots .$$

We form the quotient complexes

$$0 \to F^{\bullet k+1} \to F^{\bullet k} \to G^{\bullet k} \to 0.$$

Now consider the theory of the long exact sequence for a short exact sequence of complexes

$$0 \to A^\bullet \xrightarrow{i} B^\bullet \xrightarrow{\pi} C^\bullet \to 0.$$

One takes an element $c \in C^k$, lifts it back to some $b \in B^k$, forms db and lifts that back to some $a \in A^{k+1}$, as summarised by the 'zig-zag' picture

$$
\begin{array}{ccc}
a & \xrightarrow{\ i\ } & db \\
 & & \uparrow \\
b & \xrightarrow{\ \pi\ } & c.
\end{array}
$$

The lift back to a only exists if c is closed, and then one shows that a is closed, and has a cohomology class depending only on that of c. The map $[c] \to [a]$ is the connecting homomorphism.

Although the focus of this manoeuvre is to produce a from c, we cannot escape the fact that in reality we have produced a pair $\binom{a}{b}$ covering c by the map $\binom{a}{b} \to \pi(b)$ and satisfying

$$D \binom{a}{b} := \begin{pmatrix} -d & 0 \\ -i & d \end{pmatrix} \binom{a}{b} = \binom{0}{0}.$$

It is easy to see that $D^2 = 0$, and so we have produced a complex $\tilde{C}^\bullet$, called the mapping cone of i, whose terms are the same as those of $A^\bullet[1] \oplus B^\bullet$ but with differential D. What the construction of the connecting homomorphism really shows is that closed elements of $C^\bullet$ lift to closed elements of $\tilde{C}^\bullet$ and exact elements lift to exact elements. This proves $\tilde{C}^\bullet$ quasi-isomorphic to $C^\bullet$ by $\binom{a}{b} \to \pi(b)$. Moreover,

$$
\begin{array}{ccccccc}
0 & \to & B^\bullet & \longrightarrow & \tilde{C}^\bullet & \longrightarrow & A^\bullet[1] & \to & 0 \\
 & & b & \mapsto & \binom{0}{b}, \binom{a}{b} & \mapsto & a
\end{array}
$$

is an exact sequence making the diagram

$$
\begin{array}{ccc}
& C^\bullet & \rightsquigarrow & A^\bullet[1] \\
& {}^\pi\nearrow & \Uparrow \quad \nearrow & \\
A^\bullet \xrightarrow{\ i\ } & B^\bullet & \xrightarrow{\ \delta\ } & \tilde{C}^\bullet
\end{array}
$$

commute. The curly arrow doesn't exist as a morphism of complexes but identifying $H^\bullet(\tilde{C})$ with $H^\bullet(C)$, it induces the connecting homomorphism.

Returning to the quotient complexes of the filtered complex, we can form the diagram

$$
\begin{array}{ccccccccccc}
G^{\bullet 0} & \rightsquigarrow & F^{\bullet 1}[1] & \rightsquigarrow & G^{\bullet 1}[1] & \rightsquigarrow & F^{\bullet 2}[2] & \rightsquigarrow & G^{\bullet 2}[2] & \rightsquigarrow & F^{\bullet 3}[3] \quad \cdots \\
\Uparrow \quad \nearrow & & \searrow \quad \Uparrow \quad \nearrow & & & \searrow \quad \Uparrow \quad \nearrow & & & & & \\
F^\bullet \ \to\ \tilde{G}^{\bullet 0} & & \longrightarrow & & \tilde{G}^{\bullet 1}[1] & & \longrightarrow & & \tilde{G}^{\bullet 2}[2] & & \longrightarrow \quad \cdots \\
\binom{f'}{f} & \longmapsto & & \binom{0}{f'} & \binom{f'}{f} & \longmapsto & & \binom{0}{f'} & \binom{f'}{f} & \longmapsto & \cdots
\end{array}
$$

The lower row is clearly an exact sequence of complexes. Let us call it the double complex associated to a filtered complex. The upper row shows that at the level of cohomology, $G^{\bullet 0}$ and $\tilde{G}^{\bullet 0}$ are quasi-isomorphic, and $\tilde{G}^{\bullet k} \to \tilde{G}^{\bullet k+1}$ induces the map $H^\bullet(G^{\bullet k}) \to H^\bullet(G^{\bullet k+1})$ which is the connecting homomorphism $G^{\bullet\bullet} \to F^{\bullet\bullet}[1]$ followed by the quotient $F^{\bullet\bullet} \to G^{\bullet\bullet}$.

The story is now complete. Given the generalised de Rham complex, and a filtration by closed sets $M = Z_0 \supset Z_1 \supset Z_2 \cdots$ the Cousin de Rham complex is the double complex associated to the filtered complex

$$
\mathbb{D}^\bullet(M) \supset \mathbb{D}^\bullet_{Z_1}(M) \supset \mathbb{D}^\bullet_{Z_2}(M) \supset \cdots .
$$

The point is that we can behave as though the quotients in

$$
0 \to \mathbb{D}^\bullet_{Z_{k+1}}(M) \to \mathbb{D}^\bullet_{Z_k}(M) \to \mathbb{D}^\bullet_{Z_k \backslash Z_{k+1}}(M \backslash Z_{k+1}) \to 0
$$

form a double complex

$$
\tilde{\mathbb{D}}^\bullet_{M \backslash Z_1}(M) \to \tilde{\mathbb{D}}^\bullet_{Z_1 \backslash Z_2}(M \backslash Z_2)[1] \to \tilde{\mathbb{D}}^\bullet_{Z_2 \backslash Z_3}(M \backslash Z_3)[2] \to \cdots
$$

which computes de Rham cohomology. The complexes induced in the vertical cohomology along the rows are called Cousin complexes. Their terms consists of local cohomology spaces, and their differentials consist of the connecting homomorphism of each exact sequence followed by restriction to the complement of the next level.

4. The meromorphic Cousin-Dolbeault complex

On a complex manifold M let us consider the Dolbeault analogue of the generalised de Rham complex. Let $\Omega^{r,s}_c(U)$ denote the smooth forms of type (r, s) with compact support in the open set $U \subset M$, and define

$$
\mathbb{D}^{p,q}(U) = \Omega^{n-p,n-q}_c(U)'.
$$

Define $\mathbb{D}^{p,q}(U) \xrightarrow{\bar{\partial}} \mathbb{D}^{p,q+1}(U)$ by

$$\bar{\partial}T(\psi) = (-1)^{q+1}T(\bar{\partial}\psi)$$

and for $\omega \in \Omega^{p,q}(U)$ define

$$\hat{\omega}(\psi) = \int_U \omega \wedge \psi \qquad \psi \in \Omega_c^{n-p,n-q}(U).$$

As in the de Rham case

$$\Omega^{p,q}(U) \xrightarrow{\wedge} \mathbb{D}^{p,q}(U)$$

is a morphism of complexes. However, it is not a quasi isomorphism. In particular the kernel of $\mathbb{D}^{0,0} \xrightarrow{\bar{\partial}} \mathbb{D}^{0,1}$ is much larger than the sheaf of holomorphic functions. The duality arguments which worked for the de Rham case fail because the vector spaces involved in the Dolbeault case are infinite dimensional.

The failure is rectified if we define

$$\mathbb{D}^{p,q}(U) = \Omega_c^{n-p,n-q}(U)^*$$

where $*$ denotes the topological dual of $\Omega_c^{n-p,n-q}(U)$ when it carries the topology of uniform convergence of all derivatives of sequences whose supports lie in a given compact set. The elements of $\mathbb{D}^{p,q}(U)$ are called currents of type (p,q) and locally appear as differential forms with distribution coefficients. For the Dolbeault complex of currents, elliptic regularity theorems show that $\wedge$ is a quasi-isomorphism [4]. However, unlike the sheaves of generalised forms, the sheaves of currents are not injective, or even flabby. In particular the restriction map

$$\mathbb{D}^{p,q}(M) \to \mathbb{D}^{p,q}(M\backslash Z)$$

is not surjective in general. The definition of the Cousin Dolbeault complex still makes sense, but it does not relate in any simple way to $H_Z(M,\mathcal{O})$, defined via the right derived functor of $\Gamma_Z(M,-)$.

The Dolbeault complex of currents relates instead to a meromorphic or algebraic version of Cousin theory [10]. In order to introduce it, let us consider the concept of meromorphic sections of a sheaf with poles confined to a given complex subvariety $Z \subset M$. Classically such a section is one defined on the complement of Z, but having the property that, for any holomorphic function ϕ vanishing on Z to a sufficiently high order, ϕs has an extension across Z. To be precise, let $\mathcal{Z}$ denote the sheaf of holomorphic functions vanishing on Z and $\mathcal{Z}^n$ be the sheaf of ideals generated by products of at least n functions from $\mathcal{Z}$. Then a section s defined on the complement of Z is said to be meromorphic with poles confined to Z if, for some n, each point $z \in Z$ has a neighbourhood on which ϕs has an extension to the whole neighbourhood, for every ϕ in $\mathcal{Z}^n$ defined on the neighbourhood. In classical situations, where the sections of $\mathcal{F}$ are continuous functions, the extensions are unique, and the association of the extension of ϕs to each ϕ in $\mathcal{Z}^n$ defines a homomorphism of sheaves $\mathcal{Z}^n \to \mathcal{F}$.

These homomorphisms can be regarded as describing fractions at each point z, by producing a numerator in $\mathcal{F}_z$ to go with each denominator ϕ in $\mathcal{Z}_z^n$. The restriction of each fraction to the complement of Z reproduces the section s.

In non-classical situations, such as when $\mathcal{F}$ is a sheaf of currents, the extensions are not unique, i.e. there are sections which are supported in Z. In this case there are many homomorphisms $\mathcal{Z}^n \to \mathcal{F}$ which, considered as systems of fractions, restrict to the same section of $\mathcal{F}$ on the complement of Z, and so one cannot readily identify meromorphic sections having poles confined to Z with certain sections defined on the complement of Z. In this general case it is the fractions that we call the meromorphic sections of the sheaf, and not the sections defined on the complement of Z. We denote the sheaf of meromorphic sections of the sheaf $\mathcal{F}$ with poles along Z by $\mathcal{F}_{[M/Z]}$ and define it by

$$\mathcal{F}_{[M/Z]} = \varinjlim_n \mathcal{H}om(\mathcal{Z}^n, \mathcal{F})$$

where $\mathcal{H}om$ denotes the sheaf of homomorphisms between sheaves.

Note that the direct limit cannot be regarded simply as the union of the sheaves $\mathcal{H}om(\mathcal{Z}^n, \mathcal{F})$, because there may be some homomorphisms on $\mathcal{Z}^n$ whose restriction to $\mathcal{Z}^{n+m}$ is zero for some $m > 0$. The image of such a homomorphism on $\mathcal{Z}^n$ consists of sections of $\mathcal{F}$ which are annihilated upon multiplication by sections of $\mathcal{Z}^m$. Let us define the sheaf $\mathcal{F}_{[Z]}$ by

$$\mathcal{F}_{[Z]} = \varinjlim_n \mathcal{H}om(\mathcal{O}/\mathcal{Z}^n, \mathcal{F}).$$

Because $\mathcal{Z}^n$ coincides with $\mathcal{O}$ on the complement of Z it follows that the sections of $\mathcal{F}_{[Z]}$ have support in Z, and in fact form an $\mathcal{O}$ submodule of $\mathcal{F}_Z$. We say that these sections are *algebraically supported in* Z.

Note that a local homomorphism $\mathcal{Z}^n \to \mathcal{F}$ restricts to zero on the complement of Z if and only if its image consists of sections of $\mathcal{F}_Z$, while it corresponds to the zero section of $\mathcal{F}_{[M/Z]}$ if and only if its image consists of sections of $\mathcal{F}_{[Z]}$. Hence we recover the classical situation, in which the meromorphic sections of $\mathcal{F}$ with poles confined to Z can be identified with their restrictions to the complement of Z, if and only if $\mathcal{F}_Z$ is equal to $\mathcal{F}_{[Z]}$ over the relevant open set. It is comforting to note that this is the case if $\mathcal{F}$ is a sheaf of currents [**10**]. For example, the space of distributions supported at $0 \in \mathbb{C}^n$ is spanned by the delta function and its derivatives. These are precisely the distributions annihilated by monomials in $z_1, \ldots, z_n$ of sufficiently high order.

The sheaf of homomorphisms into $\mathcal{F}$ applied to the short exact sequences of sheaves

$$0 \to \mathcal{Z}^n \to \mathcal{O} \to \mathcal{O}/\mathcal{Z}^n \to 0$$

yields the exact sequence of sheaves

$$0 \to \mathcal{F}_{[Z]} \to \mathcal{F} \to \mathcal{F}_{[M/Z]}.$$

For any open set U, $\mathcal{F}_{[M/Z]}(U)$ can be regarded as the algebraic part of $\mathcal{F}(U\backslash Z)$, and $\mathcal{F}_{[Z]}$ is effectively defined by the exact sequence above, which should be regarded as an algebraic version of the defining sequence for $\mathcal{F}_Z$. More generally, for subvarieties $Z_2 \subset Z_1$, let us extend the definition of $\mathcal{F}_{[M/Z]}$ by putting

$$\mathcal{F}_{[Z_1/Z_2]} = \varinjlim_n \mathcal{H}om(\mathcal{Z}_2^n, \mathcal{F}_{[Z_1]}) = \varinjlim_n \mathcal{H}om(\mathcal{Z}_2^n/\mathcal{Z}_1^n, \mathcal{F}).$$

It is the sheaf of sections of $\mathcal{F}$ algebraically supported in Z_1 which are meromorphic along Z_2. There is an exact sequence

$$0 \to \mathcal{F}_{[Z_2]} \to \mathcal{F}_{[Z_1]} \to \mathcal{F}_{[Z_1/Z_2]}.$$

In [7] Kempf defines $\mathcal{F}_{[Z_1/Z_2]}$ to be the quotient of $\mathcal{F}_{[Z_1]}$ by $\mathcal{F}_{[Z_2]}$, which is slightly different to our definition, but in the end yields the same theory. Notice that, since $\mathcal{Z}_2 = \mathcal{O}$ when $Z_2 = \emptyset$, it follows that $\mathcal{F}_{[Z/\emptyset]} = \mathcal{F}_{[Z]}$.

We define the algebraic local cohomology sheaves $\mathcal{H}^\bullet_{[Z_1/Z_2]}(\mathcal{F})$ to be the right derived functors of $\underline{\Gamma}_{[Z_1/Z_2]} : \mathcal{F} \to \mathcal{F}_{[Z_1/Z_2]}$, and $H^\bullet_{[Z_1/Z_2]}(\mathcal{F})$ to be the right derived functors of $\Gamma_{[Z_1/Z_2]} : \mathcal{F} \to \mathcal{F}_{[Z_1/Z_2]}(M)$. The most important property of sheaves of currents is that they are acyclic for the functor $\underline{\Gamma}_{[Z_1/Z_2]}$, and its global counterpart $\Gamma_{[Z_1/Z_2]}$ i.e. for any closed varieties $Z_2 \subset Z_1$, and sheaf of currents $\mathcal{F}$, $\mathcal{H}^i_{[Z_1/Z_2]}(\mathcal{F}) = 0$ and $H^i_{[Z_1/Z_2]}(\mathcal{F}) = 0$ for every $i > 0$ [10]. In particular this implies that, for any line bundle $\mathcal{L}$, the Dolbeault resolution of $\mathcal{L}$ by sheaves of currents with values in $\mathcal{L}$ can be used to compute its algebraic local cohomology sheaves. In other words the cohomology of the complex of sheaves

$$0 \to \mathbb{D}^{0,0} \otimes \mathcal{L}_{[Z_1/Z_2]} \xrightarrow{\bar{\partial}} \mathbb{D}^{0,1} \otimes \mathcal{L}_{[Z_1/Z_2]} \xrightarrow{\bar{\partial}} \quad \cdots \quad \xrightarrow{\bar{\partial}} \mathbb{D}^{0,n} \otimes \mathcal{L}_{[Z_1/Z_2]} \to 0$$

is $\mathcal{H}^\bullet_{[Z_1/Z_2]}(\mathcal{L})$, and the cohomology of the complex of global sections

$$0 \to \mathbb{D}^{0,0} \otimes \mathcal{L}_{[Z_1/Z_2]}(M) \xrightarrow{\bar{\partial}} \mathbb{D}^{0,1} \otimes \mathcal{L}_{[Z_1/Z_2]}(M) \xrightarrow{\bar{\partial}} \cdots \xrightarrow{\bar{\partial}} \mathbb{D}^{0,n} \otimes \mathcal{L}_{[Z_1/Z_2]}(M) \to 0$$

is $H^\bullet_{[Z_1/Z_2]}(\mathcal{L})$. Another consequence of the acyclicity of sheaves of currents is that the sequence

$$0 \to \mathcal{F}_{[Z_2]} \to \mathcal{F}_{[Z_1]} \to \mathcal{F}_{[Z_1/Z_2]} \to 0$$

and its counterpart of global sections

$$0 \to \mathcal{F}_{[Z_2]}(M) \to \mathcal{F}_{[Z_1]}(M) \to \mathcal{F}_{[Z_1/Z_2]}(M) \to 0$$

are exact whenever $\mathcal{F}$ is a sheaf of currents. This can be taken to say that sheaves of currents are algebraically flabby.

Given a filtration of a complex manifold M by closed analytic subvarieties $M = Z_0 \supset Z_1 \supset \cdots \supset Z_n$, and a line bundle $\mathcal{L}$, we filter the Dolbeault complex of currents with values in $\mathcal{L}$ by their algebraic supports in these varieties. Let us consider the filtered complex

$$\mathbb{D}^{n,\bullet} \otimes \mathcal{L}(M) \supset \mathbb{D}^{n,\bullet} \otimes \mathcal{L}_{[Z_1]}(M) \supset \quad \cdots \quad \supset \mathbb{D}^{n,\bullet} \otimes \mathcal{L}_{[Z_n]}(M).$$

We will call the Cousin double complex associated to this filtered complex the
(meromorphic) Cousin Dolbeault complex. Consecutive terms of the filtration
give rise to the short exact sequences

$$0 \to \mathbb{D}^{n,\bullet} \otimes \mathcal{L}_{[Z_{k+1}]}(M) \xrightarrow{i_k} \mathbb{D}^{n,\bullet} \otimes \mathcal{L}_{[Z_k]}(M) \longrightarrow \mathbb{D}^{n,\bullet} \otimes \mathcal{L}_{[Z_k/Z_{k+1}]}(M) \to 0.$$

The vertical complexes of the Cousin Dolbeault complex are the mapping cones of
the injections i_k shifted down by k degrees. Their cohomology is the cohomology
of the complexes $\mathbb{D}^{n,\bullet} \otimes \mathcal{L}_{[Z_k/Z_{k+1}]}(M)[k]$. The Cousin Dolbeault complex always
computes the cohomology of $\mathbb{D}^{n,\bullet} \otimes \mathcal{L}(M)$, the Dolbeault complex of currents with
values in $\Omega^n \otimes \mathcal{L}$, where Ω^n is the line bundle of holomorphic n-forms. Since
$\mathbb{D}^{n,\bullet} \otimes \mathcal{L}$ is a resolution of $\Omega^n \otimes \mathcal{L}$, this cohomology is the sheaf cohomology
$H^\bullet(M, \Omega^n \otimes \mathcal{L})$. On the other hand, if it transpires, as with the Cousin de Rham
complex of a cell decomposition, that the vertical complexes have cohomology
only in degree zero, then the Cousin complex induced in degree zero, namely

$$0 \to H^{n,0}_{[M/Z_1]}(\mathcal{L}) \xrightarrow{\delta} H^{n,1}_{[Z_1/Z_2]}(\mathcal{L}) \xrightarrow{\delta} \quad \cdots \quad \xrightarrow{\delta} H^{n,n}_{[Z_n]}(\mathcal{L}) \to 0$$

also computes $H^\bullet(M, \Omega^n \otimes \mathcal{L})$.

5. The BGG resolution

The BGG resolution arises as an instance of a Cousin complex, namely the
meromorphic Cousin complex of a homogeneous line bundle over the flag manifold of a semisimple lie group G, endowed with the Schubert cell filtration.
$G = SL(2, \mathbb{C})$ provides the simplest example. We take the Borel subgroup B to
be the upper triangular matrices, and treating the first column of a matrix as
homogeneous co-ordinates for a point in $\mathbb{P}^1$ defines a map $SL(2, \mathbb{C}) \to \mathbb{P}^1$ which
identifies G/B with $\mathbb{P}^1$. We let ∞ denote the point $\begin{bmatrix} 1 \\ 0 \end{bmatrix}$ in $\mathbb{P}^1$. It is fixed under
the left action of B, and its complement, comprising points of the form $\begin{bmatrix} z \\ 1 \end{bmatrix}$,
constitutes another B orbit identified with $\mathbb{C}$. The Cousin complex of the single
step filtration $\mathbb{P}^1 \supset \{\infty\}$ reduces to the connecting homomorphism for the short
exact sequence of (vertical) complexes.

$$0 \to \mathbb{D}^{1,1}_{[\infty]}(\mathbb{P}^1, \mathcal{O}(n)) \longrightarrow \mathbb{D}^{1,1}(\mathbb{P}^1, \mathcal{O}(n)) \longrightarrow \mathbb{D}^{1,1}_{[\mathbb{P}^1/\infty]}(\mathbb{P}^1, \mathcal{O}(n)) \to 0$$

$$\uparrow \qquad\qquad\qquad \uparrow \qquad\qquad\qquad \uparrow$$

$$0 \to \mathbb{D}^{1,0}_{[\infty]}(\mathbb{P}^1, \mathcal{O}(n)) \longrightarrow \mathbb{D}^{1,0}(\mathbb{P}^1, \mathcal{O}(n)) \longrightarrow \mathbb{D}^{1,0}_{[\mathbb{P}^1/\infty]}(\mathbb{P}^1, \mathcal{O}(n)) \to 0$$

We shall describe in detail the cohomology and connecting homomorphism

$$H^{1,0}_{[\mathbb{P}^1/\infty]}(\mathbb{P}^1, \mathcal{O}(n)) \xrightarrow{\delta} H^{1,1}_{[\infty]}(\mathbb{P}^1, \mathcal{O}(n))$$

in this case.

Sections of $\mathcal{O}(n)$ correspond to functions $f(z_1, z_2)$ of the homogeneous co-ordinates (z_1, z_2) which are homogeneous of degree n. Thus z_2^n corresponds to a holomorphic section s which is non-vanishing on $\mathbb{C}$, and z_1^n corresponds to a holomorphic section t which is non-vanishing on a neighbourhood of ∞. A section of $\mathbb{D}^{1,0}(\mathbb{C}, \mathcal{O}(n))$ has the form θs where θ is a current of type $(1,0)$. It is meromorphic if and only if $(\frac{1}{z})^k \theta s$ extends to ∞ for some k, and holomorphic if and only if θ is holomorphic. By elliptic regularity this implies that $\theta = f dz$ for some holomorphic function f. Since $s = (\frac{1}{z})^n t$ and $dz = -z^2 d(\frac{1}{z})$ we have

$$\theta s = -(\frac{1}{z})^{n-2} d(\frac{1}{z}).f.t$$

so that $f dz s$ is meromorphic at ∞ if and only if f is, i.e. f must be a polynomial. Hence $H^{1,0}_{[\mathbb{P}^1/\infty]}(\mathbb{P}^1, \mathcal{O}(n))$ consists of sections $f dz s$ where f is a polynomial.

$H^{1,1}_{[\infty]}(\mathbb{P}^1, \mathcal{O}(n))$ consists of sections θt where θ is a current of type $(1,1)$ with support at ∞, modulo those which are $\bar{\partial}$ exact. The space of currents supported at ∞ is spanned by $\frac{\partial^{n+m}\delta}{\partial z^n \partial \bar{z}^m}, n, m = 0, \ldots$ where δ is the δ-form at ∞, i.e. $\phi \to \phi(\infty)$. Any such current involving derivatives with respect to $\bar{z}$ is $\bar{\partial}$ exact, while the others are not. We put $\partial^n \delta = \frac{\partial^n \delta}{\partial z^n}$, and identify $H^{1,1}_{[\infty]}(\mathbb{P}^1, \mathcal{O}(n))$ with the span of $\partial^n \delta \, t, n = 0, 1, \ldots$.

The connecting homomorphism can now be identified with a map of principal parts. Following through the construction of the connecting homomorphism, we take an element $f dz s$ from $H^{1,0}_{[\mathbb{P}^1/\infty]}(\mathbb{P}^1, \mathcal{O}(n))$ and lift it back to an element T in $\mathbb{D}^{1,0}(\mathbb{P}^1, \mathcal{O}(n))$. Thus T is a current such that

$$T(\phi d\bar{z}) = \int_{\mathbb{P}^1} f \phi dz \wedge d\bar{z} s$$

whenever the support of ϕ avoids ∞. The existence of such a T follows by the 'algebraic flabbiness' of the space of currents, which means effectively by the Division Theorem for distributions. Having obtained the lift T we now compute $\bar{\partial} T$. It will correspond to a current supported at ∞ since, as the theory of the connecting homomorphism shows, it can be lifted back to $\mathbb{D}^{1,0}_{[\infty]}(\mathbb{P}^1, \mathcal{O}(n))$. Considering its cohomology class amounts to restricting it to test functions ϕ which are holomorphic in a neighbourhood of ∞, i.e. such that the support of $\bar{\partial}\phi$ avoids ∞. Therefore we can calculate

$$\bar{\partial} T(\phi) = -T(\bar{\partial}\phi) = -\int_{\mathbb{P}^1} f dz \wedge \bar{\partial}\phi \, s = -\int_{C_r} f \phi dz \, s$$

where C_r is a circle of radius r about ∞ enclosing a disc which does not meet the support of $\bar{\partial}\phi$, and the last identity follows by Stokes Theorem. Changing to the co-ordinate $w = \frac{1}{z}$ centred on ∞, and the basis section t, we can express this as

$$\bar{\partial} T(\phi) = \int_{C_r} f(\frac{1}{w})\phi(w) w^{n-2} dw \, t.$$

It follows that the meromorphic current $z^k dzs$ maps via the connecting homo-morphism to $\frac{1}{(k-n+1)!}\partial^{k-n+1} t$, and this describes the connecting homomorphism completely.

The Lie algebra $\mathfrak{g} = sl(2,\mathbb{C})$ acts on sections of line bundles and on differential forms through the Lie derivative along fundamental vector fields. By duality it also acts on currents. Hence both $H^{1,0}_{[\mathbb{P}^1/\infty]}(\mathbb{P}^1, \mathcal{O}(n))$ and $H^{1,1}_{[\infty]}(\mathbb{P}^1, \mathcal{O}(n))$ are $\mathcal{U}(\mathfrak{g})$ modules where $\mathcal{U}(\mathfrak{g})$ is the universal enveloping algebra of $\mathfrak{g}$. Let X, Y and H be the standard upper triangular, lower triangular and diagonal basis for $sl(2,\mathbb{C})$. Denoting their fundamental vector fields by the same letter, we have

$$X = -\frac{\partial}{\partial z} = w^2\frac{\partial}{\partial w} \qquad Y = z^2\frac{\partial}{\partial z} = -\frac{\partial}{\partial w} \qquad H = -2z\frac{\partial}{\partial z} = 2w\frac{\partial}{\partial w}.$$

We also have

$$Xs = 0 \qquad Ys = -nzs \qquad Hs = ns$$
$$Xt = -nwt \qquad Yt = 0 \qquad Ht = -nt.$$

It follows that

$$
\begin{aligned}
H\left(z^k dzs\right) &= (Hz^k)dzs + z^k d(Hz)s + z^k dzHs &= (n-2-2k)z^k dzs \\
X\left(z^k dzs\right) &= (Xz^k)dzs + z^k d(Xz)s + z^k dzXs &= -kz^{k-1}dzs \\
Y\left(z^k dzs\right) &= (Yz^k)dzs + z^k d(Yz)s + z^k dzYs &= (-n+2+k)z^{k+1}dzs.
\end{aligned}
$$

The first line shows that the sections $z^k dzs$ are weight vectors of weight $n-2-2k$, while the second line shows that dzs is a heighest weight vector. The third line shows that, for $n < 2$, $H^{1,0}_{[\mathbb{P}^1/\infty]}(\mathbb{P}^1, \mathcal{O}(n))$ is freely generated by dzs under the action of Y, and so it is the Verma module for $sl(2,\mathbb{C})$ of highest weight $n-2$. Similarly, at ∞ we can show that

$$H\partial^k\delta = -2k\partial^k\delta \qquad X\partial^k\delta = k(k-1)\partial^{k-1}\delta \qquad Y\partial^k\delta = \partial^{k+1}\delta.$$

From which it follows that

$$H(\partial^k t) = (-n-2k)\partial^k t \qquad X(\partial^k t) = k(n-k-1)\partial^{k-1}t \qquad Y(\partial^k t) = \partial^{k+1}t.$$

It follows that δt is a highest weight vector generating $H^{1,1}_{[\infty]}(\mathbb{P}^1, \mathcal{O}(n))$ freely under the action of Y, and so it is the Verma module for $sl(2,\mathbb{C})$ of highest weight $-n$.

For $n \leq 0$ both $H^{1,0}_{[\mathbb{P}^1/\infty]}(\mathbb{P}^1, \mathcal{O}(n))$ and $H^{1,1}_{[\infty]}(\mathbb{P}^1, \mathcal{O}(n))$ are Verma modules, and the connecting homomorphism embeds $H^{1,0}_{[\mathbb{P}^1/\infty]}(\mathbb{P}^1, \mathcal{O}(n))$ in $H^{1,1}_{[\infty]}(\mathbb{P}^1, \mathcal{O}(n))$ as the maximal submodule of the latter, with quotient the irreducible finite di-mensional representation of $sl(2,\mathbb{C})$ with highest weight $-n$. Thus this one step Cousin complex has cohomology zero in degree 0 and the irreducible representa-tion of highest weight $-n$ in degree 1. By the Cousin Dolbeault theory the sheaf cohomology of $\mathcal{O}(n-2) = \mathcal{O} \otimes \Omega^1$ must be the same, and this computation can be regarded as the Cousin analogue of a Cech computation of this cohomology, although it reveals the representation structure in addition to the cohomology.

For a general semisimple Lie group G the story is much the same as for $SL(2,\mathbb{C})$. With $M = G/B$ one considers the filtration $M = Z_0 \supset Z_1 \supset \cdots \supset Z_n$

by B orbits of codimension $\geq k$. Once again the differentials of the Cousin complex arise as the connecting homomorphisms of the short exact sequences

$$0 \longrightarrow \mathbb{D}^{n,\bullet}_{[Z_{k+1}]}(M, \mathcal{L}_\lambda) \longrightarrow \mathbb{D}^{n,\bullet}_{[Z_k]}(M, \mathcal{L}_\lambda) \longrightarrow \mathbb{D}^{n,\bullet}_{[Z_k/Z_{k+1}]}(M, \mathcal{L}_\lambda) \longrightarrow 0$$

this time followed by the map induced from the next quotient

$$\mathbb{D}^{n,\bullet}_{[Z_{k+1}]}(M, \mathcal{L}_\lambda) \longrightarrow \mathbb{D}^{n,\bullet}_{[Z_{k+1}/Z_{k+2}]}(M, \mathcal{L}_\lambda)$$

where $\mathcal{L}_\lambda$ is the homogeneous line bundle determined by the character λ of B. If Z_{k+1} is defined in Z_k by the vanishing of a single function then the connecting homomorphism is once again defined by a residue computation. In order that the Cousin complex should compute the cohomology of $\mathcal{L}_\lambda$ it is necessary and sufficient that $\mathbb{D}^{n,\bullet}_{[Z_k/Z_{k+1}]}(M, \mathcal{L}_\lambda)$ have cohomology only in degree k, and this cohomology forms the k-th term of the Cousin complex.

Recall that if T is a Cartan subgroup in B then each B orbit in G/B contains a unique T fixed point wB, and w satisfies $Tw = wT$. In other words w belongs to the Weyl group of (G, T), and so we have an indexing of the B orbits in G/B by the Weyl group. As usual, let $\mathfrak{t}$ and $\mathfrak{b}$ be the lie algebras of T and B and let $\mathfrak{n}$ be the span of the root spaces of $\mathfrak{t}$ in $\mathfrak{b}$; these are called the positive roots. Then $B = NT$, where N is the subgroup of G with lie algebra $\mathfrak{n}$, and the orbit of B through wB is the same as the orbit of N, since T fixes wB. The stabiliser of wB in G is $wBw^{-1} = TwNw^{-1}$ so its stabiliser in N is $N_w = N \cap wNw^{-1}$. If C_w denotes the B orbit through wB then we have identified $C_w = N/N_w$. Its tangent space at w is identified with $\mathfrak{n}/\mathfrak{n}_w$ and hence with the span of those root spaces which lie in $\mathfrak{n}$ but not in $\mathfrak{n}_w$. These correspond to the positive roots which are not the conjugate under w of some (other) positive root, and their number gives the dimension of C_w. The number of positive roots which *are* the conjugate of a positive root under w gives the codimension of C_w and is called the length of w, which we denote $l(w)$. It follows that $Z_k \backslash Z_{k+1}$ is the disjoint union of those cells C_w with $l(w) = k$. Since $\mathbb{D}^{n,\bullet}_{[Z_k/Z_{k+1}]}(M, \mathcal{L}_\lambda)$ can be regarded as a subcomplex of $\mathbb{D}^{n,\bullet}_{[Z_k \backslash Z_{k+1}]}(M\backslash Z_{k+1}, \mathcal{L}_\lambda)$ it is not hard to show that the restriction maps

$$\mathbb{D}^{n,\bullet}_{[Z_k/Z_{k+1}]}(M, \mathcal{L}_\lambda) \longrightarrow \mathbb{D}^{n,\bullet}_{[\bar{C}_w/\partial C_w]}(M, \mathcal{L}_\lambda)$$

define an isomorphism between $\mathbb{D}^{n,\bullet}_{[Z_k/Z_{k+1}]}(M, \mathcal{L}_\lambda)$ and the direct sum of the complexes $\mathbb{D}^{n,\bullet}_{[\bar{C}_w/\partial C_w]}(M, \mathcal{L}_\lambda)$, which leads to an identity

$$H^{n,\bullet}_{[Z_k/Z_{k+1}]}(M, \mathcal{L}_\lambda) = \bigoplus_{l(w)=k} H^{n,\bullet}_{[\bar{C}_w/\partial C_w]}(M, \mathcal{L}_\lambda).$$

If w is the longest element of the Weyl group, so that C_w is the single point $*$ corresponding to B itself, then we can argue exactly as in the $SL(2, \mathbb{C})$ case that $H^{n,n}_{[*]}(M, \mathcal{L}_\lambda)$ is spanned by the delta-form and its holomorphic derivatives, tensored by a non-vanishing holomorphic section of $\mathcal{L}_\lambda$. Moreover, if $Y_1, \ldots, Y_n$ are root vectors, one from each negative root space, then $exp(z_1 Y_1 + \cdots + z_n Y_n)*$

covers a neighbourhood of $*$, on which $z_1, \ldots, z_n$ are holomorphic co-ordinates, and $\frac{\partial}{\partial z_i}(*) = Y_i(*)$ where Y_i also denotes the fundamental vector field corresponding to Y_i. Thus $H^{n,n}_{[*]}(M, \mathcal{O})$ is spanned by $Y_1^{k_1} \cdots Y_n^{k_n} \delta$ where $k_1, \ldots, k_n = 0, 1, \ldots$, which is to say that it is freely generated by the delta form δ over $\mathcal{U}(\mathfrak{n}^-)$, the universal enveloping algebra of the span $\mathfrak{n}^-$ of the negative root spaces. Since $*$ is fixed by B it is elementary that δ is annihilated by $\mathfrak{b}$, and thus a highest weight vector of weight 0. It is easy to construct a holomorphic section t_λ of $\mathcal{L}_\lambda$ over a neighbourhood of $*$ which is annihilated by $\mathfrak{n}^-$. It necessarily has weight λ, and since its derivatives under $\mathfrak{n}$ vanish at $*$ one can conclude that $\delta \otimes t_\lambda$ is a highest weight vector in $H^{n,n}_{[*]}(M, \mathcal{L}_\lambda)$ which freely generates it under $\mathcal{U}(\mathfrak{n}^-)$. Hence it is the Verma module of highest weight λ. Note that for $k < n$ $H^{n,k}_{[*]}(M, \mathcal{L}_\lambda) = 0$ by the holomorphic Poincaré Lemma.

As the $SL(2, \mathbb{C})$ example shows, the cohomology along other B orbits need not be Verma modules. In particular $H^{1,0}_{[\mathbb{P}^1/\infty]}(\mathbb{P}^1, \mathcal{O}(n))$ is not a Verma module for $n \geq 2$. Kempf shows [7] how to compute the character of the cohomology spaces in general, and concludes that $H^{n,l(w)}_{[\bar{C}_w/\partial C_w]}(M, \mathcal{L}_\lambda)$ has the same character as the Verma module of highest weight $w(\lambda + \rho) - \rho)$. An argument of Brylinski [3] can be adapted to show [10] that there are connecting homomorphisms which embed these cohomology modules into $H^{n,n}_{[*]}(M, \mathcal{L}_\lambda)$, and this proves that they are Verma modules.

However, there is the potential for more direct methods using distributions. In the case of $H^{n,l(w)}_{[\bar{C}_w/\partial C_w]}(M, \mathcal{O})$, each cell C_w carries an obvious highest weight vector which is the analogue of the Thom class in the de Rham setting. If $k = l(w)$, then $\Lambda^{n-k}(\mathfrak{n}/\mathfrak{n}_w)^*$ is 1-dimensional, and the natural action of $\mathfrak{n}_w$ is trivial, since it is a nilpotent Lie algebra. It therefore gives rise to a 1-dimensional space of N-invariant $(n-k, 0)$-forms on C_w. If ω is such an $(n-k, 0)$ form then for any $(0, n-k)$-form ϕ of compact support we define

$$\tau_w(\phi) = \int_{C_w} \omega \wedge \phi.$$

The N-invariance of ω implies the N-invariance of the $(n, n-k)$-current τ. Moreover, T acts on τ according to its action on $\Lambda^{n-k}(\mathfrak{n}/\mathfrak{n}_w)^*$, which is the sum of the weights in $\mathfrak{n}$ which are not in $\mathfrak{n}_w$. If ρ is one half the sum of the positive roots, it is well known that this can be expressed as $w\rho - \rho$. Hence the current τ is a highest weight vector with this weight. It is easy to prove that it is a closed current with support in $\bar{C}_w$, and meromorphic along ∂C_w. It remains to find a direct proof that it freely generates $H^{n,l(w)}_{[\bar{C}_w/\partial C_w]}(M, \mathcal{O})$ under the action of $\mathcal{U}(\mathfrak{n}^-)$.

References

1. I. N. Bernstein, I. M. Gelfand and S. I. Gelfand, *Differential operators on the base affine space and a study of $\mathfrak{g}$-modules*, Proceedings: Summer School on Lie Groups of Bolyai Janos Math. Soc., Halsted, New York, 1975, pp. 21–64.

2. R. Bott and L. Tu, *Differential forms in algebraic topology*, Springer, 1982.

3. J. L. Brylinski, *Differential operators on the flag varieties*, Asterisque **87-88** (1981), 43–60.

4. P. A. Griffith and J. Harris, *Principles of Algebraic Geometry*, Wiley, 1978.

5. R. Hartshorne, *Residues and Duality*, Lecture Notes in Math., vol. 20, Springer, 1966.

6. G. Kempf, *The Grothendieck-Cousin complex of an induced representation*, Adv. in Math. **29** (1978), 310–396.

7. A. T. Lundell and S. Weingram, *The Topology of CW Complexes*, Van Nostrand Reinhold, 1969.

8. B. Iversen, *Cohomology of Sheaves*, Springer, 1986.

9. B. Malgrange, *Division des distributions*, I–IV, *Séminaire Schwartz* t.4, 1959/60: Unicité du problém de Cauchy, Division des distributions, no. 21–25.

10. M. K. Murray and J. W. Rice, *Algebraic Local Cohomology and the Cousin-Dolbeault Complex*, preprint.

11. E. Witten, *Supersymmetry and Morse Theory*, J. Diff. Geom. **17** (1982), 661–692.

SCHOOL OF INFORMATION SCIENCE AND TECHNOLOGY, FACULTY OF SCIENCE AND ENGINEERING, THE FLINDERS UNIVERSITY OF SOUTH AUSTRALIA, GPO BOX 2100, ADELAIDE, SOUTH AUSTRALIA 5001

E-mail address: johnr@maths.flinders.edu.au

Contemporary Mathematics
Volume **154**, 1993

Dolbeault Cohomologies and Zuckerman Modules

H. W. WONG

Representation theorists are interested in explicit constructions for representations. Amongst others, there are two well known general methods: the algebraic constructions via Lie algebra cohomologies (Vogan-Zuckerman modules), and the analytic construction of cohomologies associated with certain vector bundles. This paper is concerned with a conjecture about the Dolbeault cohomologies of Fréchet vector bundles over certain open orbits of a generalized flag manifold: it was conjectured that they yield "essentially the same" representations as the analogous Vogan-Zuckerman constructions.

Throughout this paper, unless it is indicated otherwise, a real Lie group is denoted by a Roman upper case with "0"-subscript, its complexification, if it exists, by that without any subscript, the corresponding Lie algebras by their counterparts in lower case letters.

Let G_0 be a connected semisimple linear Lie group, K_0 a maximally compact subgroup. While the assumption on the group can be relaxed at times, it is adhered to for the sake of simplicity. The related Cartan involution, θ, yields the decomposition $\mathbf{g}_0 = \mathbf{k}_0 \oplus \mathbf{p}_0$. Let Y be a generalized flag manifold, it is the space of conjugates of a parabolic subalgebra under the action of the integral complex group; the flag manifold X is a special case. Let D be an open G_0 orbit, it must exist because there are only finitely many G_0 orbits. D carries a G_0 invariant measure when Y is X, otherwise it may not. When it does, as a homogeneous space, $Y = G_0/L_0$. Here L_0 is the centralizer of a compact torus C_0, it is reductive and connected ([**W**]). Let L_0 stabilize a θ-stable parabolic subalgebra $\mathbf{q}$, write $\mathbf{q} = \mathbf{l} \oplus \mathbf{u}$, where $\mathbf{u}$ is the nilradical of $\mathbf{q}$. For future reference,

---

1991 *Mathematics Subject Classification*. Primary 22E45; Secondary 22E30.

This paper is in final form and no version of it will be submitted for publication elsewhere.

it is recalled that $\zeta \in \mathbf{c}_0$ can be chosen so that

$$\mathbf{l} = \{X \in \mathbf{g} : ad(\zeta)(X) = 0\},$$

(1)

$$\mathbf{u} = \oplus_{\substack{\alpha \in \triangle(\mathbf{h},\mathbf{g}) \\ \alpha(i\zeta)<0}} \mathbf{g}^\alpha; \quad \bar{\mathbf{u}} = \oplus_{\substack{\alpha \in \triangle(\mathbf{h},\mathbf{g}) \\ \alpha(i\zeta)>0}} \mathbf{g}^\alpha.$$

Here $\mathbf{h}_0$ is a fundamental (i.e. maximally compact) CSA containing $\mathbf{c}_0$. Let T_0 be a maximal torus of K_0 so that $C_0 \subseteq T_0 \subseteq H_0$. $\triangle(\mathbf{h}, \mathbf{g})$ is the set of roots of $\mathbf{g}$ with respect to $\mathbf{h}$; $\mathbf{g}^\alpha$, $\alpha \in \triangle(\mathbf{h}, \mathbf{g})$, is the root space of α. Generally, a notation such as $\triangle(\mathbf{h}, \mathbf{u})$ means the set of $\mathbf{h}$ roots whose root spaces lie in $\mathbf{u}$. $\triangle^+ = \triangle^+(\mathbf{h}, \mathbf{g})$ and $\triangle_L^+ = \triangle^+(\mathbf{h}, \mathbf{l})$ are chosen so that $\mathbf{n} = \mathbf{u} \oplus \mathbf{n}_L$, here $\mathbf{n}$ and $\mathbf{n}_L$ are the direct sums of the negative root spaces ([V]). Further, let $\mathbf{s} := \mathbf{l} \cap \mathbf{p} \oplus \mathbf{u}$, hence $\mathbf{q} = \mathbf{s} \oplus \mathbf{l} \cap \mathbf{k}$, and denote by $\wp : \mathbf{q} \longrightarrow \mathbf{s}$ the obvious projection.

Let Q be the subgroup in G corresponding to $\mathbf{q}$, then $Y = G/Q$ and D inherits from it a homogeneous complex structure. This can be described intrinsically by prescribing the sheaf $\mathcal{O}$ of holomorphic functions. Let $W \subseteq D$ be open and $p : G_0 \to D = G_0/L_0$ the projection, then

$$\mathcal{O}(W) = \{f \in C^\infty(p^{-1}(W))|r(X)f = 0, X \in \mathbf{q}\}.$$

Here r is the right differentiation. Homogeneous holomorphic vector bundles over D can be similarly described. Let (π, V) be a finite rank $(L_0, \mathbf{q})$ representation. It gives rise to an induced homogeneous holomorphic vector bundle $\mathbf{V} \to D$. The sheaf $\mathcal{O}(\mathbf{V})$ of holomorphic sections can be prescribed as: $\mathcal{O}(\mathbf{V})(W) = \{f \in C^\infty(p^{-1}(W), V)|r(X)f + \pi(X)f = 0, X \in \mathbf{q}\}$ ([**TW**, §3], [**GS**]).

Notice that G_0 acts on the sheaf cohomologies of $\mathcal{O}(\mathbf{V})$. However, for them to be representations, there should be topologies. Recall that the cohomologies can be computed from the Dolbeault complex, which can be represented, as a complex of topological G_0 modules, by the following relative Lie algebra cohomology complex: $\hom(\wedge^*\mathbf{u}, C^\infty(G_0) \otimes V)^{L_0}$, together with the differentials defined by:

$$df(X_1 \wedge \ldots \wedge X_i) = \sum_{k=1}^{i}(-1)^{k+1}\gamma(X_k)\{f(X_1 \wedge \ldots \wedge \hat{X}_k \wedge \ldots \wedge X_i)\}$$
$$+ \sum_{r<s}(-1)^{r+s}f([X_r, X_s] \wedge X_1 \wedge \ldots \wedge \hat{X}_r \wedge \ldots \wedge \hat{X}_s \wedge \ldots \wedge X_i);$$

here $X_k \in \mathbf{u}, k = 1, \ldots, i$, and $f \in \hom(\wedge^i\mathbf{u}, C^\infty(G_0) \otimes V)^{L_0}$. γ is the the action of $\mathbf{u}$ on $C^\infty(G_0) \otimes V$ via $r \otimes \pi$. The following complex will be used instead because, as topological G_0 modules, it gives the same cohomologies: $A^i(V) := \hom_{L_0 \cap K_0}(\wedge^i\mathbf{s}, C^\infty(G_0) \otimes V)$ with the differentials d given by

$$df(X_1 \wedge \ldots \wedge X_i) = \sum_{k=1}^{i}(-1)^{k+1}\gamma(X_k)\{f(X_1 \wedge \ldots \wedge \hat{X}_k \wedge \ldots \wedge X_i)\}$$

(2) $$+ \sum_{r<s}(-1)^{r+s}f(\wp([X_r, X_s]) \wedge X_1 \wedge \ldots \wedge \hat{X}_r \wedge \ldots \wedge \hat{X}_s \wedge \ldots \wedge X_i)$$

where $f \in A^i(V)$ and $X_k(\in \mathbf{s})$ acts on $C^\infty(G_0) \otimes V$ via $r \otimes \pi$. The point is that $A^*(V)$ is the complex, with C^∞ coefficients and values in the pull-back bundle $\pi_0^{-1}\mathbf{V}$, on $G_0/L_0 \cap K_0$ with differentials which are de Rham along the (Euclidean) fibers of the fibration $\pi_0 : G_0/L_0 \cap K_0 \longrightarrow G_0/L_0$ and Dolbeault transversally. Therefore the Poincaré lemma comes in handy.

Note that the members of the complexes have natural Fréchet topologies; therefore the natural topologies for the cohomologies would be the induced topologies. But the induced topologies are required to be Hausdorff before the cohomologies can be studied as representations, which is the case if and only if the differentials have closed ranges. This is often far from clear and indeed a serious hurdle to surmount.

Here are some constructions whose settings are related to the one just outlined.

Borel-Weil-Bott's Theorem. Only the setting is described here. Consider the case when G_0 is compact, i.e. $G_0 = K_0$, $D = X$ compact, and $L_0 = H_0$ a maximal torus. Irreducible representations of H_0 are the characters $\mathbf{C}_\lambda$, $\lambda \in \mathbf{h}^*$ integral. Compactness of D means the cohomologies are finite dimensional and the closed range property holds automatically [1].

Discrete Series Representations. Consider the case when G_0 contains a compact CSG $L_0 = H_0$. D is an open orbit of X, generally non-compact. There is now no a priori reason why the differentials in the Dolbeault complex have closed ranges. The following is proved in ([**S1**]).

THEOREM 1. *If* $(\lambda + \rho, \alpha) << 0$ *(via Killing form) for all* $\alpha \in \triangle^+$, *let* $\mathbf{S} = \dim_{\mathbf{C}} K_0/H_0$, *then*

  (i) *The closed range property holds.*
  (ii) *Cohomologies vanish at all degrees except* $\mathbf{S}$, *and* $H^{\mathbf{S}}$ *is the discrete series representation parametrized by* λ.

The sufficiently negative condition can be relaxed provided the theorem is reformulated in a manner apparent when the conjecture is stated. The last statement can be made more precise, but a digression into globalizations is needed (see [**S2**]).

If V is a representation of G_0 (i.e. V is a locally convex, Hausdorff and complete TVS on which G_0 acts continuously) which has finite length and is admissible (i.e. all K_0 types have only finite multiplicities), its space of K_0 finite vectors $V_{(K_0)}$ then forms a $(K_0, \mathbf{g})$ module which has finite length and is admissible. These are precisely the defining properties of a Harsih-Chandra module, and all Harish-Chandra modules arise in this manner. Under this situation, $V_{(K_0)}$ is called the Harish-Chandra module of V and V a globalization of $V_{(K_0)}$.

The process $V \mapsto V_{(K_0)}$ is an exact functor between the appropriate categories. However, there does not appear to have any inverse process on first

---

[1] If a continuous map from a Fréchet space to another has finite codimesional range, then the range is closed.

glance because the same Harish-Chandra module admits infinitely many glob-
alizations, to be illustrated as follows. The spaces $C^{-\omega}(S^1)$, $C^{-\infty}(S^1)$, $L^p(S^1)$,
$C^\infty(S^1)$, $C^\omega(S^1)$ are different $SL_2(\mathbf{R})$ representations with the same Harish-
Chandra module: $\oplus_{n \in \mathbf{Z}} \mathbf{C} e^{in\theta}$. Therefore, it can be asked which globalization of
the discrete series is obtained.

The following (due to Casselman, Wallach, and Schmid) sets the framework.

PROPOSITION 2. *There are four natural globalizations attached to each Harish-Chandra module. Each of the processes forms an exact functor.*

To see concretely what these four canonical globalizations look like, let the
Harish-Chandra module be a principal series representation, so that it can be
realized as the K_0 finite sections of a vector bundle over G_0/P_0 induced from a
P_0 representation V: $L^2(G_0/P_0, V)_{(K_0)}$ (P_0 is a parabolic subgroup). Then the
four globalizations are: $C^\omega(G_0/P_0, V) \subseteq C^\infty \subseteq C^{-\infty} \subseteq C^{-\omega}$.

The outer two representations are called the minimal and maximal globalizations. The names mean what they suggest, for example, all other globalizations
inject continuously and equivariantly into the maximal globalization.

Schmid's result above can be refined: the non-vanishing cohomology is in fact
the maximal globalization of the discrete series.

The Zuckerman Modules. The proof of any closed range property is usually very hard. Partly to avoid this analytical problem, Zuckerman considered
the algebraic construction now named after him. It can be viewed as an imit-
ation of Schmid's construction, and is later systematically used by Vogan in a
wider context. To describe Zuckerman's construction, there are two functors to
be considered (see [**V**]).

Let $V \in \mathcal{M}(L_0 \cap K_0, \mathbf{q})$ (the category of $(L_0 \cap K_0, \mathbf{q})$ compatible modules,
they are here implicitly required to be locally $L_0 \cap K_0$ finite). The production
functor $pro : \mathcal{M}(L_0 \cap K_0, \mathbf{q}) \to \mathcal{M}(L_0 \cap K_0, \mathbf{g})$, $V \mapsto \hom_{\mathbf{U}(\mathbf{g})}(\mathbf{U}(\mathbf{g}), V)_{(L_0 \cap K_0)}$
turns out to be exact. There is another functor $\Gamma : \mathcal{M}(L_0 \cap K_0, \mathbf{g}) \to \mathcal{M}(K_0, \mathbf{g})$,
$V \mapsto V_{(K_0)}$.

DEFINITION 3. *The Zuckerman (ith derived functor) module $R^i(V)$ is the ith
derived functor of Γ, applied to $pro(V)$.*

To compute the Zuckerman modules, one uses the following relative Lie al-
gebra cohomology complex: $\to \hom_{\mathbf{U}(\mathbf{l} \cap \mathbf{k})}(\mathbf{U}(\mathbf{g}), V \otimes \wedge^\star(\mathbf{q}/\mathbf{l} \cap \mathbf{k})^*)_{(K_0)} \to$. See
[**Wo**, Corollary 60] for a proof. The complex is just $A^*(V)$ but with the smooth
coefficients replaced essentially by formal power series at the identity coset. One
may ask how are Dolbeault cohomologies related to the Zuckerman modules.
Call the relative Lie algebra cohomology complex $A^{FOR,\star}(V)_{(K_0)}$. The following
map turns out to be the key tool:

DEFINITION 4. *The Taylor series map*

$$\Delta \equiv \Delta_G : \{C^\infty(G_0) \otimes V \otimes \wedge^i \mathbf{s}^*\}_{(K_0)}^{L_0 \cap K_0} \longrightarrow A^{FOR,i}(V)_{(K_0)}$$

is defined as follows. For any $f \in \{C^\infty(G_0) \otimes V \otimes \wedge^i \mathbf{s}^*\}_{(K_0)}^{L_0 \cap K_0}$, *it can be written as* $f = \sum_i f_i \otimes w_i$, *satisfying* $r(X)f_i \otimes w_i + f_i \otimes X \cdot w_i = 0$, $X \in \mathbf{l} \cap \mathbf{k}$. *Define* $(\Delta f)(u) = \sum_i \ell(u)f(e) \otimes w_i$, *here* $u \in \mathbf{U}(\mathbf{g})$.

REMARK 5. *The differentials for* $A_{(K_0)}^{FOR,\star}$ *are defined by:*

$$df(u; X_1 \wedge \ldots \wedge X_{p+1}) = \sum_{k=1}^{p+1} \{ (-1)^k f(X_k \cdot u; X_1 \wedge \ldots \wedge \hat{X}_k \wedge \ldots \wedge X_{p+1})$$
$$+ (-1)^{k+1} X_k \cdot f(u; X_1 \wedge \ldots \wedge \hat{X}_k \wedge \ldots \wedge X_{p+1})\}$$

$$(3) \qquad + \sum_{k<l} (-1)^{k+l} f(u; \wp([X_k, X_l]) \wedge X_1 \wedge \ldots \hat{X}_k \ldots \hat{X}_l \ldots X_{p+1}),$$

with X_i *'s in* $\mathbf{s}$, $u \in \mathbf{U}(\mathbf{g})$, *and* $f \in A^{FOR,p}$.

Non-Open Orbits. While there is no direct bearing on the conjecture, it is nonetheless the other natural direction to generalize the case of discrete series construction on X. It is concerned with the cohomologies of a C-R complex associated with a line bundle on a (non-open) orbit. There are three well known classification schemes in representation theory: Beilinson and Bernstein's, Langlands', and Vogan and Zuckerman's. Schmid and Wolf have shown that the standard modules which are used to classify irreducible representations in each of the last two schemes are manufactured from data associated with a different set of G_0 orbits in X. The first scheme is obtained anologously but K orbits are used instead. In their language, each scheme chooses a different polarization ([**SW2**]).

Let $\rho(\bar{\mathbf{u}})$ be the half sum of roots $\alpha, \mathbf{g}^\alpha \in \bar{\mathbf{u}}$. Now the conjecture alluded at the beginning of the talk can be formulated.

CONJECTURE 6. *Let* V *be the maximal globalization of a Harish-Chandra module of* L_0. *Consider the induced "holomorphic Fréchet bundle"* $\mathbf{V} \to D$, *and the associated Dolbeault complex Dolb*.*

  (i) *The complex Dolb* has the closed range property, the cohomologies form admissible representations of finite lengths.*
  (ii) *When* V *admits an* L *infinitesimal character* $\chi_{L,\lambda}$ *and trivial* $\mathbf{u}$ *action, the cohomologies admit the* G *infinitesimal character* $\chi_{G,\lambda+\rho(\bar{\mathbf{u}})}$.
  (iii) Δ *is a quasi-isomorphism. The cohomologies are the maximal globalizations of the corresponding Harish-Chandra modules* $R^{(\star)}(V)$.

The following vanishing theorem follows from a well known result (see for example [**V**]), assuming the conjecture.

COROLLARY 7. *If V admits an L infinitesimal character $\chi_{L,\lambda}$ and trivial $\mathbf{u}$ action, and if $Re(\lambda + \rho(\bar{\mathbf{u}}), \alpha) < 0, \alpha \in \triangle^+(\mathbf{h}, \bar{\mathbf{u}})$, then all cohomologies vanish except that of degree $\mathbf{S} = dim_{\mathbf{C}}K_0/K_0 \cap L_0$, the dimension of a maximal compact subvariety $S \subseteq D$.*

THEOREM 8. *The conjecture holds when* $\dim V < \infty$.

For details, refer to [**Wo**]. The work in [**BKZ**] is of related interest. Here is a brief outline of the proof. By a standard trick of tensoring and the functoriality of the maximal globalization functor, it suffices to prove theorem 8 when V is irreducible and is "sufficiently negative". Further, "vanishing above $\mathbf{S}$" can be obtained by a general result of complex analysis ([**SW1**] and [**AG**]). It can also be proved that, under a "sufficiently negative" assumption, the closed range property holds. The point is that the fibration $G_0/L_0 \cap K_0 \to G_0/K_0$, while not holomorphic, still allows one to perform a Leray spectral sequence type argument. The spectral sequence degenerates and has $E_{(2)}^{\mathbf{S},0}$ as the only non-vanishing term. Moreover, the non-vanishing cohomology can be identified with the kernel of an order one differential operator on G_0/K_0 with real analytic coefficients. Further, the operator can be computed in an explicit way, and can be seen to be elliptic. This fact, together with looking at a filtration of the sheaf $\mathcal{O}(\mathbf{V})$ by the degree of the order of vanishing along the subvariety S, enables one to see that the cohomology is the maximal globalization of a Harish-Chandra module, which is identified to be the related Zuckerman module via the Taylor series map.

REFERENCES

[AG]   A. Andreotti and H. Grauert, *Théorèmes de finitude pour la cohomologie des espaces complexes*, Bull. Soc. Math. France **90** (1962), 193–259.

[BKZ] L. Barchini, A. W. Knapp, and R. Zierau, *Intertwining operators into Dolbeault cohomology representations*, J. Func. Anal. **107** (1992), 302–341.

[GS]   P. A. Griffiths and W. Schmid, *Locally homogeneous complex manifolds*, Acta Math. **123** (1969), 253–302.

[S1]   W. Schmid, *Homogeneous complex manifolds and representations of semisimple Lie groups*, Ph.D. dissertation, University of California, Berkeley 1967, Representation Theory and Harmonic Analysis on Semisimple Lie Groups, Math. Surveys and Monographs, vol. 31, Amer. Math. Soc., 1989, pp. 223–286.

[S2]   _____, *Boundary value problems for group invariant differential equations*, Élie Cartan et les Mathématiques d'Aujourdui, Astérisque (1985), 311–321.

[SW1] W. Schmid and J. A. Wolf, *A vanishing theorem for open orbits on complex flag manifolds*, Proc. Amer. Math. Soc. **92** (1984), 461–464.

[SW2] _____, *Geometric quantization and derived functor modules for semisimple Lie groups*, J. Func. Anal. **90** (1990), 48–112.

[TW]   J. A. Tirao and J. A. Wolf, *Homogeneous holomorphic vector bundles*, Indiana Univ. Math. J. **20** (1970/71), 15–31.

[V]   D. A. Vogan, Jr., *Representations of Real Reductive Lie Groups*, Progress in Math. vol. 15, Birkhäuser, 1981.

[W]   J. A. Wolf, *The action of a real semisimple group on a complex flag manifold I: Orbit structure and holomorphic arc components*, Bull. Amer. Math. Soc. **75** (1969), 1121–1237.

[Wo]   H. W. Wong, *Dolbeault cohomologies and Zuckerman modules associated with finite rank representations*, Ph.D. Dissertation, Harvard University, 1992.

MATHEMATICAL INSTITUTE, 24-29 ST. GILES', OXFORD OX1 3LB, ENGLAND
E-mail address: wong@maths.oxford.ac.uk

Contemporary Mathematics
Volume **154**, 1993

Unipotent Representations and
Derived Functor Modules

DAN BARBASCH

1. Introduction

The purpose of this talk is to describe some results on unipotent representations, highlighting the relations between the geometric treatment in [**ABV**] and the more representation–theoretic viewpoint in [**BV**].

Let G be a connected reductive complex group and fix an inner class of real forms. Then motivated by considerations coming from the trace formula, Arthur considers maps (see below and section 5 for the relevant definitions)

$$(1.1) \qquad \psi \; : \; \mathcal{W}_{\mathbb{R}} \times SL(2, \mathbb{C}) \longrightarrow {}^{\vee}G^{\Gamma}.$$

He then conjectures that, attached to such a map there should be a *packet* of irreducible representations (of real forms in the given inner class) satisfying certain character identities with respect to endoscopic groups (see the introduction of [**ABV**] for the detailed version).

In particular, the case when $\psi|_{\mathcal{W}_{\mathbb{R}}}$ is trivial and $\psi|_{SL(2,\mathbb{C})}$ determines an even nilpotent orbit (via the Jacobson–Morozov Theorem) is called *special unipotent*. In this case (1.1) takes on the form

$$(1.2) \qquad \psi : SL(2, \mathbb{C}) \longrightarrow {}^{\vee}G,$$

where ${}^{\vee}G$ is the (complex) dual group. Then the packet attached to such a parameter can be defined as follows.

DEFINITION. *An admissible representation of a real form $G(\mathbb{R})$ is called special unipotent if it has infinitesimal character*

$$\lambda_{\check{\mathcal{O}}} = d\psi \begin{bmatrix} 1/2 & 0 \\ 0 & -1/2 \end{bmatrix}$$

1991 *Mathematics Subject Classification.* Primary 22E46, 22E47.

Research partially supported by NSF Grant DMS 91 04117.

This paper is in final form and no version of it will be submitted for publication elsewhere.

and its annihilator in the universal enveloping algebra is maximal subject to this condition.

This is the viewpoint taken in [**BV**] where for complex groups character formulas for special unipotent representations are derived. These formulas are of the type conjectured by Arthur.

In [**ABV**], a different approach is taken. The category of Harish–Chandra modules is related to constructible or perverse sheaves on an algebraic space Y equivariant with respect to the action of an algebraic group H. Then the Arthur packets can be defined in terms of charcteristic cycles for sheaves. This approach has the advantage that it is more general in that it applies to any Langlands parameter not just an Arthur parameter, but also the conjectured character identities relevant to endoscopic groups are a consequence of Lefschetz fixed point theorems of Goresky–MacPherson.

Another feature of this setting implies that to each packet one can attach a stable combination of characters in a natural way; and the coefficients are given in terms of characters of representations of component group of the centralizer of the image of the map (1.1). In chapter 27 of [**ABV**] it is shown that for the case of special unipotent representations the two definitions coincide. However the characteristic cycles are not very easy to compute in a general setting.

A large portion of this article will be devoted to describing the geometric setting that allows us to define the Arthur packets. This is a (very incomplete) summary of [**ABV**]. Since the material is so technical we will be forced to mostly refer to [**ABV**] for details and motivation.

The classification of $(\mathfrak{g}, K)$ modules in terms of the dual group is described in section 2. In section 3 we introduce the various geometric categories relevant for our analysis and the notion of geometrically stability. Then we describe their relation to the notion of stable combinations of characters introduced by Langlands and Shelstad. Arthur parameters and the relation to Definition 1.1 are sketched in section 5. We also discuss various conjectures and relations between unipotent representations and derived functor modules. This is related to [**B2**].

As already mentioned, one of the advantages for working with characteristic cycles is that the coefficients in the stable linear combinations of characters have a natural interpretation as dimensions of representations of certain component groups. In section 6 we prove that in the complex case, these characters coincide with the results in [**BV**]. This result is joint work with D. Vogan.

2. The Classification

2.1. We review the definitions of real forms and their representations considered in [**ABV**]. For motivation and details we refer to the introduction and chapter 2 of [**ABV**].

Let G be a complex connected reductive algebraic group defined over $\mathbb{R}$, and let $\Gamma = \mathbb{Z}_2$ be the Galois group of $\mathbb{C}$ over $\mathbb{R}$. A *real form* of G is an *antiholomorphic* involutive automorphism $\sigma : G \longrightarrow G$. Two such forms σ, σ' are called *equivalent* if they are conjugate under G, in other words if there is $g \in G$ such that $\sigma' = Ad(g) \circ \sigma \circ Ad(g^{-1})$. They are called *inner* if there is $g \in G$ such that $\sigma' = Ad(g) \circ \sigma$.

An inner class gives rise to a *(weak extended)* group G^Γ which contains G as a subgroup of index 2 so that

$$1 \longrightarrow G \longrightarrow G^\Gamma \longrightarrow \Gamma \longrightarrow 1.$$

Then a *pure real form* of G is an element $\delta \in G^\Gamma - G$ satisfying $\delta^2 = 1$. A *representation* is a pair (π, δ) where δ is a pure real form and π a Harish–Chandra module for the fixed points $G(\mathbb{R}, \sigma(\delta))$, where $\sigma(\delta)$ is the antiholomorphic involutive automorphism $\sigma(\delta)(g) = \delta \cdot g \cdot \delta^{-1}$. We denote by $\Pi(G/\mathbb{R})$ the set of equivalence classes (under the action of G) of irreducible representations.

2.2. Let $^\vee G$ be the dual group. Then an inner class of real forms of G gives rise to an *L–group* $^\vee G^\Gamma$ (this is an extension of $^\vee G$ by Γ where the action is holomorphic rather than antiholomorphic).

Recall also that the *Weil group of* $\mathbb{R}$ is defined as

$$(2.2.1) \qquad \mathcal{W}_\mathbb{R} = \mathbb{C}^\times \times \{1, j\}, \quad jzj^{-1} = \bar{z}, \ j^2 = 1.$$

Then a Langlands parameter is a group homomorphism

$$(2.2.2) \qquad \phi : \mathcal{W}_\mathbb{R} \longrightarrow {}^\vee G^\Gamma$$

compatible with the maps into Γ and such that $\phi(\mathbb{C}^\times)$ is formed of semisimple elements.

Then the *pure Langlands component group of* ϕ is the component group

$$(2.2.3) \qquad A_\phi^{loc} = {}^\vee G(\phi)/{}^\vee G(\phi)_0,$$

where $^\vee G(\phi)$ is the centralizer of the image of ϕ.

The set of Langlands parameters will be denoted $P(G/\mathbb{R})$. The pairs (ϕ, τ), where $\tau \in \widehat{A_\phi^{loc}}$, are called *complete Langlands parameters* and the set of equivalence classes under the action of $^\vee G$ will be denoted $\Xi(G/\mathbb{R})$.

The relation between the sets $\Pi(G/\mathbb{R})$ and $\Xi(G/\mathbb{R})$ is as follows.

THEOREM. *There is a natural bijection*

$$\Pi(G/\mathbb{R}) \leftrightarrow \Xi(G/\mathbb{R}).$$

Again, we refer to [**ABV**] for precise statements of a more general version involving *strong real forms*.

2.3. We now describe the Langlands parameters more explicitly.

PROPOSITION. *The set of Langlands parameters can be put in 1-1 correspondence with the set of pairs*

$$\{ (y, \lambda) \in {}^\vee G \times {}^\vee \mathfrak{g} \}$$

subject to the conditions

   (1) $[\lambda, Ad(y)\lambda] = 0$,
   (2) $y^2 = exp(2i\pi\lambda)$,
   (3) $y \in {}^\vee G^\Gamma - {}^\vee G$ *is semisimple.*

An *infinitesimal character* is an orbit $\mathcal{O}_\mathbb{R} = (\mathcal{Y}, \mathcal{O})$ where

$\mathcal{O}$ is a semisimple ${}^\vee G$–conjugacy class of an element $\lambda \in {}^\vee \mathfrak{g}$

(2.3.1) $\mathcal{Y}$ is a ${}^\vee G$–conjugacy class of an element $y \in {}^\vee G^\Gamma - {}^\vee G$,

y^2 is conjugate to $e(\lambda) = exp(2i\pi\lambda)$ under ${}^\vee G$.

2.4. Given λ, let ${}^\vee \mathfrak{g}_i = \{ x : [\lambda, x] = ix \}$, and

(2.4.1) $$\mathcal{F}(\lambda) = \lambda + \sum_{i \in \mathbb{N}^+} {}^\vee \mathfrak{g}_i, \quad \mathcal{F}(\mathcal{O}) = \{\mathcal{F}(\lambda)\}_{\lambda \in \mathcal{O}}$$

Denote by $\mathcal{C}(\mathcal{O}_\mathbb{R})$ the ${}^\vee G$–conjugacy class $\mathcal{Y}^2 = e(\mathcal{O})$. Then the map $e : \lambda \mapsto exp(2i\pi\lambda)$ is well defined from $\mathcal{F}(\mathcal{O})$ to $\mathcal{C}(\mathcal{O}_\mathbb{R})$. In fact, if we give $\mathcal{F}(\mathcal{O})$ the structure of an algebraic variety, e becomes a morphism of algebraic varieties.

DEFINITION. *The set of Langlands parameters with infinitesimal character $\mathcal{O}_\mathbb{R}$ is*

$$P(\mathcal{O}_\mathbb{R}, {}^\vee G^\Gamma) = \{ \phi = (y, \lambda) : y \in \mathcal{Y}, \lambda \in \mathcal{O} \}.$$

The geometric parameter space of infinitesimal character $\mathcal{O}_\mathbb{R}$ is the fiber product

$$X(\mathcal{O}_\mathbb{R}, {}^\vee G^\Gamma) = \mathcal{Y} \times_{\mathcal{C}(\mathcal{O}_\mathbb{R})} \mathcal{F}(\mathcal{O}),$$

obtained by using the map e and the squaring map from $\mathcal{Y}$ to $\mathcal{C}(\mathcal{O}_\mathbb{R})$.

2.5. Fix $c \in \mathcal{C}(\mathcal{O}_\mathbb{R})$, $\Lambda \in \mathcal{F}(\mathcal{O})$ such that $e(\Lambda) = c$ and $y \in \mathcal{Y}$ such that $y^2 = c$. Write $P(\Lambda)$ for the stabilizer of Λ in ${}^\vee G(c)$ and $K(y)$ for the centralizer of y, a symmetric subgroup of ${}^\vee G(c)$. Let $\mathcal{P}$ be the variety of parabolic subgroups in ${}^\vee G(c)$ conjugate to $P(\Lambda)$.

PROPOSITION. *The group ${}^\vee G$ acts on $P(\mathcal{O}_\mathbb{R}, {}^\vee G^\Gamma)$ and $X(\mathcal{O}_\mathbb{R}, {}^\vee G^\Gamma)$ in a natural fashion. There is a ${}^\vee G$–equivariant isomorphism*

$$X(\mathcal{O}_\mathbb{R}, {}^\vee G^\Gamma) \cong {}^\vee G \times_{K(y)} \mathcal{P}.$$

The orbits of ${}^\vee G$ on X are in 1–1 correspondence with the orbits of $K(y)$ on $\mathcal{P}$; in particular there are only finitely many of them. The isotropy group ${}^\vee G^x$ at a point $x = (y, \Lambda) \in X$ is $K(y) \cap P(\Lambda)$.

The map

$$p : P(\mathcal{O}_\mathbb{R}, {}^\vee G^\Gamma) \longrightarrow X(\mathcal{O}_\mathbb{R}, {}^\vee G^\Gamma)$$

induces a bijection between $^\vee G$–conjugacy classes on these two spaces. The component groups of the centralizers correspond.

2.6.   Example. Let $G = Sl(2, \mathbb{R})$. Then $^\vee G^\Gamma = PSl(2) \times \{1, \tau\}$, direct product, where $\tau^2 = 1$. Take λ to be regular integral, in fact such that $e(\lambda) = 1$. Then $\mathcal{P} = \mathcal{B}$, the variety of Borel subgroups. There are two conjugacy classes of y's,

$$(2.6.1) \quad \begin{aligned} y &= \tau, & K(y) &= PSl(2, \mathbb{C}), \\ y &= \begin{bmatrix} i & 0 \\ 0 & -i \end{bmatrix} \tau, & K(y) &= \begin{bmatrix} e^{i\theta} & 0 \\ 0 & e^{-i\theta} \end{bmatrix} \cup \begin{bmatrix} e^{i\theta} & 0 \\ 0 & e^{-i\theta} \end{bmatrix} \begin{bmatrix} 0 & 1 \\ -1 & 0 \end{bmatrix}. \end{aligned}$$

Consider the second case. Identifying $\mathcal{B}$ with a sphere, there are two orbits, the union of the North and South pole, and the rest. The component group for the open orbit is $\mathbb{Z}_2$. The two characters correspond to $DS(\pm)$. The component group for the poles is trivial. The corresponding representation is the trivial representation occuring in the spherical principal $PS(sph)$. The other (irreducible) principal series $PS(sgn)$ of $Sl(2, \mathbb{R})$, matches the orbit of the *other* y.

3.   Local Systems and Characteristic Cycles

3.1. We consider the setting of a (complex) algebraic variety Y with an action of an algebraic group H with finitely many orbits. Then we can consider several categories,

(3.1.1)

$\mathcal{C}(Y, H) = $ category of H–equivariant constructible sheaves on Y,

$\mathcal{P}(Y, H) = $ category of H–equivariant perverse sheaves on Y,

$\mathcal{D}(Y, H) = $ category of H–equivariant regular holonomic D–modules on Y.

Each of these categories is abelian and objects have finite length. In each of these categories, irreducible objects are parametrized by equivariant local systems $(S_\xi, \mathcal{V}_\xi)$ on orbits. The Grothendieck groups of these categories are isomorphic. The second and third categories are isomorphic; the isomorphism is called the *Riemann–Hilbert correspondence.*

3.2. The Grothendieck group $K(Y, H)$ admits functionals defined in the following manner. Let $y \in S$, be an element in the orbit Y. Then for a constructible sheaf C, define

$$(3.2.1) \qquad \chi_S^{loc} : K(Y, H) \longrightarrow \mathbb{N}, \qquad \chi_S^{loc}(C) = dim C_y$$

On a perverse sheaf P the formula is the Euler characteristic

$$(3.2.2) \qquad \chi_S^{loc}(P) = \sum (-1)^i \, dim(H^i P)_y.$$

(The cohomology sheaves of perverse sheaves are constructible.)

DEFINITION. *A $\mathbb{Z}$–linear functional $\eta : K(Y, H) \longrightarrow \mathbb{Z}$ is called geometrically stable if it is a $\mathbb{Z}$–linear combination of χ_S^{loc}.*

3.3. The main motivation for this definition is the following Theorem of MacPherson and Kashiwara. First we introduce a few more notions.

Attached to every H–equivariant regular holonomic D–module, there is a $\mathbb{Z}$–linear combination of conormal bundles

$$(3.3.1) \qquad\qquad Ch(P) = \sum_S \chi_S^{mic}(P)\overline{T_S^*(Y)},$$

called the *Characteristic Cycle* of P. The χ_S^{mic} are integer valued and additive for short exact sequences, so define functionals on $K(Y, H)$.

THEOREM. (MacPherson–Kashiwara) *The functionals χ_S^{mic} are geometrically stable. More precisely, there are integers $c(S, S')$ such that*

$$\chi_S^{mic}(P) = \sum_{S \subset \overline{S'}} c(S, S')\chi_{S'}^{loc}(P).$$

4. Multiplicity Matrices

4.1. Recall the categories $\mathcal{C}(Y, H)$, $\mathcal{P}(Y, H)$ and $\Pi(G/\mathbb{R})$, the set of equivalence classes of irreducible representations of pure real forms. Given a parameter ξ, we write

$$(4.1.1) \quad \begin{aligned} &\pi(\xi) \text{ for the irreducible representation parametrized by } \xi, \\ &M(\xi) \text{ for the standard module parametrized by } \xi, \\ &\mu(\xi) \text{ for the irreducible constructible sheaf parametrized by } \xi, \\ &P(\xi) \text{ for the irreducible perverse sheaf parametrized by } \xi. \end{aligned}$$

The *geometric* multiplicity matrix $\{m_g(\gamma, \xi)\}$ captures the relation between the bases $\mu(\xi)$ and $P(\xi)$ in the Grothendieck group $KX(^\vee G^\Gamma)$,

$$(4.1.2) \qquad\qquad \mu(\xi) = (-1)^{dim(\xi)} \sum m_g(\gamma, \xi)P(\gamma).$$

Here we have made a direct sum over all possible orbits $\mathcal{O}$.

The representation–theoretic matrix $m_r(\gamma, \xi)$ captures the relation between $\pi(\xi)$ and $M(\xi)$,

$$(4.1.3) \qquad\qquad M(\xi) = \sum m_r(\gamma, \xi)\pi(\xi).$$

The main result about these matrices is that they are inverse transpose of each other. Precisely, the result is as follows.

THEOREM. *There is a natural perfect pairing*

$$< \,, >: K\Pi(G/\mathbb{R}) \times KX(^\vee G^\Gamma) \longrightarrow \mathbb{Z}$$

defined by

$$< M(\xi), \mu(\xi') > = e(G(\mathbb{R}, \delta(\xi))\delta_{\xi, \xi'}.$$

This pairing then has the property that

$$< \pi(\xi), P(\xi') > = (-1)^{dim(\xi)} e(G(\mathbb{R}, \delta(\xi))) \delta_{\xi, \xi'}.$$

In other words, if the pairing is set up so that $\{M\}$ and $\{\mu\}$ are dual bases, then so are $\{\pi\}$ and $\{P\}$.

In the above formula, $e(G(\mathbb{R}, \delta(\xi)))$ is Kottwitz's invariant of the real form.

COROLLARY. *Let $\overline{K}(\Pi(G/\mathbb{R})$ be the set of formal linear combinations of irreducible representations of pure real forms. Then $\overline{K}$ may be identified with the space $\mathbb{Z}$–linear functionals of $KX(^\vee G^\Gamma)$. In this identification,*

(1) $M(\xi): C \mapsto e(G(\mathbb{R}, \delta)) m(\mathcal{V}_\xi, C|_{S_\xi}),$
(2) $P(\xi): Q \mapsto (-1)^{d(\xi)} e(G(\mathbb{R}, \delta)) m(\mathcal{P}(\xi), C|_{S_\xi}).$

4.2. Let now $\eta = \sum n(\xi) \pi(\xi, \delta(\xi))$ be a formal linear combination of characters which is *locally finite*. This means that for each pure real form, only finitely many $n(\xi) \neq 0$. For such a combination we can write the *distribution character*

$$(4.2.1) \qquad\qquad \Theta(\eta, \delta) = \sum_{\delta(\xi) \cong \delta} n(\xi) \Theta(\pi(\xi)).$$

This is a well defined function on the strongly regular elements of $G(\mathbb{R}, \delta)$.

DEFINITION. *A locally finite formal character is called stable if for any two pure real forms δ, δ', and $g \in G(\mathbb{R}, \delta) \cap G(\mathbb{R}, \delta')$ strongly regular,*

$$\Theta(\eta, \delta)(g) = \Theta(\eta, \delta')(g).$$

The main result here is a recasting of Shelstad's work on stable characters.

THEOREM. *The strongly stable formal virtual characters correspond precisely to the geometrically stable linear functionals.*

In view of this we make the following definition

DEFINITION. *(18.9 in [**ABV**]) Let $\phi \in P(^\vee G^\Gamma)$ be a Langlands parameter. The stable standard representation attached to ϕ is*

$$\eta_\phi^{loc} = \sum_{\substack{\xi = (\phi, \tau), \\ \delta(\xi) = \delta}} e(G(\mathbb{R}, \delta)) M(\xi).$$

4.3. Given a parameter ϕ as above, we can attach to it a different stable combination of characters corresponding to the characteristic cycle.

DEFINITION. *(19.16 in [**ABV**]) The micro–packet attached to ϕ is the set of equivalence classes of irreducible representations*

$$\Pi(G/\mathbb{R})_\phi^{mic} = \{ (\pi(\xi), \delta(\xi)) \ : \ \chi_{S(\phi)}^{mic}(P(\xi)) \neq 0. \}$$

The corresponding strongly stable virtual character is

$$\eta_\phi^{mic} = \sum_{\pi' \in \Pi(G/\mathbb{R})_\phi^{mic}} e(\pi')(-1)^{d(\pi')-d(\phi)} \chi_{S(\phi)}^{mic}(P')\pi'$$

Here,

- *$e(\pi')$ is Kottwitz's sign $e(G(\mathbb{R}, \delta))$ attached to the real form for π',*
- *$d(\phi)$ and $d(\pi')$ are the dimensions of the orbits corresponding to ϕ, π',*
- *χ_ϕ^{mic} is defined in (2.5.1),*
- *P' is the irreducible perverse sheaf corresponding to π' via 4.1.*

4.4. Example. Consider the case of Example 2.6. The geometrically stable functionals are clear; we take a dual basis to $\mu(\xi)$ for $K\Pi$ and average with character values over the component groups.

Write $S_0 = \{N_p, S_p\}$, S_1 for the open orbit for the second y, and S_2 for the orbit coming from the first y. Then

(4.4.1)
$$\begin{aligned}
Ch(P(S_0)) &= \overline{T_{S_0}^*}, \\
Ch(P(S_1, -1)) &= \overline{T_{S_1}^*} \\
Ch(P(S_1, 1)) &= \overline{T_{S_1}^*}, \\
Ch(P(S_2)) &= \overline{T_{S_2}^*}.
\end{aligned}$$

Thus the *microlocal* stable functionals are

(4.4.2)
$$\begin{aligned}
\chi_{S_0}^{mic} &= Triv, \\
\chi_{S_1}^{mic} &= DS(+) + DS(-), \\
\chi_{S_2}^{mic} &= PS(sgn).
\end{aligned}$$

5. Arthur Parameters

5.1. Motivated by considerations coming from the trace formula, Arthur introduced the space $Q(^\vee G^\Gamma)$ of maps (definition 22.4 in [**ABV**])

(5.1.1)
$$\psi : \mathcal{W}_\mathbb{R} \times SL(2, \mathbb{C}) \longrightarrow {}^\vee G^\Gamma,$$

satisfying

(1) the restriction of ψ to $\mathcal{W}_\mathbb{R}$ is tempered, in other words, the closure of $\psi(\mathcal{W}_\mathbb{R})$ is compact in the analytic topology,
(2) the restriction of ψ to $SL(2)$ is holomorphic.

Two such parameters are called *equivalent* if they are conjugate by $^\vee G$, the set of equivalence classes being denoted $\Psi(^\vee G^\Gamma)$. Define also

(5.1.2)
$$\begin{aligned}
{}^\vee G(\psi) &= \text{ centralizer of } \psi(\mathcal{W}_\mathbb{R} \times SL(2, \mathbb{C})) \text{ in } {}^\vee G^\Gamma, \\
A_\psi &= {}^\vee G_\psi / ({}^\vee G_\psi)_0.
\end{aligned}$$

The associated Langlands parameter is defined as

$$(5.1.3) \qquad \phi_\psi(w) = \psi(w, \begin{bmatrix} |w|^{1/2} & 0 \\ 0 & |w|^{-1/2} \end{bmatrix}).$$

Arthur conjectures that there should be a family of representations attached to each such map which occur as local factors in the residual spectrum of automorphic forms. In particular these representations should be *unitary*. This family should be parametrized by some finite group akin to the component group occuring in the Langlands classification described above.

We recall the description of such parameters (before Proposition 22.9 in [**ABV**]). Write ψ_0 for the restriction of ψ to $\mathcal{W}_\mathbb{R}$ and ψ_1 for its restriction to $SL(2)$. Then ψ_0 is determined by

$$(5.1.4) \qquad (y_0, \lambda_0), \qquad y_0 \in {}^\vee G^\Gamma - {}^\vee G, \qquad \lambda_0 \in {}^\vee \mathfrak{g}.$$

For ψ_1, define

$$(5.1.5) \qquad y_1 = \psi_1 \begin{pmatrix} i & 0 \\ 0 & -i \end{pmatrix}), \qquad \lambda_{\check{\mathcal{O}}} = d\psi_1 \begin{pmatrix} 1/2 & 0 \\ 0 & -1/2 \end{pmatrix}.$$

Then the parameter ϕ_ψ is associated to the pair

$$(5.1.6) \qquad y = y_0 y_1, \qquad \lambda = \lambda_0 + \lambda_{\check{\mathcal{O}}}.$$

5.2. DEFINITION. *The Arthur packet attached to ψ is the micropacket*

$$\Pi(G/\mathbb{R}) = \Pi(G/\mathbb{R})^{mic}_{\phi_\psi}.$$

5.3. An Arthur parameter ψ is called *unipotent* if $\psi|_{\mathbb{C}^*} = Triv$. They are classified as follows.

PROPOSITION. (Corollary 27.3 in [**ABV**]) *Consider the set of parameters ψ with fixed ψ_1. Let $S = Cent({}^\vee G^\Gamma, \psi_1(SL(2))$ and*

$$S_0 = S \cap {}^\vee G, \quad S_1 = S - S_0, \quad S_1^{(2)} = \{y_0 \in S_1 : y_0^2 = 1\}.$$

Then unipotent Arthur parameters restricting to ψ_1 are in 1–1 correspondence with $S_1^{(2)}$. Equivalence classes correspond to conjugacy classes under S_0.

If the parameter ψ corresponds to y_0, then

$$A_\psi = S_0(y_0)/(S_0(y_0)_0.$$

5.4. We use the setup of the previous section; but we turn things around somewhat in that we fix an orbit $\mathcal{O}$ coming from a

(5.4.1) $$\lambda_{\check{\mathcal{O}}} = d\psi_1 \begin{pmatrix} 1/2 & 0 \\ 0 & -1/2 \end{pmatrix}, \qquad \Lambda = \mathcal{F}(\lambda_1).$$

Assume for simplicity that the nilpotent determined by ψ_1 is even, and $e(\Lambda) = \psi_1(-I) = 1$. Then let

(5.4.2) $$E = d\psi_1 \begin{pmatrix} 0 & 1 \\ 0 & 0 \end{pmatrix}.$$

The stabilizer $P(\Lambda)$ is a parabolic subgroup, and let $\mathcal{P}(\Lambda) \cong {}^\vee G / P(\Lambda)$ be the variety of parabolic subgroups conjugate to $P(\Lambda)$. The nilpotent orbit $\mathcal{Z}(\Lambda)$ is the Richardson orbit defined by the nilradical of $P(\Lambda)$, also the image of the cotangent bundle of $\mathcal{P}(\Lambda)$ under the moment map. Let

(5.4.3) $$\mathcal{I}(\Lambda) = \{\, y \in {}^\vee G^\Gamma - {}^\vee G \; : \; y^2 = 1 \,\},$$

and list the orbits under ${}^\vee G$ as $\mathcal{I}_1, \ldots \mathcal{I}_s$ with representatives $y_1, \ldots y_s$. Then let ${}^\vee K_j = {}^\vee G(y_j)$ and form

(5.4.4) $$X_j(\mathcal{O}, {}^\vee G^\Gamma) = {}^\vee G \times_{{}^\vee K_j} \mathcal{P}(\Lambda).$$

Let ${}^\vee \mathfrak{g} = {}^\vee \mathfrak{k}_j + {}^\vee \mathfrak{s}_j$ be the Cartan decomposition.

THEOREM. *The following sets are in 1–1 correspondence.*

(1) *Equivalence classes of unipotent Arthur parameters supported on* $X_j(\mathcal{O}, {}^\vee G^\Gamma)$.
(2) ${}^\vee K_j$ *conjugacy of θ_j-stable parabolic subgroups in $\mathcal{P}(\Lambda)$ whose nilradical intersected with $\mathfrak{s}_j$ meets $\mathcal{Z}(\Lambda)$.*
(3) ${}^\vee K_j$ *orbits in* ${}^\vee \mathfrak{s}_j \cap \mathcal{Z}(\Lambda)$.

Let $I_{\mathcal{P}(\Lambda)}$ be the primitive ideal attached to $\mathcal{P}(\Lambda)$, the kernel of the operator representation of $U({}^\vee \mathfrak{g})$ on $\mathcal{P}(\Lambda)$. Then a representation belongs to the Arthur packet supported on the (union of) $X'_j s$ if and only if its annihilator is the maximal ideal dual to $I_{\mathcal{P}(\Lambda)}$.

This Theorem summarizes Chapter 27 in [**ABV**].

5.5. The θ–stable parabolic subalgebras in Theorem 5.4 are closed orbits of ${}^\vee K_j$ on $\mathcal{P}(\Lambda)$. The fact that these closed orbits give unipotent representations can be generalized as follows.

Let us write $\check{\mathcal{O}}$ for the orbit $\mathcal{Z}(\Lambda)$. Fix $\theta = \theta_j$, ${}^\vee K = {}^\vee K_j$. Let $\check{\mathcal{O}}^\mathbb{R}$ be an orbit of ${}^\vee K_j$ in $\check{\mathcal{O}}^\mathbb{R} \cap {}^\vee \mathfrak{s}$.

THEOREM. *Let ${}^\vee \mathfrak{q} = {}^\vee \mathfrak{l} + {}^\vee \mathfrak{u}$ be a θ–stable parabolic subalgebra with Levi component ${}^\vee L$. Suppose $\check{\mathcal{O}}^\mathbb{R}$ satisfies the following condition.*

There is an even orbit $\check{\mathcal{O}}_{{}^\vee \mathfrak{l}}$ and a ${}^\vee K \cap {}^\vee L$ orbit $\check{\mathcal{O}}_{{}^\vee \mathfrak{l}}^\mathbb{R}$ in $\check{\mathcal{O}}_{{}^\vee \mathfrak{l}} \cap ({}^\vee \mathfrak{s} \cap {}^\vee \mathfrak{l})$ such that

$$\overline{\breve{\mathcal{O}}^{\mathbb{R}} \cap [\breve{\mathcal{O}}^{\mathbb{R}}_{\vee_{\mathfrak{l}}} + (^{\vee}\mathfrak{u} \cap {}^{\vee}\mathfrak{s})]} = \overline{\breve{\mathcal{O}}^{\mathbb{R}}_{\vee_{\mathfrak{l}}} + (^{\vee}\mathfrak{u} \cap {}^{\vee}\mathfrak{s})}.$$

Then the (appropriately defined) induction functor ${}^{\vee}K \times_{{}^{\vee}K \cap {}^{\vee}L}$ *takes representations dual to unipotent representations in the packet attached to* $\breve{\mathcal{O}}^{\mathbb{R}}_{\vee_{\mathfrak{l}}}$ *to representations dual to unipotent representations attached to* $\breve{\mathcal{O}}^{\mathbb{R}}.$

On the level of $({}^{\vee}\mathfrak{g}, {}^{\vee}K)$ modules, the functor ${}^{\vee}K \times_{{}^{\vee}K \cap {}^{\vee}L}$ corresponds to the usual derived functor construction $\mathcal{R}_{\vee \mathfrak{q}}$.

The combination of induction and duality applied to $(\mathfrak{g}, K)$ modules for G will be called *coinduction*.

CONJECTURE. *Any unipotent representation is obtained by a combination of usual unitarity preserving derived functor induction (real or complex) and coinduction.*

The coinduction part takes care of the cases when the representations are isolated. But it is not necessarily true that if the (complexification of the) WF–set of a representation is induced, that the representation itself is obtained by the unitarity preserving functors. This is false in the complex case already. But it is false even if you allow some obvious complementary series.

A special case of this conjecture is treated in [**B**]. This special case covers all the special unipotent representations attached to rigid orbits in $\mathfrak{g}$ for classical groups.

6.   CHARACTERS OF COMPONENT GROUPS

6.1. In this section we discuss the results about special unipotent representations of complex groups in light of the theory of characteristic cycles. Much of what we say is true for real groups, but we stick to complex groups for now.

First, the localization theory of Beilinson–Bernstein allows us to identify the category of regular ${}^{\vee}K$–equivariant $\mathcal{D}_{\lambda}$–modules on the flag variety with the category of $({}^{\vee}\mathfrak{g}, {}^{\vee}K)$–modules. This interchanges $\mathcal{D}$–modules $\mathcal{M}$ with $({}^{\vee}\mathfrak{g}, {}^{\vee}K)$–modules M. Recall also the notion of characteristic variety (with multiplicities) of Vogan. For a $({}^{\vee}\mathfrak{g}, {}^{\vee}K)$–module M, this is a combination

(6.1.1) $$Ch_{alg}(M) = \sum m_{alg}(\mathcal{O}, M)\overline{\mathcal{O}}.$$

Here the $\mathcal{O}$ are ${}^{\vee}K$–orbits in $({}^{\vee}\mathfrak{g}/{}^{\vee}\mathfrak{k})^*$. The characteristic cycle is of the same nature. For Ch, we grade $\mathcal{D}$ and $\mathcal{M}$ to obtain a (graded) module for regular functions on $T^*_{\mathcal{B}}$. For Ch_{alg}, we grade $U({}^{\vee}\mathfrak{g})$ and M to obtain a graded module for $S({}^{\vee}\mathfrak{g})$. In both cases we take the characteristic cycle in the sense of algebraic geometry. We will need the relation between the two. This is provided by Borho–Brylinski. We state a slightly sharper version of J.T. Chang.

Recall the moment map

$$(6.1.2) \qquad \mu \ : \ T^*(\mathcal{B}) \longrightarrow (^\vee\mathfrak{g})^*,$$
$$\mu(x,\zeta) = Ad^*(x)\zeta \in (^\vee\mathfrak{g})^*$$

(we identify $\mathcal{B}$ with cosets of elements in the group).

THEOREM. ([**C**], Corollary 2.5.6). *Suppose*

$$Ch(\mathcal{M}) = \sum m(T_Z^*(\mathcal{B}), \mathcal{M}) \cdot \overline{T_Z^*(\mathcal{B})}.$$

Then

$$m_{alg}(\mathcal{O}, M) = \sum_{\mu(\overline{T_Z^*(\mathcal{B})})=\overline{\mathcal{O}}} m(T_Z^*(\mathcal{B}), \mathcal{M})c_Z,$$

where c_Z is the Euler characteristic of the (generic) fiber of μ over $T_Z^*(\mathcal{B})$ with coefficients in $\mathcal{L}(\lambda)$.

6.2. In view of 6.1, we work with $(^\vee\mathfrak{g}, {}^\vee K)$ modules. Precisely, recall $\lambda_{\breve{\mathcal{O}}}$ from (5.1.4) satisfying $e(\lambda_{\breve{\mathcal{O}}}) = 1$. The Harish–Chandra modules on $^\vee G$ corresponding to the unipotent representations on G are irreducible modules $^\vee L$ satisfying

$$(6.2.1) \qquad \{ \ ^\vee L \ : Ann(^\vee L) \subset U(^\vee\mathfrak{g}) \text{ is the kernel of the operator}$$
$$\text{representation of } U(^\vee\mathfrak{g}) \text{ on } \mathcal{P}(\lambda_{\breve{\mathcal{O}}}). \ \}$$

These correspond to the irreducible perverse sheaves that are relevant to our discussion. They are irreducible modules whose annihilator is minimal subject to the condition that $\tau^L(^\vee L)$ and $\tau^R(^\vee L)$ contain all the simple roots satisfying $(\alpha, \lambda_{\breve{\mathcal{O}}}) = 0$. Their infinitesimal character is $\breve{\rho}$.

In the complex case, the characteristic variety is the closure of a single orbit. One of the main properties of the characteristic cycle is its behaviour with respect to coherent continuation.

THEOREM. (Joseph–King–Vogan) *Let M be irreducible and embed it in a coherent family of (virtual) representations $\{M(\eta)\}$. Let $\breve{\mathcal{O}}$ be its characteristic variety. Then viewed as a function on the (complexified) Cartan subalgebra $^\vee\mathfrak{h}$, $m_{alg}(\breve{\mathcal{O}}, \eta)$ is a harmonic polynomial of degree $|\Delta^+| - \frac{1}{2}dim(\breve{\mathcal{O}})$ The representation generated by m_{alg} is irreducible, and the degree is the lowest where it occurs in the harmonic part of $S^*(^\vee\mathfrak{h})$.*

In the setting of Theorem 6.1, the m_Z's are independent of η and the c_Z's are polynomials of the appropriate degree.

6.3. With the notation as in the previous sections, let $W_0 = W(\lambda_{\breve{\mathcal{O}}})$. For $\sigma \in \widehat{W}$, write $\breve{\sigma} = \sigma \otimes sgn$ and consider the virtual characters

$$(6.3.1) \qquad R_{\breve{\sigma}} = \frac{1}{|W_0|^2} \sum_{w \in W} tr\breve{\sigma}(w) \sum_{x,y \in W_0} sgn(x)sgn(y)^\vee M(\breve{\rho}, xwy\breve{\rho}),$$
$$R_\sigma = \frac{1}{|W_0|^2} \sum_{w \in W} tr\sigma(w) \sum_{x,y \in W_0} M(\lambda_{\breve{\mathcal{O}}}, xwy\lambda_{\breve{\mathcal{O}}})$$

where M, $^\vee M$ are the standard (principal series) modules.

PROPOSITION. $R_{\check{\sigma}}$ is zero if sgn_{W_0} does not occur in $\sigma|_{W_0}$. Similarly, $R_\sigma = 0$ unless $\sigma|_{W_0}$ contains the trivial representation.

This is the analogue of Lemma 6.2 in [BV] and is proved in the same way. Furthermore,

$$(6.3.2) \qquad\qquad < R_\sigma, R_{\check{\sigma}} >= 1.$$

As described in section 3, the pairing is between the $R_{\check{\sigma}}$ at infinitesimal character $(\check{\rho}, \check{\rho})$ and R_σ at infinitesimal character $\lambda_{\check{\mathcal{O}}}$.

Let $L_0, \ldots, L_k$ be the irreducible unipotent representations at infinitesimal character $\lambda_{\check{\mathcal{O}}}$, corresponding to $\check{\mathcal{O}}$. Let $^\vee L_0, \ldots, {}^\vee L_k$ be the *dual* representations. Let σ_i be the representations in the relevant left cell, and $\check{\sigma}_i = \sigma_i \otimes sgn$. As in [BV], there is a quotient $\overline{A}(\check{\mathcal{O}})$, of the component group of the centralizer of $\check{\mathcal{O}}$ so that the character theory of the L_i corresponds to the character theory of $\overline{A}(\check{\mathcal{O}})$. Namely

$$(6.3.3) \qquad\qquad L_i \leftrightarrow tr\pi_i, \qquad R_{\sigma_i} \leftrightarrow [x_i].$$

In this correspondence, $\pi_0 = Triv$ and $[x_0] = [1]$.

On the other hand we can write

$$(6.3.4) \qquad\qquad {}^\vee L_i = \sum_j c_{ij} R_{\check{\sigma}_j}(1) + \text{ ``other terms''}$$

To compute the c_{ij}'s we dot with the

$$(6.3.5) \qquad\qquad R_{\sigma_j} = \sum tr\pi_j(x_i) L_i.$$

The *"other terms"* transform according to other representations of W. Their dot product with R_{σ_j} vanishes, so that we get

$$(6.3.6) \qquad\qquad c_{ij} = tr\pi_j(x_i).$$

6.4. THEOREM.

$$\chi_\phi^{mic}({}^\vee L_i) = tr\pi_i(1) = dim\ \pi_i.$$

PROOF. The assertion follows from the following facts. Write $^\vee\mathfrak{p}$ for the $\theta-$ stable parabolic subalgebra $^\vee\mathfrak{p}(\Lambda) = {}^\vee\mathfrak{l} + {}^\vee\mathfrak{n}$ corresponding to the parabolic subgroup $P(\Lambda) = LN$ in 5.4.

(1) There is a character χ on $^\vee\mathfrak{p}$ such that the derived functor modules

$$\mathcal{R}_{^\vee\mathfrak{p}}^i(\chi) = \begin{cases} {}^\vee L_0 & \text{if } i = dim({}^\vee\mathfrak{k} \cap {}^\vee\mathfrak{n}), \\ 0 & \text{otherwise.} \end{cases}$$

(2) $\chi_\phi^{mic}({}^\vee L_0) = 1$.

(3) $\chi_\phi^{mic}(R_{\check{\sigma}_i}) = 0$ for $i \neq 0$.

The first fact is just 5.5 for $^\vee\mathfrak{q} = {}^\vee\mathfrak{p}$ (and the appropriate orbits). The irreducibility and vanishing follows from [**V**]. (Recall that, by the localization theory Beilinson–Bernstein, the infinitesimal character of the $^\vee L$ is $(\check\rho, \check\rho)$).

The multiplicity 1 statement follows from 1 by a version of Blattner's formula. Namely, when we grade the relevant $\mathcal{D}$–module, we get sections of the pullback of the line bundle defined by χ to the corresponding conormal bundle.

For the third fact, observe first that the orbit corresponding to ϕ is the only one which will occur in the formula for $m_{alg}(\check{\mathcal{O}})$. This is a (much simpler) special case of Theorem 27.10 in [**ABV**].

Then consider the coherent families for the $^\vee L$'s. Decompose $R_{\check\sigma_i}$ into irreducible characters. Then $\chi_\phi^{mic}(R_{\check\sigma_i})$ is the corresponding combination of the $\chi_\phi^{mic}(^\vee L_j)$. If this is not 0, then the corresponding combination of $m_{alg}(\check{\mathcal{O}}, {}^\vee L_j)$ is also not zero. As a function of the parameter η of the coherent family, this is a polynomial of degree equal to $deg\check\sigma_0 = |\Delta^+| - dim\check{\mathcal{O}}$. On the other hand, by the transformation property of $R_{\check\sigma_i}$ this polynomial must transform like $\check\sigma_i$. For $i \neq 0$, the representation $\check\sigma_i$ occurs only in strictly higher degree, so this polynomial must be identically 0. Thus, $\chi_\phi^{mic}(R_{\check\sigma_i}) = 0$ as well.

Then the value $\chi_\phi^{mic}(R_{\check\sigma_0})$ is computed from facts (2), (3) and the expression of $^\vee L_0$ in terms of the $R_{\check\sigma_i}$. Then the theorem follows from the formulas in (6.3.3) □

REFERENCES

[ABV]   J. Adams, D. Barbasch, and D. Vogan, *The Langlands Classification and Irreducible Characters for Real Reductive groups*, Birkhäuser, 1992.

[B]   D. Barbasch, *Unipotent representations for real reductive groups*, Proceedings of the International Congress of Mathematicians, 1990, vol. 2, Springer, 1991, pp. 769–777.

[BV]   D. Barbasch and D. Vogan, *Unipotent representations of complex semisimple Lie groups*, Ann. of Math. **121** (1985), 41–110.

[C]   J.T. Chang, *Asymptotics and Characteristic Cycles for Representations of Complex Groups*, preprint, 1991.

[V]   D. Vogan, *Unitarizability of certain series of representations*, Ann. of Math. **120** (1984), 141–187.

DEPARTMENT OF MATHEMATICS, CORNELL UNIVERSITY, WHITE HALL, ITHACA, NEW YORK 14853-7901, USA

E-mail address: barbasch@math.cornell.edu

Contemporary Mathematics
Volume **154**, 1993

Unitarity of Certain Dolbeault Cohomology Representations

ROGER ZIERAU

ABSTRACT. The unitary structure on Dolbeault cohomology representations is studied. The method is to use the explicit formula of an intertwining operator introduced in [**2**] to show, in certain cases, that each K-finite cohomology class is represented by an L_2 strongly harmonic form.

§1. Introduction. One of the most important areas of representation theory is the study of the irreducible unitary representations of semisimple Lie groups. Two important aspects of this are the classification of all unitary representations and the realization of these in a natural analytic and/or geometric fashion. The classification has not yet been completed, however there are many series of representations which are known to be unitary (or unitarizable). One such series of representations consists of the cohomologically induced representations which were introduced by Zuckerman as an algebraic analogue of Dolbeault cohomology of an induced bundle on an elliptic coadjoint orbit. Recently Wong ([**14**]) has shown that the cohomologically induced representations do indeed correspond to Dolbeault cohomology (in the sense that (i) the Dolbeault cohomology is a continuous admissible representation and (ii) its Harish-Chandra module is a cohomologically induced representation). The point of this article is to show how one can, in certain cases, unitarize the Dolbeault cohomology representations by producing a natural G-invariant Hermitian form on an appropriate space of L_2 harmonic forms. The main tool is the intertwining operator studied in [**1**] and [**2**] which maps a principal series representation into the harmonic forms. This gives an explicit enough formula that one can for example show square integrability of harmonic forms in some special cases. It is hoped that this method will work

1991 *Mathematics Subject Classification.* Primary 22E45; Secondary 22E46.
Partially supported by National Science Foundation Grant DMS 89 02352.
This paper is in final form and no version of it will be submitted for publication elsewhere.

in broad generality. Other methods have been used to study this question in various special cases. See for example [7], [10], [8], [13], and [3].

More precisely, we let G be a connected linear semisimple Lie group and $G_{\mathbb{C}}$ its complexification. Let X be a generalized flag variety for $G_{\mathbb{C}}$ and let $\mathcal{D} = G \cdot x_0$ be an open orbit in X. We assume $\mathcal{D}$ is measurable so, by Theorem 6.3 in [12], $\mathcal{D} \simeq G/H$ where H is the centralizer of a compact torus in G. Note that any elliptic coadjoint orbit is of this form. The holomorphic tangent space at the identity coset may be considered as an invariant complex polarization. Let χ be a unitary character of H and $\mathcal{L}_\chi \to \mathcal{D}$ the associated Hermitian homogeneous line bundle. The representations we wish to unitarize are the Dolbeault cohomology $H^s(G/H, \mathcal{L}_\chi)$ spaces of the line bundles $\mathcal{L}_\chi$. This may be considered as the natural setting for geometric quantization for such orbits. The open orbit $\mathcal{D}$ carries two metrics of interest. One is an indefinite invariant hermitian metric defined in terms of the Killing form and the other is positive definite but not invariant (see §5). The latter is used to define the notion of a square integrable $\mathcal{L}_\chi$-valued differential form. The invariant metric serves two purposes. It defines both the formal adjoint $\overline{\partial}^*$ of the Dolbeault operator $\overline{\partial}$ and an indefinite invariant global Hermitian form on a space of L_2 harmonic forms.

DEFINITION 1.1.

(a) $\mathcal{H}^s(G/H, \mathcal{L}_\chi)$ will denote the space of $\mathcal{L}_\chi$-valued differential forms of type $(0, s)$ which are strongly harmonic in the sense that $\overline{\partial}\omega = 0$ and $\overline{\partial}^*\omega = 0$.

(b) $\mathcal{H}^s_{(2)}(G/H, \mathcal{L}_\chi)$ will denote the space of $(0, s)$ forms which are both strongly harmonic and square integrable.

Since $\mathcal{H}^s_{(2)}(G/H, \mathcal{L}_\chi)$ consists of closed forms there is a natural map q from $\mathcal{H}^s_{(2)}(G/H, \mathcal{L}_\chi)$ into the Dolbeault cohomology $H^s(G/H, \mathcal{L}_\chi)$. There is (see §5) a natural G-invariant global Hermitian form defined on $\mathcal{H}^s_{(2)}(G/H, \mathcal{L}_\chi)$.

A satisfactory geometric quantization for the elliptic coadjoint orbits might consist of verifying the following statements:

(A) Every K-finite cohomology class is represented by a square integrable strongly harmonic form.

(B) The global invariant form is positive semi-definite on $\mathcal{H}^s_{(2)}(G/H, \mathcal{L}_\chi)$.

(C) The nullspace of this form is the kernel of q.

One would then conclude that an L_2 inner product is defined on the K-finite vectors in $H^s(G/H, \mathcal{L}_\chi)$. We will establish (A)-(C) as a theorem under the following conditions:

(1) A natural negativity condition on $\mathcal{L}_\chi$, see (2.1).

(2) The real ranks of G and H are equal.

(3) H is the fixed point set of an involution, so $\mathcal{D} \simeq G/H$ is a semisimple symmetric space with an invariant complex structure.

In part I we will give an account of the construction of the intertwining map which, as mentioned above, is the main tool. Everything in part I is contained in

[2] however the point of view here is a little different. We are a bit less technical than in [2] and it is hoped that the idea of the construction will be a little clearer here. The proofs will be a bit sketchier. We assume for part I that the real ranks of G and H are the same. This is the assumption in [2]. In part II the intertwining operator discussed in part I will be used to prove that (A)-(C) are true when conditions (1)-(3) hold. The two significant facts which must be established are the square integrability of forms in the image of the intertwining operator and the positive semi-definiteness of the global invariant form. The proof of the former requires condition (3) in order to obtain some estimate of the growth of the forms. An integration formula on $\mathcal{D}$ is also needed. The proof of the latter, in some sense, does not use condition (3). We will also give an example which will indicate how to calculate the operator and will also suggest that the unitarity results given here should hold in more generality.

Part I

§2. The intertwining map. As in the introduction, G is a semisimple linear Lie group, $G_{\mathbb{C}}$ its complexification, $\mathfrak{g}_0$ the Lie algebra of G and $\mathfrak{g}$ its complexification. Similar notation holds for other Lie groups and Lie algebras. Fix a Cartan involution θ. The corresponding maximal compact subgroup will be denoted by K. We assume that rank(G)=rank(K). Thus, there is a Cartan subalgebra $\mathfrak{t}$ of $\mathfrak{g}$ contained in $\mathfrak{k}$. The elliptic coadjoint orbits are of the form $Ad^*(G) \cdot \lambda \simeq G/H$ for $\lambda \in \mathfrak{t}^* \subset \mathfrak{g}^*$, $H \equiv$ centralizer of λ in G.

Let $\Delta(\mathfrak{t}, \mathfrak{g})$ be the roots of $\mathfrak{t}$ in $\mathfrak{g}$, then the set of $\mathfrak{t}$-roots in $\mathfrak{h}$ is

$$\Delta(\mathfrak{t}, \mathfrak{h}) = \{\alpha \in \Delta(\mathfrak{t}, \mathfrak{g}) : \langle \lambda, \alpha \rangle = 0\}.$$

A θ-stable parabolic subalgebra $\mathfrak{h} + \mathfrak{q}_+$ is determined by its $\mathfrak{t}$-roots:

$$\Delta(\mathfrak{t}, \mathfrak{h} + \mathfrak{q}_+) = \{\alpha \in \Delta(\mathfrak{t}, \mathfrak{g}) : \langle \lambda, \alpha \rangle \geq 0\}.$$

The unipotent radical $\mathfrak{q}_+$ is determined by $\Delta(\mathfrak{t}, \mathfrak{q}_+) = \{\alpha \in \Delta(\mathfrak{t}, \mathfrak{g}) : \langle \lambda, \alpha \rangle > 0\}$. The complex conjugate $\bar{\mathfrak{q}}_+$ of $\mathfrak{q}_+$ with respect to the real form $\mathfrak{g}_0$ has roots $\Delta(\mathfrak{t}, \bar{\mathfrak{q}}_+) = -\Delta(\mathfrak{t}, \mathfrak{q}_+)$ and is denoted by $\mathfrak{q}_-$. Fix any positive system $\Delta^+ = \Delta^+(\mathfrak{t}, \mathfrak{g})$ in $\Delta(\mathfrak{t}, \mathfrak{g})$ so that $\Delta(\mathfrak{q}_+) \subset \Delta^+$ (this amounts to any positive system in $\Delta(\mathfrak{t}, \mathfrak{h})$ along with $\Delta(\mathfrak{q}_+)$). Let $H_{\mathbb{C}}$, Q_+ and Q_- be the connected subgroups of $G_{\mathbb{C}}$ with Lie algebras $\mathfrak{h}, \mathfrak{q}_+$ and $\mathfrak{q}_-$. One can easily see that $\mathcal{D} = G/H$ is an open orbit in the flag variety $G_{\mathbb{C}}/H_{\mathbb{C}}Q_-$. Thus, $\mathcal{D}$ is a complex manifold with holomorphic and anti-holomorphic tangent spaces at the identity coset given by $\mathfrak{q}_+$ and $\mathfrak{q}_-$.

We will assume throughout that $rank_{\mathbb{R}}(H) = rank_{\mathbb{R}}(G)$ (this is condition (2) of the introduction), that is, there is a maximal abelian subspace $\mathfrak{a}_0 \subset \mathfrak{h}_0 \cap \mathfrak{p}_0$ which is maximal abelian in $\mathfrak{p}_0$. We fix such an $\mathfrak{a}_0$. We choose a positive system $\sum^+(\mathfrak{a}, \mathfrak{g})$ of $\mathfrak{a}$ roots in $\mathfrak{g}$ and let $\sum^+(\mathfrak{a}, \mathfrak{h})$ be $\sum(\mathfrak{a}, \mathfrak{h}) \cap \sum^+(\mathfrak{a}, \mathfrak{g})$.

Let $A = \exp(\mathfrak{a}_0)$ and let $\mathfrak{n} = \displaystyle\sum_{\alpha \in \Sigma^+(\mathfrak{a},\mathfrak{g})} \mathfrak{g}^{(\alpha)}$ (where $\mathfrak{g}^{(\alpha)}$ is the rootspace for the $\mathfrak{a}$-root α). Set $N = \exp(\mathfrak{n}_0)$. Then $G = KAN$ is an Iwasawa decomposition of G and $P = MAN$ is a minimal parabolic subgroup where $M = Z_K(\mathfrak{a})$, the centralizer of $\mathfrak{a}$ in K. We fix a Cartan subalgebra $\mathfrak{t}_M \subset \mathfrak{t}$ of $\mathfrak{m}$. Note that $\mathfrak{m} = \mathfrak{m} \cap \mathfrak{h} + \mathfrak{m} \cap \mathfrak{q}_+ + \mathfrak{m} \cap \mathfrak{q}_-$.

Let $\mathbb{C}_\chi$ be a 1-dimensional unitary representation of H, $\chi \in i\mathfrak{t}_0^*$ the differential of $\mathbb{C}_\chi|_T$. A holomorphic homogeneous line bundle $\mathcal{L}_\chi \to G/H$ is determined. We assume that the following negativity condition holds:

$$(2.1) \qquad\qquad \langle \chi + \rho_{\mathfrak{k}}, \gamma \rangle < 0 \text{ for all } \gamma \in \Delta(\mathfrak{k} \cap \mathfrak{q}_+)$$

where $\rho = \rho(\Delta^+(\mathfrak{t}, \mathfrak{g}))$ is half the sum of the roots in $\Delta^+(\mathfrak{t}, \mathfrak{g})$. We will construct a nonzero intertwining operator from a principal series representation into the space $A^s(G/H, \mathcal{L}_\chi)$ of $\mathcal{L}_\chi$-valued C^∞ forms on G/H of type (o, s). Here $s = \dim_{\mathbb{C}}(K/K \cap H)$, the dimension of a maximal compact subvariety of G/H. Note that $A^s(G/H, \mathcal{L}_\chi) = \{C^\infty(G) \otimes \wedge^s(\mathfrak{q}_+) \otimes \mathbb{C}_\chi\}^L = \{\omega : G \to \wedge^s(\mathfrak{q}_+) \otimes \mathbb{C}_\chi : \omega(gl) = l^{-1} \cdot \omega(g)\}$.

The following notation will be used for principal series representations. We have fixed a minimal parabolic subgroup $P = MAN$ of G. If $(\delta_M \otimes e^\nu \otimes 1, W)$ is a representation of P we set $I(W) = \{\varphi : G \to W : \varphi$ is C^∞ and $\varphi(gman) = \delta(m^{-1}) e^{-(\nu+\rho)}(a)\varphi(g)\}$. We also denote this by $C^\infty(G/P, W)$ the C^∞ sections of the homogeneous bundle $W \otimes \mathbb{C}_\rho$ with fiber $W \otimes \mathbb{C}_\rho$ at eP. The same notation will be used when considering representations of H induced from the parabolic subgroup $H \cap P$, that is, the shift by $\rho_{\mathfrak{h}}$ does not appear in the notation.

The strategy for constructing the intertwining map is as follows. If V is a smooth representation of G and $C^\infty(G/H, \mathcal{E})$ is the space of C^∞ sections of a finite dimensional homogeneous vector bundle $\mathcal{E} \to G/H$ with fiber E at eH then

$$(2.2) \qquad\qquad \operatorname{Hom}_G(V, C^\infty(G/H, \mathcal{E})) \simeq \operatorname{Hom}_H(V, E).$$

The correspondence between intertwining operators $A \in \operatorname{Hom}_G(V, C^\infty(G/H, \mathcal{E}))$ and $a \in \operatorname{Hom}_H(V, E)$ is $A(v)(g) = a(g^{-1}v)$ if a is given (and $a(v) = A(v)(e)$ if A is given). Our goal is to find a nonzero element of $\operatorname{Hom}_H(V, E)$ where $E = \wedge^s(\mathfrak{q}_+) \otimes \mathbb{C}_\chi$ and V is an appropriate principal series representation. This is of course equivalent (by taking duals) to finding a nonzero element of $\operatorname{Hom}_H(E^*, \mathcal{D}'(G/P, W^*))$ where $\mathcal{D}'(G/P, W^*) = \{C^\infty(G/P, W)\}'$, the continuous dual. Thus, $H \cdot eP \subset G/P$ is a closed H-orbit.

It seems to be very difficult to say much about when a finite dimensional representation E^* of H occurs in the space of distributions $\mathcal{D}'(G/P, W^*)$. However, one can give some useful information about H-finite distributions which are supported in a closed H-orbit in G/P. Since $\mathfrak{a} \subset \mathfrak{h} \cap \mathfrak{p}$, $H = K \cap H \cdot A \cdot N \cap H$ is an Iwasawa decomposition of H and $P \cap H = M \cap H \cdot A \cdot N \cap H$ is a minimal parabolic of H.

LEMMA 2.3. *The distributions in $\mathcal{D}'(G/P, \mathcal{W}^*)$ which restrict to smooth functions on $H \cdot eP$ may be identified with $C^\infty(H/P \cap H, \mathcal{W}^* \otimes \mathbb{C}_{-\rho+\rho_\mathfrak{h}})$.*

PROOF. In general, if S is a closed submanifold of M and $\mathcal{V}$ is a vector bundle on M then the space of $\mathcal{V}$-valued distributions on M is $\mathcal{D}'(M, \mathcal{V}) = C_0^\infty(M, \mathcal{V}^* \otimes \wedge^{top}(T^*M))'$. This contains $\mathcal{D}'(S, \mathcal{V} \otimes \wedge^{top}(TM/TS))$, and this in turn contains the space of smooth distributions $C^\infty(S, \mathcal{V} \otimes \wedge^{top}(TM/TS))$. The inclusion of $C^\infty(S, \mathcal{V} \otimes \wedge^{top}(TM/TS))$ into $\mathcal{D}'(M, \mathcal{V})$ is given by integrating over S. In the present situation $M = G/P$, $S = H \cdot eP \simeq H/H \cap P$, $\mathcal{V} = \mathcal{W}^* \otimes \mathbb{C}_\rho$, $\wedge^{top}(T^*M) = \mathbb{C}_{2\rho}$ and $\wedge^{top}(TM/TS) = \mathbb{C}_{-2\rho+2\rho_\mathfrak{h}}$, taking into account our ρ-shifts we get

$$(2.4) \qquad C^\infty(H/H \cap P, \mathcal{W}^* \otimes \mathbb{C}_{-\rho+\rho_\mathfrak{h}}) \subset \mathcal{D}'(G/P, \mathcal{W}^*),$$

and this inclusion is given by integrating over $H \cap K$.

The principal series parameters will now be determined. Let δ_M be the finite dimensional irreducible representation of M with extreme weight $\chi + 2\rho(\mathfrak{k} \cap \mathfrak{q}_+)$ restricted to $\mathfrak{t}_M$. This is dominant for $\Delta^+(\mathfrak{t}_M, \mathfrak{m} \cap \mathfrak{h}) \cup \Delta(\mathfrak{m} \cap \bar{\mathfrak{q}}_+)$ where $\Delta^+(\mathfrak{t}_M, \mathfrak{m} \cap \mathfrak{h})$ is any positive system for $\mathfrak{m} \cap \mathfrak{h}$. Also, $\delta_M|_{M \cap H}$ contains a copy of the one dimensional representation $\wedge^s(\mathfrak{k} \cap \mathfrak{q}_+) \otimes \mathbb{C}_\chi|_{M \cap H}$. We let $W \in \widehat{MAN}$ be given by

$$(2.5) \qquad \delta_M \otimes e^{\rho_\mathfrak{h}} \otimes 1, \quad \rho_\mathfrak{h} = \frac{1}{2} \sum_{\alpha \in \Sigma^+(\mathfrak{a}, \mathfrak{h})} \alpha.$$

Then $V = I(W) = C^\infty(G/P, \mathcal{W})$ is the (normalized) principal series representation of interest.

PROPOSITION 2.6. *For $E = \wedge^s(\mathfrak{q}_+) \otimes \mathbb{C}_\chi$ and W as in (2.5)*

$$Hom_H(E^*, \mathcal{D}'(G/P, \mathcal{W}^*)) \neq 0.$$

PROOF. We look for distributions transforming as vectors in E^* which are supported in the closed orbit $H \cdot eP$ and are smooth on this orbit. So we need to construct a nonzero element of

$$Hom_H(E^*, C^\infty(H/P \cap H, \mathcal{W}^* \otimes \mathbb{C}_{-\rho+\rho_\mathfrak{h}}))$$
$$\simeq Hom_{P \cap H}(E^*, W^* \otimes \mathbb{C}_{-\rho+2\rho_\mathfrak{h}})$$
$$\simeq Hom_{P \cap H}(W \otimes \mathbb{C}_{\rho-2\rho_\mathfrak{h}}, E)$$
$$(2.7) \qquad \simeq Hom_{M \cap H \cdot A}(\delta_M \otimes \mathbb{C}_{\rho-\rho_\mathfrak{h}}, E^{\mathfrak{n} \cap \mathfrak{h}}).$$

We use the following lemma.

LEMMA 2.8. *Let G_1 be any connected reductive Lie group, K_1 a maximal compact subgroup and $P_1 = M_1 A_1 N_1$, $(M_1 = Z_{K_1}(\mathfrak{a}))$ a minimal parabolic subgroup. Let τ be a one dimensional representation of K_1. If a finite dimensional* *irreducible representation* (π, F) *of* G_1 *contains the* K_1 *type* τ *then* M_1 *acts by* $\tau|_{M_1}$ *on* $F^{\mathfrak{n}_1}$.

PROOF OF LEMMA. Let $v \in F$ be such that $\pi(k)v = \tau(k)v$, for all $k \in K_1$. Decompose v into $\mathfrak{a}$-weight vectors; $v = v_0 + v_1 + v_2 + \cdots$ with $v_0 \in F^{\mathfrak{n}_1}$. To see that $v_0 \neq 0$, let $\overline{P}_1 = M_1 A_1 \overline{N}_1$ be the opposite parabolic subgroup. As $G_1 = \overline{N}_1 A_1 K_1$, $\pi(G_1)v = \pi(\overline{N}_1 A_1)v$ spans F. Let λ be the highest $\mathfrak{a}$ weight of F. If $v_0 = 0$ then $\pi(\overline{N}_1)v = \sum_{\lambda > \mu} F_\mu$ ($F_\mu = \mu$- weight space of F), but this is a contradiction to $\pi(G_1)v$ spanning F. Since $F^{\mathfrak{n}_1}$ is M_1-irreducible, $F^{\mathfrak{n}_1} = C \cdot v_0$. The lemma is proved.

Apply this lemma to H, $P \cap H = M \cap H \cdot A \cdot N \cap H$, $\tau = \wedge^s(\mathfrak{k} \cap \mathfrak{q}_+) \otimes \mathbb{C}_\chi$ and any H-subrepresentation F of $\wedge^s(\mathfrak{q}_+) \otimes \mathbb{C}_\chi$ with $Hom_{K \cap H}(\tau, F) \neq 0$. We conclude that $Hom_{M \cap H}(\wedge^s(\mathfrak{k} \cap \mathfrak{q}_+) \otimes \mathbb{C}_\chi, F^{\mathfrak{n} \cap \mathfrak{h}}) \neq 0$ and so $Hom_{M \cap H}(\delta_M, F^{\mathfrak{n} \cap \mathfrak{h}}) \neq 0$. We must show that some such F has highest $\mathfrak{a}$-weight $\rho - \rho_\mathfrak{h}$. Recall that $\mathfrak{a} + \mathfrak{t}_M$ is a Cartan subalgebra and that we may write $\mathfrak{a} + \mathfrak{t}_M$-roots as pairs $(\alpha, \beta) \in \mathfrak{a}^* \times \mathfrak{t}_M^*$ and we write $X_{\alpha,\beta}$ for a root vector for (α, β). Note that if $X_{\alpha,\beta} \in \mathfrak{n} \cap \mathfrak{q}_+$ then $\theta(X_{\alpha,\beta})$ is a root vector for $X_{-\alpha,\beta}$ and $X_{\alpha,-\beta} \in \mathfrak{n} \cap \mathfrak{q}_-$. One can check that

(1) $\mathfrak{k} \cap \mathfrak{q}_+$ is spanned by $\{X_{\alpha,\beta} + \theta(X_{\alpha,\beta}) : X_{\alpha,\beta} \in \mathfrak{n} \cap \mathfrak{q}_+\} \cup \{X_{0,\beta}\}$

(2) $\mathfrak{p} \cap \mathfrak{q}_+$ is spanned by $\{X_{\alpha,\beta} - \theta(X_{\alpha,\beta}) : X_{\alpha,\beta} \in \mathfrak{n} \cap \mathfrak{q}_+\}$.

Set $U = \sum F$, with F ranging over all irreducible H-subrepresentations of $\wedge^s(\mathfrak{q}_+) \otimes \mathbb{C}_\chi$ with $Hom_{H \cap K}(\tau, F) \neq 0$. Now decompose U into $\mathfrak{a}$-weight spaces. The vector which spans $\wedge^s(\mathfrak{k} \cap \mathfrak{q}_+) \otimes \mathbb{C}_\chi$ is

$$u = (X_{\alpha,\beta} + \theta(X_{\alpha,\beta})) \wedge \cdots \wedge X_{0,\beta'} \wedge \cdots \otimes 1 \in \wedge^s(\mathfrak{k} \cap \mathfrak{q}_+) \otimes \mathbb{C}_\chi,$$

the exterior product of the s basis elements of $\mathfrak{k} \cap \mathfrak{q}_+$ listed in (1) above. Note that u is in U. Writing u as a sum of weight vectors we get

$$u = \sum X_{\alpha,\beta} \wedge \cdots \wedge \theta(X_{\alpha',\beta'}) \wedge \cdots \wedge X_{0,\beta''} \wedge \cdots \otimes 1$$

(the sum is over (α, β), $(0, \beta)$ as in (1)). The highest possible weight in this expression has no $\theta(X_{\alpha,\beta})$ terms. This highest weight vector is $X_{\alpha,\beta} \wedge \cdots \wedge X_{0,\beta'} \wedge \cdots$ (with (α, β), $(0, \beta')$ as in (1)). The highest $\mathfrak{a}$-weight occurring is

$$\sum_{X_{\alpha,\beta} \in \mathfrak{n} \cap \mathfrak{q}_+} \alpha = \frac{1}{2} \sum_{X_{\alpha,\beta} \in \mathfrak{n} \cap \mathfrak{q}_+ + \mathfrak{n} \cap \mathfrak{q}_-} \alpha = \rho - \rho_\mathfrak{h} \text{ (since } \mathfrak{n} = \mathfrak{n} \cap \mathfrak{h} + \mathfrak{n} \cap \mathfrak{q}_+ + \mathfrak{n} \cap \mathfrak{q}_-\text{).}$$

The proposition is proved.

Thus, a nonzero intertwining map $\mathcal{S} : I(W) \to A^s(G/H, \mathcal{L}_\chi)$ exists. The proof of Proposition 2.6 gives a nonzero $t \in Hom_{P \cap H}(W \otimes \mathbb{C}_{\rho - 2\rho_\mathfrak{h}}, E)$, see (2.7), thus a nonzero $t^* \in Hom_{H \cap P}(E^*, W^* \otimes \mathbb{C}_{-\rho + 2\rho_\mathfrak{h}})$ is determined. There corresponds a homomorphism $\tilde{T} \in Hom_H(E^*, C^\infty(H/H \cap P, \mathcal{W}^* \otimes \mathbb{C}_{-\rho + \rho_\mathfrak{h}}))$ given by $\tilde{T}_{e^*}(h) = t^*(h^{-1} \cdot e^*)$. By (2.4) we get distributions T_{e^*}:

$$T_{e^*}(\varphi) = \int_{H \cap K} \langle t^*(l^{-1} \cdot e^*), \varphi(l) \rangle dl$$

(2.9)
$$= \int_{K \cap H} \langle l^{-1} \cdot e^*, t(\varphi(l)) \rangle dl,$$

for $\varphi \in I(W)$, $e^* \in E^*$.

THEOREM 2.10. *Assume* $rank_R(H) = rank_R(G)$. *Let* χ *be a unitary character of* H *satisfying* (2.1). *Let* $W \in \hat{P}$ *be* $\delta_M \otimes e^{\rho_\mathfrak{h}} \otimes 1$, *with* $\delta_M =$ *the irreducible* M-*representation with extreme weight* $\chi + 2\rho(\mathfrak{k} \cap \mathfrak{q}_+)$. *Then there exists a nonzero intertwining map* $\mathcal{S}: I(W) \to A^s(G/H, \mathcal{L}_\chi)$. *An explicit formula is*

$$(\mathcal{S}\varphi)(g) = \int_{K \cap H} \pi(l) t(\varphi(gl)) \, dl$$

$$= \int_{K \cap H} \pi(l) \left(t(\varphi(\kappa(gl))) \right) e^{(-\rho_\mathfrak{h} - \rho)H(gl)} dl,$$

where $g = \kappa(g) \exp(H(g)) N_g$ *is the Iwasawa decomposition of* g *with respect to* KAN, $\varphi \in C^\infty(G/P, W)$ *and* π *is the representation of* H *on* $\wedge^s(\mathfrak{q}_+) \otimes \mathbb{C}_\chi$. *Alternatively, we may fix* $\psi_0 \in W$ *of* $\mathfrak{t}_M$-*weight* $\chi + 2\rho(\mathfrak{k} \cap \mathfrak{q}_+)$ *and noting that the image of* t *is one dimensional, spanned by a vector* $\omega_0 \in E^{\mathfrak{n} \cap \mathfrak{h}}$, *we may write*

$$(\mathcal{S}\varphi)(g) = \int_{K \cap H} \langle (\varphi(\kappa(gl))), \psi_0 \rangle e^{(-\rho_\mathfrak{h} - \rho)H(gl)} \pi(l)(\omega_0) \, dl.$$

PROOF. $T: E^* \to \mathcal{D}'(G/P, W)$ corresponds to the map $a: C^\infty(G/P, W) \to E$ given by

$$T_{e^*}(\varphi) = \langle e^*, a(\varphi) \rangle.$$

The intertwining operator $\mathcal{S}: C^\infty(G/P, W) \to C^\infty(G/H, \mathcal{E})$ corresponding to a by (2.2) is given by:

$$\langle e^*, \mathcal{S}(\varphi)(g) \rangle = \langle e^*, a(g^{-1} \cdot \varphi) \rangle$$

$$= T_{e^*}(g^{-1} \cdot \varphi)$$

$$= \int_{K \cap H} \langle l^{-1} \cdot e^*, t(\varphi(gl)) \rangle dl$$

$$= \int_{K \cap H} \langle e^*, \pi(l) t(\varphi(gl)) \rangle dl$$

$$= \langle e^*, \int_{K \cap H} \pi(l) t(\varphi(gl)) dl \rangle.$$

The other formulas follow easily.

REMARK 2.11. We could have simply written down this formula then checked that it is a well defined intertwining map (using an integral formula on page 198 of [2]). If one takes this approach then it must be verified that $\mathcal{S}$ is not identically zero. This is essentially contained in Theorem 4.2.

REMARK 2.12. The $\rho_\mathfrak{h}$ appearing here is not unique, that is the original positive system $\Sigma^+(\mathfrak{a}, \mathfrak{g})$ was somewhat arbitrary. It takes some effort, but one can show that for an appropriate choice of positive system $\rho_{\mathfrak{h}}$ will be dominant. This is contained in [2, Proposition 11.1].

REMARK 2.13. The proof of Proposition 2.6 shows that if ω_0 is decomposed under the $H \cap K$ type decomposition of $\wedge^s(\mathfrak{q}_+) \otimes \mathbb{C}_\chi$ there is a nonzero component of type $\chi + 2\rho(\mathfrak{k} \cap \mathfrak{q}_+)$.

REMARK 2.14. There is a general principle explained in [5] from which one can construct intertwining operators between induced representations. One can use this principle to construct $\mathcal{S}$.

§3. **Strongly Harmonic.** In this section we show that the image of $\mathcal{S}$ consists of strongly harmonic forms. $\mathcal{D} = G/H$ has an indefinite G-invariant Hermitian metric defined as follows. Let k denote the Killing form and set

$$(\xi, \eta) = k(\xi, \overline{\eta})$$

for $\xi, \eta \in \mathfrak{q}$ (conjugation is with respect to the real form $\mathfrak{g}_0$), now translate to an arbitrary tangent space. Letting $\overline{\partial}$ denote the operator of the Dolbeault complex we use this invariant metric to define the formal adjoint $\overline{\partial}^*$ of $\overline{\partial}$. Thus $\overline{\partial}$ and $\overline{\partial}^*$ are G-invariant differential operators. Recall from the introduction that an $\mathcal{L}_\chi$-valued differential form ω on G/H is strongly harmonic means $\overline{\partial}\omega = 0$ and $\overline{\partial}^* \omega = 0$.

THEOREM 3.1. *Image* $(\mathcal{S}) \subset \ker \overline{\partial} \cap \ker \overline{\partial}^*$, *in other words,* $\mathcal{S}$ *maps* $I(W)$ *into* $\mathcal{H}^s(G/H, \mathcal{L}_\chi)$, *the space of strongly harmonic forms.*

The strategy of the proof is to note that if we assume $Im(\mathcal{S}) \not\subset \ker \overline{\partial}$ then $eval_e \cdot \overline{\partial} \cdot \mathcal{S} \neq 0$ (where $eval_e$ means evaluation at eH). This is because if $\overline{\partial}\mathcal{S}\varphi \neq 0$ then $(\overline{\partial}\mathcal{S}\varphi)(g) \neq 0$ for some g; thus $eval_e \overline{\partial}\mathcal{S}(g^{-1} \cdot \varphi) = \overline{\partial}\mathcal{S}(g^{-1}\varphi)(e) = (g^{-1} \cdot \overline{\partial}\mathcal{S}\varphi)(e) = \overline{\partial}\mathcal{S}\varphi(g) \neq 0$. Furthermore $eval_e \cdot \overline{\partial} \cdot \mathcal{S}$ is an H-homomorphism $C^\infty(G/P, W) \to \wedge^{s+1}(\mathfrak{q}_+) \otimes \mathbb{C}_\chi$. Thus $Hom_L((\wedge^{s+1}(\mathfrak{q}_+) \otimes \mathbb{C}_\chi)^*, \mathcal{D}'(G/P, W^*)) \neq 0$. We will see that $(\wedge^{s+1}(\mathfrak{q}_+) \otimes \mathbb{C}_\chi)^*$ necessarily maps to distributions supported in the closed orbit $H \cdot eP$ in G/P. However, the following lemma gives enough information about the possible H-types of $\mathcal{D}'(G/P, W^*)$ which are supported in the closed orbit to exclude $(\wedge^{s+1}(\mathfrak{q}_+) \otimes \mathbb{C}_\chi)^*$ from mapping (nonzero) into distributions supported in $H \cdot eP$. A similar argument applies to show $\overline{\partial}^* \mathcal{S} = 0$.

LEMMA 3.2. *Let* G, H, P *be as in* §2. *Let* E_0 *be a finite dimensional irreducible representation of* H *and let* W *be an irreducible representation of* P *with* $W|_A = e^{-\nu}$, $\nu \in \mathfrak{a}_0^*$. *If* $E_0^* \subset \mathcal{D}'(G/P, W^*)$ *is supported in the closed orbit* $H \cdot eP$ *in* G/P *then the* $\mathfrak{a}$-*weights of* E_0 *are of the form* $\nu - \rho - \sum n_\alpha \alpha$ *with* $n_\alpha \geq 0$ *and* $\alpha \in \sum(\mathfrak{a}, \mathfrak{n} \cap \mathfrak{q}_+)$. *Furthermore, if all* $n_\alpha = 0$ *then the elements of* E_0^* *restrict to smooth distributions on* $H \cdot eP$.

PROOF. Consider $\mathcal{D}'(G/P, W^*) = C^\infty(G/P, W)'$ with W an irreducible f.d. representation of P. The distributions supported in $H/P \cap H$ which are smooth on $H/P \cap H$ are integration against functions in $C^\infty(H/P \cap H, W^* \otimes \mathbb{C}_{-\rho + \rho_{\mathfrak{h}}})$. Letting (π_0, E_0) be a finite dimensional irreducible representation of H with

$E_0^* \hookrightarrow C^\infty(H/P \cap H, W^* \otimes \mathbb{C}_{-\rho+\rho_\mathfrak{h}})$ we see that $Hom_{P \cap H}(W \otimes \mathbb{C}_{\rho-2\rho_\mathfrak{h}}, E_0)$ contains a nonzero t. Letting t^* be dual to t, the corresponding distributions are given by

$$T_v(\varphi) = \int_{K \cap H} \langle \varphi(l),\, t^*(\pi_0(l^{-1})v) \rangle \, dl$$

for $v \in E_0^*$ and $\varphi \in C_0(G/P, W)$, see (2.9). Since $H = K \cap H \cdot A \cdot N \cap H$ is an Iwasawa decomposition, we may rewrite this using an integration formula ([9, page 198]) as

$$T_v(\varphi) = \int_{\overline{N} \cap H} \langle \varphi(\kappa(\overline{n})),\, t^*(\pi_0(\kappa(\overline{n})^{-1})v) \rangle \, e^{<2\rho_\mathfrak{h}, H(\overline{n})>} \, d\overline{n},$$

where $\overline{N} \cap H$ is the Lie subgroup of G with Lie algebra $\displaystyle\sum_{\alpha \in \Sigma^+(\mathfrak{a}, \mathfrak{n} \cap \mathfrak{h})} \mathfrak{g}^{-\alpha}$. Now take v to be v_-, a lowest $\mathfrak{a}$-weight vector in E_0^*. Thus $t^*(\pi_0(\kappa(\overline{n})^{-1})v_-) = t^*(\pi_0(e^{H(\overline{n})} N_{\overline{n}} \overline{n}^{-1})v_-) = t^*(\pi_0(e^{H(\overline{n})} N_{\overline{n}})v_-) = e^{<\nu-\rho+2\rho_\mathfrak{h}, H(\overline{n})>} t^*(v_-)$. We obtain:

$$T_v(\varphi) = \int_{\overline{N} \cap H} \langle \varphi(\kappa(\overline{n})),\, t^*(v_-) \rangle \, e^{<\nu-\rho, H(\overline{n})>} \, d\overline{n}.$$

Note that $\overline{N} \approx \overline{N} \cdot eP$ is an open (dense) submanifold of $G/P \approx K/M$ and $\overline{N} \cap H$ is open and dense in the closed orbit $H \cdot eP$. Also, we may restrict a distribution T on K/M to a distribution T' on $\overline{N}$. Thus:

$$T'_{v_-}(\varphi) = \int_{\overline{N} \cap H} <\varphi(\overline{n}),\, t^*(v_-)> \, d\overline{n}.$$

This formula defines an $\overline{N} \cap H$-invariant distribution on $\overline{N}$ supported in $\overline{N} \cap H$. In fact, if we replace $t^*(v_-)$ by any $w^* \in W^* \otimes \mathbb{C}_{-\rho+2\rho_\mathfrak{h}}$ we obtain distributions T_{w^*} which are $\overline{N} \cap H$ invariant and supported in $\overline{N} \cap H$. Furthermore, any $\overline{N} \cap H$ invariant W^*-valued distribution supported in $\overline{N} \cap H$ is a normal derivative of T_{w^*} for some $w^* \in W^* \otimes \mathbb{C}_{-\rho+2\rho_\mathfrak{h}}$.

Let T be an H-finite distribution supported in the closed orbit and T' its restriction to $\overline{N}$. Assume T is the lowest weight vector so T' is $\overline{N} \cap H$- invariant. So $T' = u \cdot T'_{w^*}$ for some $w^* \in W^* \otimes \mathbb{C}_{-\rho+2\rho_\mathfrak{h}}$ and some u in the enveloping algebra of $\mathfrak{g}$ which is in the span of $\{X_1 X_2 \cdots X_j : x_i \in \mathfrak{n} \cap \mathfrak{q}\}$. We now determine the $\mathfrak{a}$-weight of T'. Let $\eta \in \mathfrak{a}_0$.

$$(\exp(\eta) \cdot T')(\varphi) = (\exp(\eta) \cdot T)'(\varphi)$$
$$= \int_{\overline{N} \cap H} \langle ((\exp(\eta)^{-1} u \cdot \varphi)(\overline{n}), w^* \rangle \, d\overline{n}$$
$$= \int_{\overline{N} \cap H} \langle Ad(\exp \eta)u \cdot \varphi(Ad(\exp \eta)\overline{n} \cdot \exp(\eta)), w^* \rangle \, d\overline{n}$$
$$= \int_{\overline{N} \cap H} e^{<\nu-\rho-\sum n_\alpha \alpha, \eta>} \langle \varphi(Ad(\exp \eta)\overline{n}), w^* \rangle \, d\overline{n}$$

$$= \int_{\overline{N} \cap H} e^{<\nu - \rho - \sum n_\alpha \alpha, \eta>} \langle \varphi(\overline{n}), \, w^* \rangle \, d\overline{n}$$

$$= e^{<\nu - \rho - \sum n_\alpha \alpha, \eta>} T'(\varphi),$$

where the summations are over certain $n_\alpha \geq 0$.

PROOF OF THE THEOREM. Let $E_0 = \wedge^{s+1}(\mathfrak{q}_+) \otimes \mathbb{C}_\chi$ and assume $eval_e \cdot \overline{\partial} \cdot S$ is nonzero. Thus, there is a nonzero H homomorphism $i : E_0^* \to \mathcal{D}'(G/P, W^*)$ (the dual of $eval_e \cdot \overline{\partial} \cdot S$). Then $i(e_0^*)(\varphi) = e_0^*(\overline{\partial}S(\varphi)(eH))$ for $e_0^* \in E_0^*$. It follows from the formula for S in Theorem 2.10 and the form of $\overline{\partial}$ that this is supported in the closed orbit $H \cdot eP$.

In order to show $Im(S) \subset \ker \overline{\partial}$ it is enough to show that the $\mathfrak{a}$-weights of E_0 are not of the form $\rho_\mathfrak{h} - \rho - \sum_{\alpha \in \Sigma(\mathfrak{a}, \mathfrak{n} \cap \mathfrak{q}_+)} n_\alpha \alpha$, $n_\alpha \geq 0$. Using the same basis for $\mathfrak{q}_+$ used in section 2 we can write basis elements of $\wedge^{s+1}(\mathfrak{q}_+)$ in the form:

$$(X_{\alpha, \beta} + \theta(X_{\alpha, \beta})) \wedge \cdots \wedge (X_{\alpha', \beta'} - \theta(X_{\alpha', \beta'})) \wedge \cdots \wedge X_{0, \beta''} \wedge \cdots$$

(the exterior product of $s+1$ terms from $\mathfrak{k} \cap \mathfrak{q}_+$ and $\mathfrak{p} \cap \mathfrak{q}_+$). The $\mathfrak{a}$-weight vectors are thus of the form

$$X_{\alpha, \beta} \wedge \cdots \wedge \theta(X_{\alpha', \beta'}) \wedge \cdots \wedge X_{0, \beta''} \wedge \cdots.$$

The lowest $\mathfrak{a}$-weight vector is $\theta(X_{\alpha, \beta}) \wedge \cdots \wedge \theta(X_{\alpha', \beta'}) \wedge X_{0, \beta''} \wedge \cdots \wedge X_{0, \beta'''} \wedge X_{\alpha, \beta}$. The first s terms have used all root vectors $X_{\alpha, \beta} \in \mathfrak{n} \cap \mathfrak{q}_+$ and all root vectors $X_{0, \beta''} \in \mathfrak{m} \cap \mathfrak{q}_+$. Thus the $\mathfrak{a}$-weight is $\rho_\mathfrak{h} - \rho + \alpha$, for some $\alpha \in \sum(\mathfrak{a}, \mathfrak{n} \cap \mathfrak{q}_+)$. Since all other $\mathfrak{a}$-weights are higher, no $\mathfrak{a}$-weight is of the allowable form. We conclude that $Im(S) \subset \ker \overline{\partial}$.

We next show that $Im(S) \subset \ker \overline{\partial}^*$. The $\mathfrak{a}$-weight vectors in $\wedge^{s-1}(\mathfrak{q}_+)$ are of the form

$$X_{\alpha, \beta} \wedge \cdots \wedge \theta(X_{\alpha', \beta'}) \wedge \cdots \wedge X_{0, \beta''} \wedge \cdots,$$

with $s-1$ terms. The lowest $\mathfrak{a}$-weight is $\rho_\mathfrak{h} - \rho$ or $\rho_\mathfrak{h} - \rho + \alpha$ (the second possibility only when no $X_{0, \beta'}$ exists, i.e., when $\mathfrak{m} \cap \mathfrak{q}_+ = 0$). The second possibility, $\rho_\mathfrak{h} - \rho + \alpha$ is never of the allowable form $\rho_\mathfrak{h} - \rho - \sum n_\alpha \alpha$. However, $\rho_\mathfrak{h} - \rho$ may be allowable. In this case all $n_\alpha = 0$ and the image of E_0^* in $\mathcal{D}'(G/P, W^*)$ is smooth. Thus, we need to show $Hom_H((\wedge^{s-1}(\mathfrak{q}_+) \otimes \mathbb{C}_\chi)^*, \, C^\infty(H/P \cap H, \, W^* \otimes \mathbb{C}_{-\rho + \rho_\mathfrak{h}})) = 0$, that is $Hom_{P \cap H}(W \otimes \mathbb{C}_{\rho - 2\rho_\mathfrak{h}}, \, \wedge^{s-1}(\mathfrak{q}_+) \otimes \mathbb{C}_\chi) = 0$. With a view to contradiction, suppose $Hom_{P \cap H}(W \otimes \mathbb{C}_{\rho - 2\rho_\mathfrak{h}}, \, \wedge^{s-1}(\mathfrak{q}_+) \otimes \mathbb{C}_\chi) \neq 0$. Then there is an irreducible H subrepresentation E_1 of E_0 so that $Hom_{M \cap H \cdot A}(W \otimes \mathbb{C}_{\rho - 2\rho_\mathfrak{h}}, \, E_1^{\mathfrak{n} \cap \mathfrak{h}}) \neq 0$. Since the $\mathfrak{a}$-weights in $E_1^{\mathfrak{n} \cap \mathfrak{h}}$ must be $\rho - \rho_\mathfrak{h}$, the $\mathfrak{a}$-weight vectors are all of the form

$$(3.3) \qquad\qquad X_{\alpha, \beta} \wedge \cdots \wedge X_{\alpha', \beta'} \wedge X_{0, \beta''} \wedge \cdots \wedge X_{0, \beta''}$$

where all $X_{\alpha, \beta} \in \mathfrak{n} \cap \mathfrak{q}_+$ appear and all $X_{0, \beta''}$ except one appear. Now note that the $\mathfrak{t}_M$-weights of W are of the form $\chi + 2\rho(\mathfrak{k} \cap \mathfrak{q}_+) + \sum_{\alpha \in \Delta(\mathfrak{t}_M, \mathfrak{m} \cap \mathfrak{q}_+)} m_\alpha \alpha$, $m_\alpha \geq 0$, but the $\mathfrak{t}_M$-weights of (3.3) are $\chi + 2\rho(\mathfrak{k} \cap \mathfrak{q}_+) - \beta$ for some $\beta \in \Delta(\mathfrak{t}_M, \mathfrak{m} \cap \mathfrak{q}_+)$. The theorem is proved.

§4. **Cohomology.** In this section we show that the image of $\mathcal{S}$ in cohomology is nonzero. As χ satisfies (2.1) the lowest K type of $H^s(G/H, \mathcal{L}_\chi)$ has highest weight $\mu_0 = w_0(\chi + \rho_\mathfrak{k}) - \rho_\mathfrak{k}$, where w_0 is the element of the Weyl group of K so that $w_0(\Delta^+(\mathfrak{l} \cap \mathfrak{k})) \subset \Delta^+$ and $w_0(\Delta^+(\mathfrak{h} \cap \mathfrak{q}_+)) \subset -\Delta^+$. The principal series representation $I(W)$ with W as in (2.5) also contains this K type, call it (μ_0, F_0). We will show that $\mathcal{S}(F_0) \not\subset Im\overline{\partial}$, hence $Im(\mathcal{S})$ is nonzero in the cohomology space $H^s(G/H, \mathcal{L}_\chi)$.

Consider the Dolbeault complexes on G/H and $K/K \cap H$:

$$\{C^\infty(G/H, \wedge^\bullet(\mathfrak{q}_+) \otimes \mathbb{C}_\chi), \overline{\partial}_{G/H}\}$$

and

$$\{C^\infty(K/K \cap H, \wedge^\bullet(\mathfrak{k} \cap \mathfrak{q}_+) \otimes \mathbb{C}_\chi), \overline{\partial}_{K/K \cap H}\}.$$

There is a "restriction" map

$$r : C^\infty(G/H, \wedge^\bullet \mathfrak{q}_+ \otimes \mathbb{C}_\chi) \to C^\infty(K/K \cap H, \wedge^\bullet(\mathfrak{k} \cap \mathfrak{q}_+) \otimes \mathbb{C}_\chi)$$

defined by restricting a form $\omega : G \to \wedge^\bullet \mathfrak{q}_+ \otimes \mathbb{C}_\chi$ to K, then projecting the values to $\wedge^\bullet(\mathfrak{k} \cap \mathfrak{q}_+) \otimes \mathbb{C}_\chi$ (orthogonal projection). This is a map of complexes, that is $r(\overline{\partial}_{G/H}\omega) = \overline{\partial}_{K/K \cap H} r(\omega)$.

PROPOSITION 4.1. $r\mathcal{S}(F_0) \neq 0$ in $C^\infty(K/K \cap H, \wedge^s(\mathfrak{k} \cap \mathfrak{q}_+) \otimes \mathbb{C}_\chi)$ and represents a nonzero cohomology class on $K/K \cap H$.

PROOF. It is enough to see that $r\mathcal{S}(\phi_0)(e) \neq 0$ for some $\phi_0 \in F_0$. The proof consists of showing how to choose such a ϕ_0.

Recall that the operator $\mathcal{S}$ is constructed from $t \in Hom_{P \cap H}(W \otimes \mathbb{C}_{\rho - 2\rho_\mathfrak{h}}, E)$, $E = \wedge^s(\mathfrak{q}_+) \otimes \mathbb{C}_\chi$, and $(\mathcal{S}\phi)(g) = \int_{K \cap H} l \cdot t(\phi(\kappa(gl))) e^{(-\rho_\mathfrak{h} - \rho)H(gl)} dl$. Consider a basis $\{X_i, Y_j\}$ of consisting of a basis $\{X_i\}$ of $\mathfrak{k} \cap \mathfrak{q}_+$ and a basis $\{Y_j\}$ of $\mathfrak{p} \cap \mathfrak{q}_+$. We get a basis $\{e_k\}$ of E with each e_k consisting of s wedges from $\{X_i, Y_j\}$, also there is a dual basis $\{e_k^*\}$. Index the e_k's so that $e_0 = X_1 \wedge X_2 \cdots \wedge X_s$. Now,

$$\mathcal{S}\phi(g) = \sum_i \left(\int_{K \cap H} \langle l^{-1} \cdot e_i^*, t(\phi(\kappa(gl))) \rangle e^{(-\rho_\mathfrak{h} - \rho)(H(gl))} dl \right) e_i.$$

Hence,

$$r(\mathcal{S}\phi)(e) = \int_{K \cap H} \langle e_0^*, l \cdot t(\phi(l)) \rangle dl \cdot e_0,$$

for $\phi \in I(W) = C^\infty(K/M, W) = \{C^\infty(K) \otimes W\}^M = \sum_{\mu \in \hat{K}} F_\mu \otimes Hom_M(F_\mu, W)$.

Note that the dimension of $Hom_{K \cap H}(\wedge^s(\mathfrak{k} \cap \mathfrak{q}_+) \otimes \mathbb{C}_\chi, F_0)$ is one (because the $\mathfrak{k}$-weight of $\wedge^s(\mathfrak{k} \cap \mathfrak{q}_+) \otimes \mathbb{C}_\chi$ is $\chi + 2\rho(\mathfrak{k} \cap \mathfrak{u}) = w_0^{-1}(w_0(\chi + \rho_\mathfrak{k}) - \rho_\mathfrak{k})$ and this is an extreme weight of F_0). We thus let f_0 be the weight vector in F_0 of weight $\chi + 2\rho(\mathfrak{k} \cap \mathfrak{q}_+)$. Also, $Hom_M(F_0, W)$ contains nonzero element f^*. Set $\phi_0(k) = f^*(k^{-1} \cdot f_0)$, so

$$r(\mathcal{S}\phi_0)(e) = \int_{K \cap H} \langle e_0^*, l \cdot t(f^*(l^{-1} f_0)) \rangle dl \cdot e_0 = e_0^*(t(f^*(f_0))) e_0$$

because l^{-1} acts on f_0 and e_0 by the same scalar.

Claim: $\langle e_0^*, t(f^*(f_0)) \rangle \neq 0$. Note that $f^*(f_0) \in W$ is the weight vector for the extreme weight $\chi + 2\rho(\mathfrak{k} \cap \mathfrak{q}_+)$. However, the image of this weight vector under t has a component in $\wedge^s(\mathfrak{k} \cap \mathfrak{q}_+) \otimes \mathbb{C}_\chi$ (see proof of Proposition 2.6).

Now we show $rS(\phi_0) \notin Im(\overline{\partial}_{K/K \cap H})$. $C^\infty(K/K \cap H, \wedge^s(\mathfrak{k} \cap \mathfrak{q}_+) \otimes \mathbb{C}_\chi) \simeq \{C^\infty(K) \otimes \wedge^s(\mathfrak{k} \cap \mathfrak{q}_+) \otimes \mathbb{C}_\chi\}^{K \cap H} \simeq \sum_{\mu \in \hat{K}} F_\mu \otimes Hom_{K \cap H}(F_\mu, \wedge^s(\mathfrak{k} \cap \mathfrak{q}_+) \otimes \mathbb{C}_\chi)$. Since μ_0 occurs in $H^s(K/K \cap H, \mathcal{L}_\chi)$ (by the Bott-Borel-Weil theorem) it is enough to show that the dimension of $Hom_{K \cap H}(F_{\mu_0}, \wedge^s(\mathfrak{k} \cap \mathfrak{q}_+) \otimes \mathbb{C}_\chi) = 1$. But this is clear since $\chi + 2\rho(\mathfrak{k} \cap \mathfrak{q}_+)$ is an extreme weight of F_{μ_0} (hence of multiplicity one).

THEOREM 4.2. *$Im(S)$ is nonzero in $H^s(G/H, \mathcal{L}_\chi)$.*

PROOF. If $S\phi = \overline{\partial}_{G/H}\omega$ then $rS\phi = r(\overline{\partial}_{G/H}\omega) = \partial_{K/K \cap H} r(\omega)$. However the proposition says that this is not the case (for all ϕ).

REMARK 4.3. By Wong's thesis ([**14**]) $H^s(G/H, \mathcal{L}_\chi)$ is an irreducible admissible representation which is a maximal globalization. At the K-finite level, the map from the principal series to cohomology is onto, thus S can be extended to an intertwining map from $\mathcal{B}(G/P, W)$, the W-valued hyperfunctions on G/P, onto $H^s(G/H, \mathcal{L}_\chi)$ See [**14**], [**11**]. We may conclude:

COROLLARY 4.4. *Every K-finite cohomology class is represented by a strongly harmonic form.*

Part II

§**5.** We now assume conditions (1), (2) and (3) of the introduction hold, that is, the negativity condition (2.1) holds, the real ranks of G and H are equal and G/H is a semisimple symmetric space. Thus H is the fixed point set of an involution σ. We may assume that the Cartan involution commutes with σ. The lie algebra $\mathfrak{g}_0$ decomposes into the ± 1 eigenspaces of σ,

$$\mathfrak{g}_0 = \mathfrak{h}_0 + \mathfrak{q}_0,$$

also,

$$\mathfrak{g}_0 = \mathfrak{k}_0 \cap \mathfrak{h}_0 + \mathfrak{k}_0 \cap \mathfrak{g}_0 + \mathfrak{p}_0 \cap \mathfrak{h}_0 + \mathfrak{p}_0 \cap \mathfrak{q}_0.$$

Similarly for the complexification $\mathfrak{g}$ of $\mathfrak{g}_0$. The tangent space at $x_0 = eH$ of $G/H = \mathcal{D} \subset X = G_\mathbb{C}/H_\mathbb{C}Q_-$ decomposes as

$$\mathfrak{q} = \mathfrak{q}_+ \oplus \mathfrak{q}_-.$$

Note that $\mathfrak{q}_+$ corresponds to the holomorphic tangent space.

The G-invariant Hermitian metric $(\ ,\ )$ was defined in §3. Note that $(\ ,\ )$ is positive definite on $\mathfrak{p} \cap \mathfrak{q}$, negative definite on $\mathfrak{k} \cap \mathfrak{q}$ and these two spaces are orthogonal. The second metric we use is positive definite and not invariant. We use the following decomposition of G,

$$(5.1) \qquad\qquad\qquad G = KBH$$

where B is $\exp(\mathfrak{b}_0)$ and $\mathfrak{b}_0$ is a maximal abelian subspace of $\mathfrak{p}_0 \cap \mathfrak{q}_0$. Define a form $\langle\ ,\ \rangle$ on the tangent space $T_{eH}(G/H) = \mathfrak{q}_0$ by switching the sign of $(\ ,\ )$ on $\mathfrak{k}_0 \cap \mathfrak{q}_0$. Therefore $\langle\ ,\ \rangle$ is positive definite. Now define $\langle\ ,\ \rangle$ on an arbitrary tangent space by setting $\langle\tau_{kb}(\xi), \tau_{kb}(\eta)\rangle = \langle\xi, \eta\rangle$ (τ_g is left translation by g on G/H). This is well defined as $kbh = k'b'h'$ implies $k \in k'z$ where z is in the centralizer of B in $H \cap K$ and $\langle\ ,\ \rangle$ is $H \cap K$-invariant.

As in part I, $\mathbb{C}_\chi$ is a unitary character of H of weight χ. The bundle corresponding to χ is $\mathcal{L}_\chi \to G/H$ and the space of $\mathcal{L}_\chi$-valued differential forms of type $(0, s)$ is denoted by $A^s(G/H, \mathcal{L}_\chi)$. Let $\#$ (and $\tilde{\#}$) denote the Hodge-Kodaira orthocomplementation operator with respect to the invariant metric (and the positive definite metric, respectively). Let $\overline{\wedge}$ (and $\tilde{\overline{\wedge}}$) denote wedge product followed by contraction with respect to the invariant metric (and positive definite metric, respectively). There are two global Hermitian inner products on G/H, one G-invariant,

$$\langle\omega, \eta\rangle_{G/H} = \int_{G/H} \omega\overline{\wedge}\#\eta$$

and one positive definite,

$$\langle\omega, \eta\rangle_{\text{pos}} = \int_{G/H} \omega\tilde{\overline{\wedge}}\tilde{\#}\eta$$

(provided these integrals exist).

DEFINITION 5.2. *We say* $\omega \in A^s(G/H, \mathcal{L}_\chi)$ *is* L_2 *if* $\|\omega\|^2_{\text{pos}} \equiv \langle\omega, \omega\rangle_{\text{pos}} < \infty$. *Recall from the introduction that* $\overline{\partial}^*$ *is the formal adjoint of* $\overline{\partial}$ *with respect to the invariant metric and*

$$\mathcal{H}^s(G/H, \mathcal{L}_\chi) = \left\{ A^s(G/H, \mathcal{L}_\chi) : \overline{\partial}\omega = 0 \text{ and } \overline{\partial}^*\omega = 0 \right\}$$

$$\mathcal{H}^s_{(2)}(G/H, \mathcal{L}_\chi) = \left\{ \omega \in A^s(G/H, \mathcal{L}_\chi) : \overline{\partial}\omega = 0, \overline{\partial}^*\omega = 0 \text{ and } \omega \text{ is } L_2 \right\}.$$

The preceding setup and the following proposition are contained in [8].

PROPOSITION 5.3.
(a) *If* ω, η *are* L_2 *then* $\langle\omega, \eta\rangle_{G/H} < \infty$,
(b) $\mathcal{H}^s_{(2)}(G/H, \mathcal{L}_\chi)$ *is* G-*invariant.*

We consider $\langle\ ,\ \rangle_{G/H}$ as a G-invariant indefinite "global" Hermitian form on $\mathcal{H}^s_{(2)}(G/H, \mathcal{L}_\chi)$. The integer s is the complex dimension of $K/K \cap H$.

§6. **Square Integrability.** In this section it is proved that the forms in the image of $\mathcal{S}$ are square integrable. To accomplish this $\mathcal{S}\varphi$ is bounded by a joint eigenfunction of the invariant differential operators on G/K. The eigenfunctions are the ones studied by Flensted-Jensen and Oshima and Matsuki in [4], [6], and their growth behavior is known. In addition it is important that there is a convenient integration formula for G/H.

The key observation is that

$$|\mathcal{S}\varphi(g)| \leq \int_{H\cap K} |\langle\varphi(\kappa(gl)),\phi_0\rangle|\, e^{-\langle\rho+\rho_\mathfrak{h},H(bl)\rangle}|l\cdot\omega_0|dl$$

(6.1)
$$\leq C\int_{H\cap K} e^{-\langle\rho+\rho_\mathfrak{h},H(g)\rangle}dl$$

where $g \in G$ and $C = \max\{|\langle\varphi(k),\phi_0\rangle l\cdot\omega_0| : k \in K,\ l \in L\cap K\}$. Thus we may focus our attention on showing that

$$\int_{H\cap K} e^{-\langle\rho+\rho_\mathfrak{h},H(g)\rangle}dl$$

decays quickly enough to give square integrability of $|\mathcal{S}\varphi(g)|$. To do this we will follow the analysis on G/K given in [9]. This analysis was developed to study the discrete series on a semisimple symmetric space for some group "dual" to G. We need not consider this dual semisimple symmetric space here (which is not G/H).

THEOREM 6.2. *If*

(1) $\langle\chi+\rho(\mathfrak{k}),\gamma\rangle < 0, \forall\gamma \in \triangle(\mathfrak{k}\cap\mathfrak{q}_+)$,
(2) $Rank_R G = Rank_R H$,
(3) G/H *is symmetric*,

then every K-finite cohomology class in $H^s(G/H,\mathcal{L}_\chi)$ is represented by a strongly harmonic L_2 form.

PROOF. The existence of $\mathcal{S}$ is guaranteed by (1) and (2). Let $\mathfrak{b}_0$ be as in section 5. Fix any positive system of roots $\Sigma^+(\mathfrak{b}_0,\mathfrak{k}_0\cap\mathfrak{h}_0+\mathfrak{p}_0\cap\mathfrak{q}_0)$ and let $\mathfrak{b}_0^+$ be the corresponding positive Weyl chamber. There is a decomposition of G given by

$$G = K\overline{B^+}H.$$

Equation (6.1) becomes

(6.3)
$$|\mathcal{S}(\varphi)(kb)| \leq C\int_{H\cap K} e^{-\langle\rho+\rho_\mathfrak{h},H(b)\rangle}dl.$$

We consider the "fundamental" functions ψ_λ on G/K defined in equation 7.16 of [9] ($w = e$ in the notation of [9]). Comparing the setup in §2 of this paper with 7.3 of [9] we see that $\mathfrak{a}$ and certain positive root systems for $\mathfrak{a}$ have been chosen so that the Iwasawa H function here is as in [9]. Thus the right hand side of (6.3) is $\psi_\lambda(b^{-1}K)$ with $\lambda = \rho_\mathfrak{h}$. So we may rewrite (6.3) as

(6.4)
$$|\mathcal{S}(\varphi)(kb)| \leq C\psi_{\rho_\mathfrak{h}}(b^{-1}K)$$

We will use the following integration formula which is given in 8.1.1 of [9].

(6.5)
$$\int_{G/H} f(gH)dg = \int_K\int_{\mathfrak{b}_0^+} f(k\exp(\xi)H)\delta(\xi)\,d\xi dk$$

where dk is Haar measure on K and $d\xi$ is Lebesgue measure on $\mathfrak{b}_0$ ($\delta(\xi)$ is given in 8.1.1.). The crucial estimate is as follows (see 7.6.4 of [9]). On chambers $\mathfrak{b}_1^+$ in $\mathfrak{b}$ defined by positive systems for $\Sigma(\mathfrak{b}_0, \mathfrak{g}_0)$ we have, for dominant $\lambda \in \mathfrak{a}^*$,

$$\psi_\lambda(\exp(\xi)K) \leq Ce^{-(1+\epsilon)\langle \hat{\rho}, \xi \rangle}, \text{ for } \xi \in \mathfrak{b}_1^+,$$

where $\hat{\rho}$ corresponds to $\mathfrak{b}_1^+$. By Remark 2.12 we may assume that $\rho_\mathfrak{b}$ is dominant. Furthermore, $\delta(\xi) \leq e^{\langle \hat{\rho}, \xi \rangle}$ for all chambers $\mathfrak{b}_1^+$. If we let $\mathfrak{b}_1^+$'s fill out $\mathfrak{b}_0^+$ we see that

$$(6.6) \qquad \int_K \int_{\mathfrak{b}_0^+} \left| \psi_{\rho_\mathfrak{b}}(\exp(\xi)K) \right|^2 \delta(\xi)\, d\xi\, dk < \infty.$$

However it is $\psi_{\rho_\mathfrak{b}}(\exp(\xi)^{-1}K)$ (note the inverse) which bounds $|\mathcal{S}(k\exp(\xi))|$. The following argument will complete the proof.

Let w_0 be the element of the Weyl group of $\Sigma(\mathfrak{b}_0, \mathfrak{k}_0 \cap \mathfrak{h}_0 + \mathfrak{p}_0 \cap \mathfrak{q}_0)$ which sends $\mathfrak{b}_0$ to $-\mathfrak{b}_0$. We consider w_0 as an element of $K \cap H$ in the following calculation.

$$\int_K \int_{\mathfrak{b}_0^+} |\mathcal{S}\varphi(k\exp(\xi))|^2 \delta(\xi)\, d\xi\, dk$$

$$= \int_K \int_{\mathfrak{b}_0^+} \left| \mathcal{S}\varphi(kw_0^{-1}\exp(Ad(w_0)\xi)w_0) \right|^2 \delta(\xi)\, d\xi\, dk$$

$$= \int_K \int_{\mathfrak{b}_0^+} \left| w_0^{-1} \cdot \mathcal{S}\varphi(kw_0^{-1}\exp(Ad(w_0\xi))) \right|^2 \delta(\xi)\, d\xi\, dk$$

$$= \int_K \int_{\mathfrak{b}_0^+} \left| \mathcal{S}\varphi(k\exp(Ad(w_0)\xi)) \right|^2 \delta(\xi)\, d\xi\, dk$$

$$= \int_K \int_{\mathfrak{b}_0^+} \left| \mathcal{S}\varphi(k\exp(-\xi)) \right|^2 \delta(-Ad(w_0)\xi)\, d\xi\, dk,$$

$$\text{by the change of variables } \xi \to -Ad(w_0)\xi \text{ on } \mathfrak{b}_0^+,$$

$$= \int_K \int_{\mathfrak{b}_0^+} \left| \mathcal{S}\varphi(k\exp(-\xi)) \right|^2 \delta(\xi)\, d\xi\, dk, \text{ since } \delta(-Ad(w_0)\xi) = \delta(\xi),$$

$$\leq \int_K \int_{\mathfrak{b}_0^+} \left| \psi_{\rho_\mathfrak{b}}(\exp(\xi)K) \right|^2 \delta(\xi)\, d\xi\, dk, \text{ by (6.4)},$$

$$< \infty, \text{ by (6.6)}.$$

Thus the square integrability of $\mathcal{S}\varphi(g)$ is established.

REMARK. A little more has been shown: if $\varphi \in C^\infty(G/P, \mathcal{W})$ then $\mathcal{S}(\varphi)$ is L_2. Thus, $\mathcal{S}$ is an intertwining map from $I(W)$ into $\mathcal{H}_{(2)}^s(G/H, \mathcal{L}_\chi)$.

§7. **Unitarity.** We prove the following proposition from which our main result, Theorem 7.9, will follow.

PROPOSITION 7.1. *Suppose that conditions* (1) *and* (2) *of Theorem* 6.2 *hold, then*

$$0 \le \int_{H\cap K} \int_{H\cap K} \langle \varphi(\kappa(gl)), \phi_0 \rangle \overline{\langle \varphi(\kappa(gl')), \phi_0 \rangle}$$

(7.2)
$$e^{-\langle \rho + \rho_{\mathfrak{h}}, H(gl) + H(gl') \rangle} \langle l\omega_0, l'\omega_0 \rangle \, dl \, dl'$$

for every $g \in G$. *Furthermore, there is a* $\varphi \in I(W)$ *so that the right hand side of* (7.2) *is not* 0.

REMARK 7.3. The right hand side of (7.2) is the integrand of the invariant global form $\langle \mathcal{S}\varphi, \mathcal{S}\varphi \rangle_{G/H}$.

REMARK 7.4. Here it is not assumed that G/H is symmetric. Thus if $\mathcal{S}\varphi$ can be shown to be L_2 in this more general case then Theorem 7.9 will hold.

PROOF. Replacing φ by $g^{-1} \cdot \varphi$ we may assume $g = e$. We may also assume that φ is K-finite. It will be shown that

(7.5)
$$0 \le \int_{H\cap K} \int_{H\cap K} \langle \varphi(l), \phi_0 \rangle \overline{\langle \varphi(l'), \phi_0 \rangle} \langle l\omega_0, l'\omega_0 \rangle \, dldl'$$

for $\varphi \in I(W)_{K\text{-finite}}$.

The K-finite functions in $I(W) = C^\infty(K/M, \mathcal{W})$ are, by the Peter-Weyl theorem, linear combinations from $F \otimes \mathrm{Hom}_M(F, W)$ with F ranging over irreducible K-representations. For $u \otimes v \in F \otimes \mathrm{Hom}_M(F, W)$ the corresponding section of $\mathcal{W}$ is

$$\varphi_{u\otimes v}(k) = v(k^{-1}u).$$

Let $v^* \in \mathrm{Hom}_M(W^*, F^*)$ be dual to v, now (7.5) is

(7.6)
$$\int_{H\cap K} \int_{H\cap K} \langle l^{-1}u, v^*(\phi_0) \rangle \overline{\langle l'^{-1}u, v^*(\phi_0) \rangle} \langle l\omega_0, l'\omega_0 \rangle \, dldl'.$$

We now verify that this is ≥ 0 by a matrix coefficient calculation. Let $\{U_\mu\}$ be the irreducible representations of $H \cap K$ (parameterized by some μ's). Assume $\varphi = u \otimes v$ and write the following decompositions into $H \cap K$-types,

$$u = \sum u_{\mu'}$$
$$v^*(\phi_0) = \sum v_{\mu''}$$
$$\omega_0 = \sum \omega_{\mu'''}$$

Now (7.6) becomes,

(7.7)
$$\sum_{\mu',\mu'',\mu'''} \sum_{\overline{\mu}',\overline{\mu}'',\overline{\mu}'''} \int_{H\cap K} \int_{H\cap K} \langle u_{\mu'}, lv_{\mu''} \rangle \langle u_{\overline{\mu}'}, l'v_{\overline{\mu}''} \rangle \langle l\omega_{\mu'''}, l'\omega_{\overline{\mu}'''} \rangle \, dldl'.$$

Using the orthogonality of matrix coefficients for the compact group $H \cap K$ we see that (7.7) is equal to

$$\sum_{\mu',\overline{\mu}',\mu'''} \int_{H\cap K} \int_{H\cap K} \langle u_{\mu'}, lv_{\mu'} \rangle \overline{\langle u_{\mu^{-1}}, l'v^{\overline{\mu}'} \rangle} \langle l\omega_{\mu'''}, l'\omega_{\overline{\mu}'''} \rangle \, dl dl'$$

$$= \sum_{\mu',\overline{\mu}',\mu'''} \int_{H\cap K} \langle u_{\mu'}, lv_{\mu'} \rangle \frac{\overline{\langle u_{\overline{\mu}'}, l\omega_{\mu'''} \rangle} \langle v_{\overline{\mu}'}, \omega_{\mu'''} \rangle}{\dim(U_{\overline{\mu}'})} \delta_{\overline{\mu}',\mu'''} \, dl$$

$$= \sum_{\mu',\overline{\mu}',\mu'''} \frac{\langle \omega_{\mu'''}, v_{\mu'} \rangle \overline{\langle u_{\overline{\mu}'}, u_{\mu'} \rangle} \langle v_{\overline{\mu}'}, \omega_{\mu'''} \rangle}{\dim(U_{\mu'}) \dim(U_{\overline{\mu}'})} \delta_{\mu',\mu'''} \cdot \delta_{\overline{\mu}',\mu'''}$$

$$= \sum_{\mu} \frac{|\langle v_\mu, \omega_\mu \rangle|^2 \|u_\mu\|^2}{(\dim(U_\mu))^2}$$

(7.8) $\geq 0.$

To see that the right hand side of (7.2) is not always zero it is enough to show (7.7) is nonzero for an appropriate choice of $\varphi = u \otimes v$. As noted in the proof of Proposition 4.1 the irreducible K-representation F_{μ_0} with extreme weight $\chi + 2\rho(\mathfrak{k} \cap \mathfrak{q}_+)$ occurs in $H^s(G/H, \mathcal{L}_\chi)$. Note that $\dim \operatorname{Hom}_M(F_{\mu_0}, W) = 1$, so we may choose v so that $v^*(\phi_0)$ is a lowest weight vector in F_{μ_0}, which is a lowest weight vector in an $H \cap K$ type U_0. By Remark 2.13 ω_0 has a component in U_0. So (7.8) is nonzero for v as above and any $u \in F_{\mu_0}$.

THEOREM 7.9. *Under the same hypothesis as Theorem 6.2 the global invariant form is defined and positive definite on $H^s(G/H, \mathcal{L}_\chi)_{K\text{-finite}}$.*

PROOF. The only thing that remains to prove is that a form $\mathcal{S}\varphi$ in the image of $\overline{\partial}$ is in the null space of the invariant form. This follows by modifying the proof of Proposition 7.7 in [13].

§8. Example. Consider $\mathcal{D} = G/H = SO(2n,1)/SO(2)^m \times SO(2(n-m),1)$, for $m = 1, 2, \ldots, n-1$. This is an example of an elliptic orbit which (for $m \geq 2$) is not a semisimple symmetric space. The real ranks of G and H are equal. There is some overlap between this example and [15].

Fix the Cartan subgroup

$$T = \exp(\mathfrak{t}_0) = \left\{ \exp \begin{pmatrix} 0 & \theta_1 & & & & \\ -\theta_1 & 0 & & & & \\ & & \ddots & & & \\ & & & 0 & \theta_m & \\ & & & -\theta_m & 0 & \\ & & & & & 1 \end{pmatrix} \right\}.$$

The weights will be denoted by $(a_1, \ldots, a_n)$ which corresponds to $\sum_{j=1}^m i a_j \theta_j$. The parabolic subgroup $H_{\mathbb{C}} Q_+$ is determined, as in §2, by

$$\lambda = (m, m-1, \ldots, 2, 1, 0, \ldots, 0).$$

So

$$\chi = (a_1, \ldots, a_m, 0, \ldots, 0)$$

defines a one dimensional unitary representation of H and χ satisfies the negativity condition (2.1) when

$$a_1 < a_2 - 1 < a_3 - 2 < \ldots < a_m - (m-1) < -2n + m.$$

Fix

$$\mathfrak{a}_0 = \left\{ \begin{pmatrix} \ddots & & & \\ & \ddots & & \\ & & \ddots & \\ & & & 0 & t \\ & & & t & 0 \end{pmatrix} : t \in \mathbb{R} \right\},$$

this is maximal abelian in $\mathfrak{h}_0 \cap \mathfrak{p}_0$ and in $\mathfrak{p}_0$. We use the positive system of $\mathfrak{a}$-roots $\sum^+(\mathfrak{a}, \mathfrak{g}) = \{t\}$. Note that $\rho = n - \frac{1}{2}$ and $\rho_{\mathfrak{h}} = n - m - \frac{1}{2}$. A (minimal) parabolic subgroup $P = MAN$ is determined, $M = Z_K(\mathfrak{a}_0) \simeq SO(2n-1)$. Following (2.5) we let δ_M be the irreducible representation of M with extreme weight $\chi + 2\rho(\mathfrak{k} \cap \mathfrak{q}_+)$. The intertwining operator is

$$\mathcal{S}\varphi(g) = \int_{H \cap K} \langle \varphi(\kappa(gl)), \psi_0 \rangle e^{-(2n-m-1)H(gl)} \pi(l) \omega_0 \, dl.$$

In order to calculate $\langle \mathcal{S}\varphi, \mathcal{S}\varphi \rangle_{\text{pos}}$ one needs an integration formula for G/H. We do the following. There is a chain of subgroups

$$H = H_m \subset H_{m-1} \subset \ldots \subset H_2 \subset H_1 \subset H_0 = G$$

so that,

(1) H_j/H_{j+1} is semisimple symmetric and,
(2) $H_j \simeq SO(2)^j \times SO(2(n-j), 1)$.

Integration is given by,

$$\int_{G/H} f(gH) dg = \int_{G/H_1} \int_{H_1/H_2} \cdots \int_{H_{m-1}/H_m} f(gh_1 \ldots h_{m-1} H) dh_{m-1} \ldots dh_1 \, dg$$

where all of the measures are the left invariant measures on the appropriate homogeneous spaces. For each H_j/H_{j+1} we may use the integration formula of

(6.5). The corresponding subgroup $B_j \subset H_j$ consists of

$$b_{j,s_j} = \exp \begin{pmatrix} & & & & & 0 \\ & & & & & \vdots \\ & & & & & s_j \\ & & & & & \vdots \\ 0 & \cdots & s_j & \cdots & & 0 \end{pmatrix}$$

with $s_j \in \mathbb{R}$ in the $(2j+1, 2n+1)$ and $(2n+1, 2j+1)$ slots. Also, $\delta_j(s_j) \leq Ce^{2(n-j-\frac{1}{2})s_j}$. Using the decomposition (5.1) for each $H_j \subset H_{j+1}$, write

$$g = k_0 b_{0,s_0} k_1 b_{1,s_1} \ldots k_{m-1} b_{m-1,s_{m-1}}$$

with $k_j \in H_j \cap K$ and $b_{j,s_j} \in B_j$. Since each B_j commutes with each $H_i \cap K$ for $i > j$, we may write

$$g = k_0 k_1 \ldots k_{m-1} b_{0,s_0} b_{1,s_1} \ldots b_{m-1,s_{m-1}}.$$

Therefore for $l \in H \cap K$

(8.1) $$H(gl) = H(b_{0,s_0} b_{1,s_1} \ldots b_{m-1,s_{m-1}} l).$$

Each element of $H \cap K \simeq SO(2)^m \times SO(2(n-m))$ decomposes as

$$l = l_1 l_2$$

with $l_1 \in SO(2)^m$ and $l_2 \in SO(2(n-m))$. Since $l_1 \in M$ (and $l_1 l_2 = l_2 l_1$), it follows from the form of $S\varphi$ given in Theorem 2.10 that integration over the $SO(2)^m$ part of $H \cap K$ may be omitted. Also, since l_2 commutes with B_j, $j = 0, 1, \ldots, m-1$ we see that

$$S\varphi(g) = \int_{SO(2(n-m))} \langle \varphi(k_0 \ldots k_m l_2 \kappa(b_{0,s_0} \ldots b_{m-1,s_{m-1}})), \psi_0 \rangle$$
$$e^{-(2n-m-1)H(b_{0,s_0}\ldots b_{m-1,s_{m-1}})} \pi(l_2)\, \omega_0\, dl_2.$$

An easy calculation gives

$$e^{H(b_{0,s_0}\ldots b_{m-1,s_{m-1}})} = \cosh(s_0) \ldots \cosh(s_{m-1}).$$

Let M be the maximum of

$$\frac{|\langle \varphi(k_0 \ldots k_m l_2 \kappa(b_{0,s_0} \ldots b_{m-1,s_{m-1}})), \psi_0 \rangle \cdot}{\langle \varphi(k_0 \ldots k_m l_2' \kappa(b_{0,s_0} \ldots b_{m-1,s_{m-1}})), \psi_0 \rangle|}$$

for $k_j \in H_j \cap K$, $l_2, l_2' \in H \cap K$ and $b_{j,s_j} \in B_j$.

For g as in equation (8.1) we get:

$$|\mathcal{S}\varphi(g)|^2 \le M \left(\prod_{j=0}^{m-1} \cosh(s_j) \right)^{-2(2n-m-1)}.$$

So,

$$\langle \mathcal{S}\varphi, \mathcal{S}\varphi \rangle_{\text{pos}} \le \int_{G/H} |\mathcal{S}\varphi(g)|^2 \, dg$$

$$\le M \int_{\mathbb{R}^+} \cdots \int_{\mathbb{R}^+} \prod_{j=0}^{m-1} (\cosh(s_j))^{-2(2n-m-1)} \delta_j(s_j) ds_1 \ldots ds_{m-1}$$

$$\le \int_{\mathbb{R}^+} \cdots \int_{\mathbb{R}^+} \prod_{j=0}^{m-1} (\cosh(s_j))^{-2(2n-m-1)} e^{2(n-j-\frac{1}{2})s_j} ds_1 \ldots ds_{m-1}$$

$$\le \int_{\mathbb{R}^+} \cdots \int_{\mathbb{R}^+} \prod_{j=0}^{m-1} e^{(-2(n-m)-2j+1)s_j} ds_1 \ldots ds_{m-1} < \infty.$$

Thus $\mathcal{S}\varphi$ is L_2 and we may conclude that Theorem 7.9 holds for this example.

REFERENCES

1. L. Barchini, *Szego mappings, harmonic forms and Dolbeault cohomology*, J. Functional Analysis (to appear).
2. L. Barchini, A. W. Knapp and R. Zierau, *Intertwining operators into Dolbeault cohomology representations*, J. Functional Analysis **107** (1992), 302–341.
3. E. G. Dunne and M. G. Eastwood, *The twistor transform*, Twistors in Mathematics and Physics, Cambridge University Press, 1990, pp. 110–128.
4. M. Flensted-Jensen, *Discrete series for semisimple symmetric spaces*, Annals of Math. **111** (1980), 253–311.
5. A. W. Knapp, *Imbedding discrete series in $L^2(G/H)$*, Harmonic Analysis on Lie Groups (Sandbjerg Estate, August 26-30, 1991), Report Series, No. 3, Copenhagen University Mathematics Institute, 1991, pp. 27–29.
6. T. Oshima and T. Matsuki, *A description of discrete series of semisimple symmetric spaces*, Advanced Studies in Mathematics, Vol. 4, North-Holland, Amsterdam, New York, Oxford, 1984.
7. R. Penrose and M. A. H. MacCallum, *Twistor theory: an approach to the quantization of fields and space-time*, Phys. Repts. **6C** (1972), 241–315.
8. J. W. Rawnsley, W. Schmid and J. A. Wolf, *Singular unitary representations and indefinite harmonic theory*, J. Functional Analysis **51** (1983), 1–114.
9. H. Schlichtkrull, *Hyperfunctions and Harmonic Analysis on Noncompact Symmetric Spaces*, Birkhäuser, 1984.
10. W. Schmid, *L^2-cohomology and the discrete series*, Annals of Math. **103** (1976), 375–394.
11. W. Schmid, *Boundary value problems for group invariant differential equations*, Soc. Math. France Astérisque, Numero hors series (1985), 311–321.
12. J. A. Wolf, *The action of a real semisimple Lie group on a compact flag manifold I: Orbit structure and holomorphic arc components.*, Bulletin Amer. Math. Soc. **75** (1969), 1121–1237.
13. J. A. Wolf, *Geometric realizations of discrete series representations in a nonconvex holomorphic setting*, Bulletin de la Sociétié Mathématique de Belgique **42** (1990), 797–812.
14. H.-W. Wong, *Dolbeault cohomologies and Zuckerman modules associated with finite rank representations*, PhD. Thesis, Harvard University (1991).

15. R. Zierau, *Harmonic forms and certain unitary representations of $SO_e(2n, 1)$*, Nova Journal of Algebra and Geometry (to appear).

MATHEMATICS DEPARTMENT, OKLAHOMA STATE UNIVERSITY, STILLWATER, OK 74078, USA

E-mail address: zierau@math.okstate.edu

Recent Titles in This Series

(Continued from the front of this publication)

(See the AMS catalog for earlier titles)